Springer Atmospheric Sciences

The Springer Atmospheric Sciences series seeks to publish a broad portfolio of scientific books, aiming at researchers, students, and everyone interested in this interdisciplinary field. The series includes peer-reviewed monographs, edited volumes, textbooks, and conference proceedings. It covers the entire area of atmospheric sciences including, but not limited to, Meteorology, Climatology, Atmospheric Chemistry and Physics, Aeronomy, Planetary Science, and related subjects.

Vasubandhu Misra

An Introduction to Large-Scale Tropical Meteorology

 Springer

Vasubandhu Misra
Department of Earth, Ocean and
Atmospheric Science
Florida State University
Tallahassee, FL, USA

ISSN 2194-5217   ISSN 2194-5225   (electronic)
Springer Atmospheric Sciences
ISBN 978-3-031-12889-9   ISBN 978-3-031-12887-5   (eBook)
https://doi.org/10.1007/978-3-031-12887-5

This Springer imprint is published by the registered company Springer Nature Switzerland AG
The registered company address is: Gewerbestrasse 11, 6330 Cham, Switzerland

Vasubandhu Misra

An Introduction to Large-Scale Tropical Meteorology

 Springer

Vasubandhu Misra
Department of Earth, Ocean and
Atmospheric Science
Florida State University
Tallahassee, FL, USA

ISSN 2194-5217   ISSN 2194-5225   (electronic)
Springer Atmospheric Sciences
ISBN 978-3-031-12889-9   ISBN 978-3-031-12887-5   (eBook)
https://doi.org/10.1007/978-3-031-12887-5

This Springer imprint is published by the registered company Springer Nature Switzerland AG
The registered company address is: Gewerbestrasse 11, 6330 Cham, Switzerland

Preface

I regard Tropical Meteorology as germane to "essential" Meteorology. The tropics are often referred as the driver of the atmospheric general circulation, exporting the excess energy it receives to higher latitudes. Further, the largest and one of the most impactful natural variations of climate, the El Niño and the Southern Oscillation, occur in the deep tropics. The largest seasonal variation of the hydroclimate is associated with the monsoons, which are deeply embedded within the tropics. And the ubiquity of Thunderstorms, the organization of convection, equatorial waves, and the occasional genesis of tropical cyclones make Tropical Meteorology a very interesting subject.

I would imagine that most graduate or undergraduate meteorological programs around the world offer some component of Tropical Meteorology as part of their curriculum given its overarching influence on anything related to Meteorology. I happen to be fortunately a part of one such department that has a cherished legacy in Tropical Meteorology. I have been a beneficiary of this institutional excellence both as a student in the department and as a faculty.

I joined Florida State University (FSU) as a graduate student, primarily inspired by my PhD supervisor Prof. T. N. Krishnamurti. Prof. Krishnamurti (popularly known as Krish) is considered the father of modern Tropical Meteorology for his pioneering work on monsoons, tropical cyclones, and numerical weather prediction. It was indeed a privilege for me to watch this academic giant from such close quarters and learn from him as much as I could.

This textbook is a humble rendering of the lessons I learned from those early days of my graduate education to the present. Of course, you grow wiser with experience as I was getting a more complete perspective on the concepts I was exposed to in my early graduate days, which inspired me to write this textbook. I have taught this course on large-scale Tropical Meteorology at Florida State University for nearly 15 years, which is reflected in this textbook.

In this day and age of self-taught pandits over the Internet and free wares, I was conflicted in writing a textbook like this. But then I recollected from my own experience that my initial impression and understanding of the field has a large fingerprint on my current understanding of the subject. Furthermore, given the broad span of topics and the breadth of the current research in Tropical Meteorology, it is not unimaginable for a new student to be lost in the "woods." Therefore, I believe that there is always a scope for guidance and help in steering such and other students through a textbook such as this.

This textbook is written at the level of senior undergraduate or first-year graduate students who require an initial introduction to tropical climate studies. The book covers a wide range of topics that include old, well-founded theories (e.g., CISK and quasi-equilibrium) to more recent developments (e.g., weak temperature gradient theory, moisture mode theory, and "attribution science").

I start the book by giving a synopsis of tropical circulation using the zonally symmetric model (Chap. 1). However, as useful and as reductionist as the zonally symmetric model is to describe the tropical climate, it has many limitations. Foremost, is the fact that many of the features in the tropics are zonally asymmetric. Chapter 2 introduces the Monsoon phenomenon, one of the most zonally asymmetric features. The chapter dwells largely on the Asian Monsoon but also briefly covers the other regional monsoons. The description of the tropical Indian Ocean is also included within this chapter as it is integral to the Asian Monsoon. The Indian Ocean is also briefly touched upon in Chap. 5, in the context of the Western Pacific Warm Pool. In Chap. 3, the concept of atmospheric convection and its role in the general circulation is developed following which the Inter-tropical Convergence Zone is discussed in Chap. 4. The theories on significance and maintenance of the Western Pacific Warm Pool are introduced in Chap. 5. The Madden-Julian Oscillation and its summer counterpart (the intraseasonal variations in the Asian Monsoon) are extensively discussed in Chap. 6 followed by the interannual variations of the El Niño and the Southern Oscillation in Chap. 7. The theories for these variations and their limitations are also presented in these chapters. Tropical Atlantic variations at interannual scales are discussed in Chap. 8. In Chap. 9, we discuss the large-scale controls on extreme weather. We also introduce the concept of "attribution science" in this chapter, which has been actively pursued by the community to understand the role of global warming on some high-impact weather events. Finally, we round off the textbook with a discussion on climate change in Chap. 10 that goes and revisits many of the topics in the earlier chapters in the context of a changing climate.

There are however some omissions from the book. For example, the land-atmosphere interactions in the tropics are not discussed. Similarly, the role of the variations in the stratosphere, like the quasi-biennial oscillation, is also not mentioned. Furthermore, the other regional monsoons besides the Asian monsoon are not adequately covered in the book. The topic of decadal variations is overlooked. I hope to include them in future editions of the book. But for now, the matter of this textbook came to me naturally as I have been teaching this material to my students at FSU and some variations of it for the past decade and more.

This textbook is a primer that merely scratches the surface of tropical meteorology. But it should provide sufficient background for a student to pursue deeper into the field. As Krish would say, "Tropical Meteorology has tremendous scope for exploration because the degrees of freedom are far too many to constrain." This is true. As the world moves to faster and more efficient technology, it is my firm belief that we will observe and model even better than ever to enlighten and further refine our understanding of Tropical Meteorology.

I have so many to thank for this book. First and foremost, my wife, Charu, my daughter, Mallika, and my father, Prof. P. K. Misra, for their unyielding support and constant encouragement. They have been pillars of my support system through the thick and thin of my travails in writing this book. My colleagues in the Department of Earth, Ocean, and Atmospheric Science, and the Center for Ocean-Atmospheric Prediction Studies who have, from their work, attracted some of the most enthusiastic students to our program, which has made my profession of teaching and research a pleasure at FSU. I would like to thank several of my MS and PhD students, and my post-docs and research scientists, like Dr. Amit Bhardwaj, Dr. Akhilesh Mishra, Dr. Nirupam Karmakar, Dr. C. B. Jayasankar, Dr. Steven Chan, Dr. Sarvesh Dubey, Dr. Satish Bastola, Dr. Haiqin Li, and Mr. Steven DiNapoli who have significantly contributed to my understanding of the subject. I also want to thank the vast number of anonymous reviewers of many of my journal publications and of this book, who have provided so much invaluable guidance even when some of their comments were unvarnished! Last but not the least, I would like to acknowledge support from NASA, NOAA, NSF, USGS, and the Ministry of Earth Sciences, India for their support of my work.

Tallahassee, FL, USA   Vasubandhu Misra

Constants

Latent heat of vaporization $(L_v) = 2.5 \times 10^6$ J/Kg
Latent heat of fusion $(L_f) = 3.3 \times 10^5$ J/Kg
Latent heat of sublimation $(L_s = L_f + L_v) = 2.8 \times 10^6$ J/Kg
Planetary albedo $(\alpha) = 0.3$
Density of air $= 1$ kg/m^3
Density of water $(\rho_w) = 1000$ kg/m^3
Density of seawater $(\rho_{sw}) = 1022.4$ kg/m^3
Radius of the Earth $(R) = 6370$ km
Specific heat capacity of water (Cw) $= 4184$ J/Kg/K
Specific heat capacity of seawater $(C_{SW}) = 3994$ J/Kg/K
Specific heat capacity of dry air $(c_{pd}) = 1004$ J/Kg/K
Stefan Boltzman constant $(\sigma) = 5.67 \times 10^{-8}$ Wm^{-2}K^{-4}
Solar constant $(S_o) = 1388$ Wm^{-2}

Contents

Acronyms

ABF	Angola Benguela Front
AEJ	African Easterly Jet
AEWs	African Easterly Waves
AGCM	Atmospheric General Circulation Model
AMM	Atlantic Meridional Mode
AMO	Atlantic Multi-decadal Oscillation
AMOC	Atlantic Meridional Overturning Circulation
APE	Available Potential Energy
ATLAS	Autonomous Temperature Line Acquisition System
BSISO	Boreal Summer Intra-Seasonal Oscillation
CAPE	Convectively Available Potential Energy
CEPEX	Central Equatorial Pacific Experiment
CFSR	Climate Forecast System Reanalysis
CIN	Convective Inhibition
CISK	Conditional Instability of the Second Kind
CLIVAR	CLImate VARiability and Predictability
CMIP3	Coupled Model Intercomparison Project Version 3
CMIP5	Coupled Model Intercomparison Project Version 5
CP	Central Pacific
DJF	December-January-February
ECMWF	European Center For Medium Range Weather Forecasting
EDJ	Eddy-Driven Jet
EL	Equilibrium Level
ENSO	El Niño and the Southern Oscillation
EOF	Empirical Orthogonal Function
EP	Eastern Pacific
ERSSTv5	Extended Reynolds SST Version 5
EUC	Equatorial Under Current
GARP	Global Atmospheric Research Program
GATE	GARP Atlantic Tropical Experiment
GHGs	Green House Gases

GOALS	Global Ocean-Atmosphere-Land System
HadISST	Hadley Center Sea Ice and Sea Surface Temperature
IMR	Indian Summer Monsoon Rainfall
IOD	Indian Ocean Dipole
ISM	Indian Summer Monsoon
ISO	Intra-Seasonal Oscillation
ITCZ	Inter Tropical Convergence Zone
ITF	Indonesian Through Flow
IWM	Indian Winter Monsoon
JJA	June-July-August
JAS	July-August-September
JJAS	June-July-August-September
JMA	Japanese Meteorological Agency
LCL	Lifting Condensational Level
LFC	Level of Free Convection
LHS	Left Hand Side
LIS	Lightning Imaging Sensor
MAM	March-April-May
MISO	Monsoon Intra-Seasonal Oscillation
MJO	Madden-Julian Oscillation
MMC	Mean Meridional Circulation
NAO	North Atlantic Oscillation
NASH	North Atlantic Subtropical High
NCEP-NCAR	National Centers for Environmental Prediction-National Center for Atmospheric Research
NEC	North Equatorial Current
NECC	North Equatorial Counter Current
NH	Northern Hemisphere
NOAA	National Oceanic and Atmospheric Administration
NWP	Numerical Weather Prediction
OGCM	Ocean General Circulation Model
OISSTv2	Optimally Interpolated Sea Surface Temperature version 2
OLR	Outgoing Longwave Radiation
OND	October-November-December
PBL	Planetary Boundary Layer
PDO	Pacific Decadal Oscillation
PMEL	Pacific Marine Environmental Laboratory
QG	Quasi-Geostrophy
RHS	Right Hand Side
RMM	Real-time Multivariate MJO
S2S	Sub-seasonal to Seasonal
SACZ	South Atlantic Convergence Zone
SE	Standing Eddies
SEC	South Equatorial Current
SH	Southern Hemisphere

SICZ	South Indian Convergence Zone
SLP	Sea Level Pressure
SODA	Simple Ocean Data Assimilation
SOI	Southern Oscillation Index
SON	September-October-November
SPCZ	South Pacific Convergence Zone
SRES	Special Report on Emissions Scenario
SST	Sea Surface Temperature
SSTA	SST Anomalies
STJ	Sub-Tropical Jet
TAO	Tropical Atmosphere Ocean
TBO	Tropospheric Biennial Oscillation
TC	Tropical Cyclone
TE	Transient Eddies
TEJ	Tropical Easterly Jet
TOA	Top of the Atmosphere
TOGA	Tropical Ocean-Atmosphere
TOGA COARE	TOGA Coupled Ocean-Atmosphere Response Experiment
TPOS	Tropical Pacific Observing System
TRITON	Triangle Trans-Ocean Buoy Network
TRMM	Tropical Rainfall Measuring Mission
TMI	TRMM Microwave Imager
VIRS	Visible and Infrared Scanner
WISHE	Wind-Induced Surface Heat Exchange
WRF	Weather and Research Forecast Model
XBT	eXpendable BathyThermograph

Chapter 1
A Synopsis of the Tropical Climate

Abstract This chapter gives a brief overview of the large-scale tropical atmospheric circulation using a two-dimensional framework of latitudes and pressure levels.

Keywords Atmospheric general circulation · Hadley circulation · Sub-tropical jet · Tropical easterly jet · Tropopause · Thermal wind · Geostrophy · Angular momentum · Buoyancy · Tibetan high · Clausius Clapeyron equation · Equivalent potential temperature · Hypsometric equation · Level of free convection · Virtual temperature · Mass continuity equation · Streamfunction · Velocity potential · Inertial stability · Hadley Cell/Circulation · Ferrel Cell · Polar Cell · Sawyer Eliassen equation · Ocean heat content

1.1 Observational Features in a Zonally Symmetric Framework

A reductionist or a simplistic system of modeling climate is usually the first step and, in many instances, a desired adaptation to gain a fundamental understanding of the phenomenon in question. The early theory of atmospheric general circulation started in a two-dimensional framework (Hadley 1735). This two-dimensional framework involves the meridional and the vertical dimension of the mass and circulation fields. For a variable (A) on a sphere at a given pressure level, the zonally averaged field is obtained as

$$[A] = \frac{1}{2\pi} \oint_0^{2\pi} A \, d\lambda \qquad (1.1)$$

where $A \in \{T, u, v$ or any other meteorological variable$\}$ and the square brackets on A denote the zonal average around a latitude circle and λ is the longitude. In dealing with climate, it is natural to conduct temporal averaging (represented by overbar), which leads us to

© Springer Nature Switzerland AG 2023
V. Misra, *An Introduction to Large-Scale Tropical Meteorology*, Springer
Atmospheric Sciences, https://doi.org/10.1007/978-3-031-12887-5_1

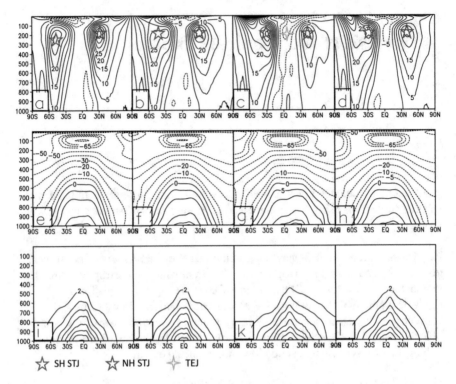

☆ SH STJ   ☆ NH STJ   ✦ TEJ

Fig. 1.1 The climatological (1979–2010) seasonal average of the zonal mean cross sections of (**a,
b, c, d**) zonal wind [ms^{-1}; contour interval is 5 ms^{-1}], (**e, f, g, h**) temperature [°C; contours of
−80 °C, −70 °C, −65 °C, −60 °C, −50 °C, −40 °C, −30 °C, −20 °C, −10 °C, −5 °C, 0 °C,
5 °C, 10 °C, 20 °C, 25 °C, 27 °C, and 30 °C are plotted], and (**i, j, k, l**) specific humidity [g kg^{-1};
contour interval is 2 g kg^{-1}] for (**a, e, i**) DJF, (**b, f, j**) MAM, (**c, g, k**) JJA, and (**d, h, l**) SON seasons
from CFSR. The ordinate in all panels is pressure in hPa

$$[A] = \frac{1}{T} \int_0^T [A] \, dt \tag{1.2}$$

Therefore, if the two operations as described in Eqs. 1.1 and 1.2 are computed at
discrete pressure levels, then zonally symmetric fields are produced as a function
of latitude (ϕ) and pressure levels. An illustration of the seasonally and zonally
averaged fields of zonal wind, temperature, and specific humidity is illustrated in
Fig. 1.1.

This two-dimensional framework allows a simple conceptualization of the
general circulation with several features of the real world that can be identified from
Fig. 1.1. For example, the following describe some of the most prominent features of
the atmospheric general circulation in a two-dimensional pressure-latitude plane:

- STJ of the middle latitudes (Fig. 1.1a–d).
- The TEJ (Fig. 1.1c).

- The easterly trade winds in the tropical latitudes (Fig. 1.1a–d).
- The contrasting lapse rates in the troposphere and stratosphere.
- The contrasting depth of the tropopause between the tropical and higher latitudes (Fig. 1.1e–h).
- The contrast in the meridional temperature gradients in the tropics (relaxed) and higher latitudes (tighter; Fig. 1.1e–h).
- The moist tropics and the comparatively drier higher latitudes (Fig. 1.1q–l).

The seasonal variations of some of these features are also very revealing, for example, the variations in the STJ across the seasons, with the winter hemisphere exhibiting the strongest display of the STJ. This is apparent in Fig. 1.1a in the NH and Fig. 1.1c in the SH. In the first order, the strong meridional temperature gradients in the winter hemisphere (Fig. 1.1e, g) set up the strong STJ by way of the *thermal wind* relationship. The seasonal variations in the strength of the STJ arise partly from the corresponding seasonal variations in the meridional gradients of the temperature. The hemispheric asymmetry with the STJ in NH winter being much stronger than the STJ in the SH winter reflects the hemispheric asymmetry in the land/ocean distribution. As there is more land mass in the NH compared to the SH, it leads to stronger meridional temperature gradients from the fact that the land surface has less heat capacity than oceans and therefore responds faster to impinging heat flux changes. Furthermore, the STJ in the NH also displays larger seasonal changes in strength and location than its SH counterpart for the same reasons. Similarly, the appearance of the TEJ in the JJA season in Fig. 1.1c reflects the buildup of the meridional land-ocean thermal contrast that sets up the Tibetan High with TEJ emanating from the southern flank of the anticyclonic flow of the high.

These time-averaged zonal mean cross sections (Fig. 1.1) not only serve as a quick, simple, and insightful introduction to the general circulation of the atmosphere but also provide a greater insight into the features that arise from the deviations about this zonal mean structure (or zonally asymmetric features).

Given the distribution of the temperature (Fig. 1.1e–h) and humidity (Fig. 1.1q–l), it is evident that the thickness of a given pressure layer is higher in the tropics than in the higher latitudes (consider the *hypsometric equation* that shows that thickness is an exclusive function of the average *virtual temperature* in the layer). As a result, a pressure gradient force gets established in a time-averaged sense with lower pressures in the tropics and higher pressures in the upper latitudes, which gets reversed in the stratosphere. Such a meridional-pressure gradient force results in an easterly flow (trade winds) in the lower latitudes of the lower troposphere following *geostrophy*. Consistent with the *mass continuity equation* and using the tropopause as a *rigid lid*, the lower tropospheric converging easterly trades and the divergent flow of the subtropics imply upper-level divergence and convergence in the tropics and the subtropics, respectively. This caricature of the circulation implies a meridional overturning circulation, with rising motion in the tropics and corresponding subsidence in the subtropics, that comprises the Hadley Cell in a two-dimensional pressure-latitude plane.

1.2   A Reductionist Model

It is tempting to overemphasize the simplicity of this two-dimensional (latitude-pressure) framework to trivialize the generation of a figure like Fig. 1.1. However, in the absence of global, gridded atmospheric reanalysis fields, the complexity of generating Fig. 1.1 from in situ observations (e.g., radiosondes launched from observing stations spread unevenly across the world) can compound very easily and can also lead to potentially erroneous conclusions as discussed in Sect. 1.4.

The construction of the temporally averaged circulation field on a latitude-pressure framework follows from defining a stream function on a sphere as

$$\frac{\partial \left[\overline{\psi}\right]}{\partial p} = [\overline{v}]\,\frac{2\pi r \cos\phi}{g} \tag{1.3}$$

$$\frac{\partial \left[\overline{\psi}\right]}{r\partial \phi} = -[\overline{\omega}]\,\frac{2\pi r}{g} \tag{1.4}$$

where $[\overline{v}]$ and $[\overline{\omega}]$ are the zonally and temporally averaged meridional wind and omega, respectively. The above two equations are defined such that they satisfy the mass continuity equation for zonally and temporally averaged flow on a sphere as given below:

$$-\frac{\tan\phi\,[\overline{v}]}{r} + \frac{1}{r}\frac{\partial\,[\overline{v}]}{\partial\phi} + \frac{1}{\cos\phi}\frac{\partial\,[\overline{\omega}]}{\partial p} = 0 \tag{1.5}$$

The streamlines defined by ψ are tangents to the flow velocity vector. In the meridional cross-section as presented in a zonally symmetric framework, $\left[\overline{\psi}\right]$ represents the meridional and the vertical motion fields so that (from rearranging Eqs. 1.3 and 1.4)

$$[\overline{v}] = \frac{g}{2\pi r \cos\phi}\frac{\partial \left[\overline{\psi}\right]}{\partial p} \tag{1.6a}$$

$$[\overline{\omega}] = -\frac{g}{2\pi r^2}\frac{\partial \left[\overline{\psi}\right]}{\partial\phi} \tag{1.6b}$$

The usefulness of the stream function lies in the fact that the flow velocity components are given by the partial derivatives of the ψ at a given spatial point and at a given instant of time. Therefore, using Eq. 1.3 with zonally and temporally averaged meridional winds on a sphere, we can compute $\left[\overline{\psi}\right]$ from the top to the bottom of the atmosphere using the boundary condition of $\left[\overline{\psi}\right] = 0$ at the top of the atmosphere and integrating to the surface, giving

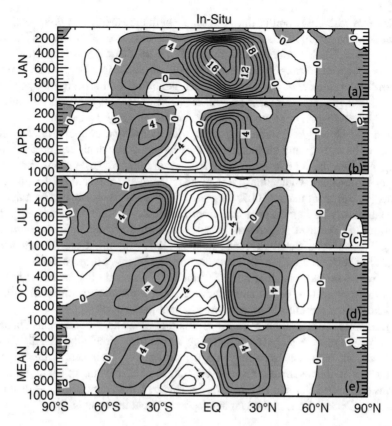

Fig. 1.2 Climatological (1968–1989) monthly mean, zonally averaged stream function ($[\bar{\psi}]$: units: 10^{10} kg/s with a contour interval of 2×10^{10} kg/s for months of (**a**) January, (**b**) April, (**c**) July, and (**d**) October, and (**e**) annual mean. The shaded values are positive values of $[\bar{\psi}]$, suggesting a clockwise circulation in the meridional-pressure plane. (Reproduced from Waliser et al. 1999)

$$[\bar{\psi}] = \int \frac{2\pi r \cos \phi}{g} [\bar{v}] \, dp \qquad (1.7)$$

Equation 1.7 provides for mass transport in kg/s in a zonal cross section shown in Fig. 1.2. In Fig. 1.2, we can identify three distinct closed, meridional overturning circulations that span the tropical, the middle, and the polar latitudes in either hemisphere. The tropical circulation is the Hadley Circulation, the middle latitude circulation is the Ferrel Cell, and that in the polar latitudes is the Polar Cell. All these circulations, by virtue of their variations in the meridional-pressure dimensions, acquire a sign, which is arbitrary but needs to be acknowledged. For example, in Fig. 1.2, the clockwise circulation in the meridional-pressure plane can be considered positive and is shaded, while the anticlockwise circulation is considered negative and appears with only contour lines of $[\bar{\psi}]$. In both the Hadley and Polar Cells, the ascent is at lower (and warmer) latitudes, and descent is over higher (and colder) latitudes. The Hadley and Polar Cells are therefore thermally direct circulations with the ascent of warm air and descent of cold air that imply heat transport along the temperature gradient. On the other hand, the Ferrel Cell is a thermally indirect circulation involving the ascent of cold air and the descent of warm air with the implied heat transport opposed to the prevalent temperature gradient.

Held and Hou (1980) indicate that *angular momentum* is approximately conserved in the overturning circulation of the Hadley Cell, which can produce very strong jets like the STJ in their poleward extremities. However, conservation of angular momentum would imply jets of near 130 ms^{-1} rather than the observed 40 ms^{-1}. But Hartmann (2007) suggests that the large-scale eddies take the momentum surplus in the upper branch of the Hadley Circulation and transport it poleward and downward to sustain surface westerlies of midlatitudes (Hartmann 2007). Furthermore, Schneider and Lindzen (1977) highlight the importance of cumulus friction in addition to latent heating in obtaining a more realistic zonal mean cross section. Schneider and Lindzen (1977) indicate that cumulus friction mixes the momentum, which strengthens the zonal mean overturning circulation and weakens the meridional temperature gradient in the tropics.

An important observation from Fig. 1.2 is that in the solstitial seasons (January and July), there is a singular overturning meridional circulation (Hadley Cell) in the tropics with the ascending and the descending branches in the summer and the winter hemispheres, respectively. On the other hand, in the equinoctial seasons (April and October in Fig. 1.2), the Hadley Circulation splits into two near symmetrical circulations in the two hemispheres about the equator.

1.3   Theory on the Maintenance of the Zonally Symmetric Circulation

Using the zonal momentum equation, geostrophic balance for the meridional momentum equation, hydrostatic approximation, mass continuity equation, and the first law of thermodynamics on a ϕ-p plane for steady-state flow, a second-order partial differential equation in $[\bar{\psi}]$ can be derived as

$$\frac{A}{r^2}\frac{\partial^2 [\bar{\psi}]}{\partial \phi^2} + \frac{2B}{r}\frac{\partial^2 [\bar{\psi}]}{\partial \phi \partial p} + C\frac{\partial^2 [\bar{\psi}]}{\partial p^2} = \cos\phi \left(\underbrace{\frac{\partial \overline{[M]}}{\partial p}}_{\text{Term 1}} - \underbrace{\frac{1}{r}\frac{\partial \overline{[H]}}{\partial \phi}}_{\text{Term 2}} \right) \tag{1.8}$$

The readers are directed to Krishnamurti et al. (2013) for a formal derivation of Eq. 1.8. Equation 1.8 is more commonly known as the Kuo-Eliassen equation and is like the *Sawyer-Eliassen equation* used in the context of frontal circulations. In a broad sense, A, B, and C in Eq. 1.8 refer to the static stability (vertical gradient of potential temperature), vertical shear of the zonal wind (or baroclinicity), and to the sum of the Coriolis Force and meridional gradient of zonal wind (or *inertial stability*), respectively. The RHS of Eq. 1.8 has essentially two forcing terms, with Term 1 essentially being the vertical gradient of the meridional gradient of momentum flux and Term 2 being the meridional gradient of diabatic heating. Equation 1.8 can be solved numerically as it generally turns out to be an elliptic equation (like a Laplacian equation) to obtain $\left[\bar{\psi}\right]$.

Equation 1.8 offers an elegant and direct way to understand the modulation of the meridional overturning circulations described by $\left[\bar{\psi}\right]$ in terms of its forcing terms. Limiting ourselves to the meridional overturning circulation in the tropics (so-called Hadley Cell), Eq. 1.8 can be rewritten after recognizing that term B is very small in the tropics (as horizontal temperature gradients are weak), and we could also substitute Eqs. 1.3 and 1.4 in Eq. 1.8 to obtain

$$-A\frac{\partial\overline{[\omega]}}{\partial\phi} + C\frac{\partial\overline{[v]}}{\partial p} = \cos\phi\left(\frac{\partial\overline{[M]}}{\partial p} - \frac{1}{r}\frac{\partial\overline{[H]}}{\partial\phi}\right) \tag{1.9}$$

Recall that the Hadley Circulation has ascent in the deep tropics and descent in the subtropics that prescribes poleward and equatorward flow in the upper and the lower troposphere, respectively. Therefore, $\frac{\partial\omega}{\partial\phi} > 0$ and $\frac{\partial v}{\partial p} < 0$. For the most part given that A and C are generally positive definite, the LHS of Eq. 1.9 is going to be negative. Therefore, the forcing on the RHS must be negative to satisfy Eq. 1.9:

$$\text{Term1}: \frac{\partial\overline{[M]}}{\partial p} \approx \frac{1}{r}\frac{\partial}{\partial\phi}\frac{\partial\left\{\left[\overline{u^*v^*}\right] + \left[\overline{u'v'}\right]\right\}}{\partial p} \tag{1.10}$$

where * and ′ refer to deviations about the zonal and the time mean, respectively. If Eq. 1.10 is discretized, for example, in the following manner to understand the forcing of the Hadley Circulation, then

$$\frac{1}{r}\frac{\partial}{\partial\phi}\frac{\partial\left\{\left[\overline{u^*v^*}\right] + \left[\overline{u'v'}\right]\right\}}{\partial p}$$

$$= \frac{1}{r}\frac{\partial\left\{\left[\overline{u^*v^*}\right]_{200\,\text{hPa}} - \left[\overline{u^*v^*}\right]_{850\,\text{hPa}}\right\} + \left\{\left[\overline{u'v'}\right]_{200\,\text{hPa}} - \left[\overline{u'v'}\right]_{850\,\text{hPa}}\right\}}{\partial\phi \times (200 - 850)} \tag{1.11}$$

although practically it is advised to avoid finite differencing over such large pressure intervals as done in Eq. 1.11! Given that $\left[\overline{u^*v^*}\right]_{200\,\text{hPa}} \gg \left[\overline{u^*v^*}\right]_{850\,\text{hPa}}$ and $\left[\overline{u'v'}\right]_{200\,\text{hPa}} \gg \left[\overline{u'v'}\right]_{850\,\text{hPa}}$, then Eq. 1.11 can be rewritten as

$$\frac{\partial M}{\partial p} \approx -\frac{1}{r} \frac{\partial \left\{ [\overline{u^* v^*}]_{200\ hPa} + [\overline{u'v'}]_{200\ hPa} \right\}}{\partial \phi \times (650\ hPa)} \tag{1.12}$$

Discretizing Eq. 1.12 further, we can, for example, rewrite the equation as

$$\frac{\partial M}{\partial p} \approx \frac{-1}{r} \frac{\left\{ [\overline{u^* v^*}]_{30N} - [\overline{u^* v^*}]_{10N} \right\} + \left\{ [\overline{u'v'}]_{30N} - [\overline{u'v'}]_{10N} \right\}}{20° \times (650\ hPa)} \tag{1.13}$$

Equation 1.13 implies $\frac{\partial M}{\partial p} < 0$ for net eddy flux of westerly momentum at the interface of subtropics and tropics. This is because the prevailing westerly momentum in the subtropic invariably implies $[\overline{u^* v^*}]_{30N} > 0$ and $[\overline{u'v'}]_{30N} > 0$ and the easterly momentum in the tropics invariably implies $[\overline{u^* v^*}]_{10N} < 0$ and $[\overline{u'v'}]_{10N} < 0$, which would ensure $\frac{\partial M}{\partial p} < 0$, following Eq. 1.13. By having $\frac{\partial M}{\partial p} < 0$, this becomes a positive forcing to the $[\overline{\psi}]$ in Eq. 1.8 because the negative value of $\frac{\partial M}{\partial p}$ satisfies the required negative forcing of Eq. 1.9 for Hadley Cell.

Similarly, Term 2 of Eq. 1.8 is going to be invariably a positive forcing to the Hadley Cell (or have a negative value to satisfy the LHS of Eq. 1.9) when the diabatic heating in the deep tropics (e.g., 10°N) is much stronger than in the subtropics (e.g., 30°N). The convective heating in the ITCZ and radiative cooling in the subtropics contribute to this meridional gradient of diabatic heating. Obviously both the forcing terms of Eq. 1.9 have a strong seasonality, which consequently also explains the seasonality of the meridional circulation of the Hadley Cell observed in Fig. 1.2.

Alternatively, Held and Hou (1980) proposed a two-level zonally symmetric model on a sphere to explain the zonally symmetric circulations. Their model used the conservation of angular momentum and thermal wind balance to estimate the strength and width of the Hadley Circulation. Although the Held and Hou model provides a reasonable width of the Hadley Cell, the strength is overestimated. Lindzen and Hou (1988) extended the Held and Hou (1980) model further by introducing asymmetric heating about the equator. They showed that as the latitude of maximum heating moved away from the equator, the winter hemisphere Hadley Cell became wider and carried much more mass flux than the summer hemisphere Hadley Cell. This variation was extremely nonlinear with a shift of the latitude of maximum heating of only 2° from the equator rendering the winter hemisphere Hadley Cell to be three times wider and carrying nearly an order of magnitude mass flux more than the summer hemisphere Hadley Cell.

1.4 Application of the Concepts of Zonally Symmetric Circulation

As mentioned earlier, the zonally symmetric framework provides a summarized view of the atmospheric general circulation features including that of the tropical features, which could be used to understand the fidelity of model simulations, atmospheric reanalysis, and also variations in climate. Waliser et al. (1999) did an enlightening study to understand the uncertainty of computing cross sections of $\left[\bar{\psi}\right]$ from atmospheric reanalysis and in situ observations. Although atmospheric reanalysis is regarded as somewhat of a revolutionary concept that ushered in extensive climate analysis, there is still abundant skepticism on the fidelity of the reanalysis. Some of this skepticism stems from the uncertainties of the data assimilation methodology used in the reanalysis, the fidelity of the model used in the reanalysis, the spatiotemporal variations in the observational density around the globe, and the changing observing systems over time and space, which can all contribute to unrealistic variations of features like the zonally symmetric circulations represented by $\left[\bar{\psi}\right]$. In contrast, in situ observations (e.g., radiosondes), while considered as a gold standard in observing the atmosphere, suffer from the varying density of such observations across spatiotemporal scales that make the rendition of fields like $\left[\bar{\psi}\right]$ from using only in situ observations very uncertain.

Waliser et al. (1999) computed $\left[\bar{\psi}\right]$ separately from the NCEP-NCAR reanalysis (Kalnay et al. 1996), monthly in situ observations, and sub-sampled reanalysis. The sub-sampled reanalysis was generated by considering grid points in the reanalysis that have at least one monthly radiosonde observation (which is made of ten or more daily observations), while all other grid points are flagged missing. The sub-sampling was done separately at each pressure level. The idea here is that by comparing the rendition of $\left[\bar{\psi}\right]$ between the NCEP-NCAR reanalysis and the sub-sampled reanalysis, one could assess the impact of in situ data sparseness on the zonally symmetric circulation. Likewise, by comparing the $\left[\bar{\psi}\right]$ field developed from in situ observations only and the sub-sampled reanalysis, we could assess the bias of the reanalysis in the zonally symmetric circulation. The broad conclusions from these comparisons were as follows:

(i) Most of the qualitative differences between NCEP-NCAR reanalysis and in situ observations can be attributed to sparse sampling and a simplified interpolation scheme to develop the gridded analysis from in situ observations. These differences largely arise from the sparse in situ observations over the oceanic and polar regions and include the following:

(a) In the equinoctial seasons, the southern Hadley Cell (with subsidence south of the equator) is shallow in the subsampled reanalysis and in the in situ observed analysis, while in the NCEP-NCAR reanalysis, it consistently extends up to 200 hPa.

(b) Greater variability in the latitude of subsidence of the Hadley Cell was found in the sub-sampled reanalysis and in the in situ observational analysis compared to the NCEP-NCAR reanalysis.

(c) Greater variability in the Ferrel and Polar Cells in the SH was observed in the in situ observed analysis and sub-sampled reanalysis compared to the NCEP-NCAR reanalysis.

(d) The in situ and the sub-sampled reanalysis overestimate the influence of ENSO on the Hadley Circulation because of their reliance on sparsely distributed observations in the oceanic regions.

(ii) The comparison between the sub-sampled and the NCEP-NCAR reanalysis revealed that the former produces a (~20%) stronger northern Hadley Cell and a (~20–60%) weaker southern Hadley Cell, reflecting a bias in the sub-sampled reanalysis, owing to poor observational coverage over the oceans.

(iii) The comparison between the in situ and the sub-sampled estimates of the meridional circulations revealed that the latter produced slightly stronger Hadley Circulations, which sometimes exceeded by ~20–30% in strength relative to the former. These differences suggest a bias in the NCEP-NCAR reanalysis to produce more energetic Hadley Circulations than what the in situ observations support.

There are several studies that are devoted to assessing the fidelity of the zonally symmetric circulations in climate model simulations (Johanson and Fu 2009; Stachnik and Schumacher 2011; Lucas et al. 2014). In the case of the climate model simulations, where the model output is available, the Kuo-Eliassen equation can be used to understand the role of its forcing terms in generating the bias in the zonally symmetric circulations of the models.

The zonally symmetric circulations can also be applied to compute the meridional transport including interhemispheric transports of the various quantities in this two-dimensional framework. The meridional flux of a quantity A is given by vA, where v is the meridional wind. Splitting v and A into time-invariant and time-varying quantity and then further splitting them into zonally averaged and zonally asymmetric components, one could deduce the following expression for temporally and zonally averaged meridional flux $\left[\overline{vA}\right]$:

$$\left[\overline{vA}\right] = \underbrace{[\overline{v}]\,[\overline{A}]}_{\text{MMC}} + \underbrace{\overline{[v']\,[A']}}_{\text{TE}} + \underbrace{[\overline{v^* A^*}]}_{\text{SE}} \tag{1.14}$$

Equation 1.14 then describes the meridional flux by temporally and zonally averaged circulation in terms of three components: MMC, TE, and SE. MMC, TE, and SE in Eq. 1.14 are Hadley Cell, Ferrel Cell (eddy flux transport by deviations of the time-invariant flow), and Polar Cell (eddy flux transport by deviations from the zonally averaged flow), respectively. Several studies have shown that the bulk of the meridional transport (e.g., of heat, momentum) from the tropics to the subtropics is carried out by the Hadley Cell, with the Ferrel Cell and the Polar Cell playing a significant role in transporting from the subtropics to the middle latitudes and from the middle latitudes to the polar latitudes, respectively. Given the cross-equatorial nature of the MMC in the solstitial seasons, it is quite interesting to understand the role of these circulations in interhemispheric transport. For example, consider the interhemispheric transport of moisture. Moisture, as measured, say, by specific humidity, resides largely in the lower troposphere. As a consequence, the direction of the lower branch of the MMC dictates the direction of the interhemispheric moisture transport. Therefore, examining the structure of the MMC for, say, boreal winter (Fig. 1.2a) or boreal summer (Fig. 1.2c) months yields the fact that the moisture is transported from the winter to the summer hemisphere. Similarly, kinetic energy, a positive definite quantity that has its largest value in the upper troposphere, gets transported by the MMC from the summer hemisphere to the winter hemisphere. Likewise, easterly momentum, which increases with height in the tropical troposphere, gets transported from the summer to the winter hemisphere by the MMC. In the equinoctial seasons, the interhemispheric transport by the MMC is not as obvious when it splits into two symmetric cells as in the solstitial seasons when the MMC acquires a single cell structure.

1.5 Limitations of the Zonally Symmetric Framework

At any point, when averaging is conducted either in space or in time or in both (e.g., $[\bar{\psi}]$), then the concept becomes abstract and deviates from realism. After all, averaging is a statistical construct. This is the reason for the commonly used adage 'climate is what you expect, weather is what you get'. Likewise, zonally symmetric features are fundamentally statistical constructs that one would not be able to directly measure, nor does it completely represent the atmospheric general circulation. However, zonally symmetric features not only provide a quick insight into the atmospheric general circulation but also give an appreciation of the zonally asymmetric features. Almost all the weather and climate phenomena that will be discussed in this book, hereafter, will invariably deal with some aspect of the zonally asymmetric feature.

  Both the sets of zonally symmetric models (Held-Hou/Lindzen-Hou, and Kuo-Eliassen), however, make use of *geostrophic balance* approximation. But at best, geostrophic balance is a crude approximation in the tropics, which fails at the equator. Another major limitation of these zonally symmetric models is that their implied thermodynamic profile of *equivalent potential temperature* (θ_e) is unsupported by observations. The mean θ_e profile in the tropics has a characteristic minimum around 600 hPa with the mean tropical atmosphere being conditionally stable above (Ooyama 1969). But the large-scale, uniform ascent at lower latitudes and poleward descent, transporting heat from the equator to poles as implied by the zonally symmetric models, will lead to an increase in θ_e in the lower troposphere and a decrease in the upper troposphere, which is contrary to observations. This type of circulation will not preserve the heat balance of the equatorial zone. However, in reality, ascending motion of the Hadley Circulation takes place in the deep convection of the relatively narrow ITCZ. The ascent of saturated air parcels in the cumulonimbus clouds of the ITCZ occurs along the pseudo-adiabats, where, θ_e is conserved and parcels arrive in the upper troposphere with moderate positive *buoyancy*. Thus, the heat balance of the equatorial zone is maintained.

The 200 hPa climatological maps of wind show several zonally asymmetrical features both in the boreal winter (Fig. 1.3a) and boreal summer (Fig. 1.3b) seasons. In the boreal winter, the major tropical asymmetric features at 200 hPa are the high pressure over northern Australia, Bolivian High, and mid-oceanic troughs in the southern subtropical Pacific and southern Atlantic Oceans (Fig. 1.3a). In the boreal summer season, the major asymmetric features at 200 hPa are the Tibetan High, TEJ, Mexican High, and mid-oceanic troughs over the northern subtropical Pacific and Atlantic Oceans (Fig. 1.3a). Similarly, at 850 hPa, the subtropical high pressure systems over the subtropical oceans and the cross-equatorial monsoonal flows are apparent asymmetrical features in both seasons (Fig. 1.3c, d). The Somali Jet off the east coast of Africa is another zonally asymmetrical feature that appears only in the boreal summer season (Fig. 1.3d). In many of these instances, the zonally asymmetrical distribution of land and ocean and topographic features (e.g.,

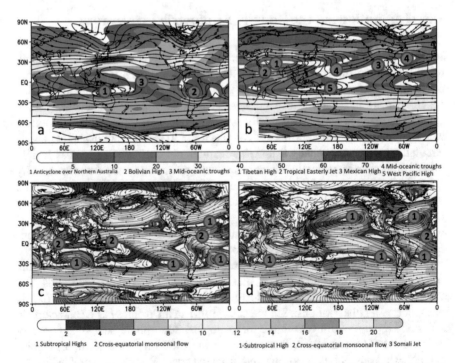

Fig. 1.3 The climatological seasonal mean (1979–2010) winds (ms^{-1}) at 200 hPa for (**a**) DJF and (**b**) JJA seasons. Similarly, the climatological seasonal mean winds (ms^{-1}) at 850 hPa for (**c**) DJF and (**d**) JJA seasons. Several asymmetrical features in the tropics are labeled beside each panel

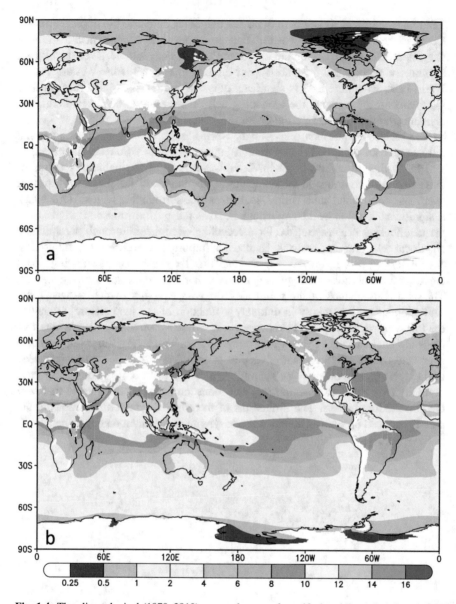

Fig. 1.4 The climatological (1979–2010) seasonal mean of specific humidity (g/kg) for (**a**) DJF and (**b**) JJA seasons at 850 hPa from CFSR

mountains, land cover) are the primary reasons for the zonally asymmetric features of the atmospheric general circulation. A prime example is that of the Tibetan High, which appears because of the uniquely high altitude of the Tibetan Plateau that also serves as an elevated heat source producing a huge dome of high pressure, with the

TEJ emanating from its southern flank and extending across to the tropical Atlantic Ocean in the boreal summer season. There are many other asymmetrical features of the general circulation that do not appear in Fig. 1.3. For example, a very significant asymmetric feature that does not appear directly in Fig. 1.3 is the Western Pacific Warm Pool, where the SST is significantly warm (>27 °C) with significant *ocean heat content* and sustains the ascending cell of the east-west Walker circulation residing over the Tropical Pacific Ocean.

The regional monsoons (e.g., Asian, North, South American, Australian, and African monsoons) are other important asymmetric features. Some of their features will be further discussed in Chap. 2. However, one of the defining features of these regional monsoons is the comparative abundance of the low-level moisture over these continental areas (Fig. 1.4a, b) from the surrounding regions. Obviously, a significant fraction of this moisture is advected by the monsoon winds from the neighboring oceans. But the local recycling of moisture through precipitation and local evaporation feedback is also an important feature of the monsoons. An interesting aspect, however, is the fact that the mean specific humidity over these continental monsoon regions is much higher than in the surrounding regions. This is a consequence of the *Clausius-Clapeyron equation*, which indicates that saturation specific humidity is uniquely a function of the temperature. Owing to the low heat capacity of the land surface, the excessive summer radiative heat flux heats the land surface to very warm temperatures. These very warm temperatures over land overlaid by constant advection of moisture-laden monsoon winds lead to the development of this very zonally asymmetric feature of the high, low-level humidity over these continental, regional monsoon regions compared to that over the surrounding oceans. The limitations of the zonally symmetric framework are clearly exposed by such an abundance of asymmetry prevalent on our planet.

Chapter 2
The Monsoon

Abstract This chapter introduces the monsoon with emphasis on the Asian Monsoon. The chapter motivates the importance of the Asian Monsoon on the atmospheric general circulation. The tropical Indian Ocean variability in reference to the seasonal cycle of the Asian Monsoon is discussed and the coupled ocean-atmosphere phenomenon of the monsoon is detailed. The role of the Tibetan Plateau on the Asian Monsoon variability is discussed followed by the description of the ENSO-monsoon teleconnection. Finally, the other regional monsoons are briefly discussed.

Keywords Monsoon · Mascarene High · Great Whirl · Somali Jet · Wyrtki Jet · Mixed layer depth · Ekman layer · Cross-equatorial flow · Arabian Sea · Bay of Bengal · Flux · Intraseasonal oscillation · Wind stress · Biennial oscillation · Indian Ocean Dipole · Inter-tropical convergence zone · Virtual temperature · Rossby waves · Tibetan Plateau · Sensible heat flux

2.1 The Asian Monsoon

The etymology of the word 'monsoon' suggests that its origins are from the Arabic word 'mausim', which refers to anything that appears once a year. It was the name given to the periodic reversal of winds in the Indian Ocean by the Arab sailors a few hundred of years ago. It is now established that this name *mausim* is external to indigenous ideas and only considers the seafarer's concept of periodic reversal of winds. This came about from the commercial interests that sustained active trade of spices, silk, and other commodities, which were dependent on sailships entering and exiting the Indian Ocean during favorable periods of the monsoon winds. Sanskrit and folk literature, however, acknowledge monsoons as periodic reversals associated with sun and rain but not winds. In any case, this brief introduction to the etymology of monsoon is to suggest that for the longest time, monsoon referred to the Indian Monsoon. Over the last several decades, the word 'monsoon' has gathered other flavors with several regional monsoons being identified e.g., the West African Monsoon, the Australian Monsoon, the Southeast Asian Monsoon,

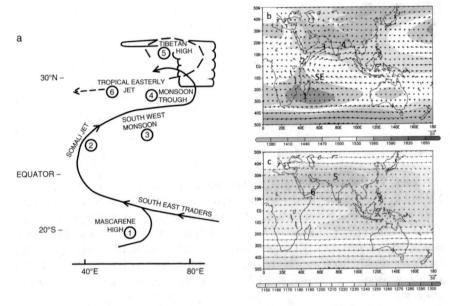

Fig. 2.1 (**a**) A schematic of the characteristic circulation features of the ISM and the corresponding climatological mean JJA (**b**) 850 hPa and (**c**) 200 hPa wind circulation and geopotential height from CFSR. The wind speed is in m/s and 850 hPa and 200 hPa heights are in meters and decameters, respectively. The numbers overlaid in panels (**b**) and (**c**) correspond to those in panel (1) and SE in panel (**b**) is Southeast Trades. (Panel (**a**) is reproduced from Krishnamurti et al. 2013)

the South American Monsoon, the North American Monsoon, etc.). A persuasive case for calling each of these as a monsoon has been made in the relevant articles that ascribe to some form of seasonal reversal of winds, the strong seasonal cycle of rainfall, and the associated circulation features with the local topography playing a critical role in the seasonal evolution as an elevated heat source, although this latter feature of the role of topography is being rigorously debated (Chou et al. 2001; Boos and Kuang 2010, 2013).

In this chapter, we will largely dwell on the Asian Monsoon as some of its features have a global imprint (Fig. 2.1). A large fraction of this chapter is dedicated to understanding the influence of the Asian Monsoon on the global climate variations. It is this global influence of the monsoon that endears itself as part of any introductory textbook in tropical meteorology.

Although the planetary scale features of the Asian Monsoon are easily recognizable, Asian Monsoon is also classified as part of two subsystems: the Indian and East Asian Monsoons (Ding and Chang 2005). Other studies classify the Asian Monsoon into three subsystems: Indian, East Asian, and West North Pacific Monsoons (Wang and LinHo 2002). If you were to consider the full seasonal cycle, then the Asian Monsoon is combined with the Australian Monsoon to be called the Asian-Australian Monsoon system. Regardless of these inhomogeneities and the distinctions therein, the Indian Monsoon is regarded to be a pronounced feature of the Asian-Australian Monsoon system (Kirtman and Shukla 2000). Furthermore, the variability of the Indian Monsoon can be thought of as the variability of the broader Asian-Australian Monsoon system.

To get a sense of the global implication of the Indian Monsoon, consider the following statistics for the All-India average:

The average amount of rainfall during the ISM = 887.36 mm (Noska and Misra 2016)

The average length of ISM in days = 124 (Noska and Misra 2016)

Therefore, average daily rain rate of the ISM= $\frac{887.36}{124}$ = 7.2 mm/day or 0.0072 m/day

Surface area of India =3.287 $\times$ 10^{12} m^2

So the amount of latent heat energy generated (from condensation) per day from the ISM

$\approx$ L$_v$ $\times$ average rain rate $\times$ surface area of India $\times$ density of water

= 2.5 $\times$ 10^6 $\times$ 0.0072 $\times$ 3.287 $\times$ 10^{12} $\times$ 1000

= 5.9 $\times$ 10^{19} joules/day = 6.8 $\times$ 10^{14} joules/s = 6.8 $\times$ 10^{14} watts

To give a perspective of this energy generation from one typical season of the ISM, compare it with the global electricity consumption in 2019, for example, which was about 22.85 $\times$ 10^{12} watts (https://en.wikipedia.org/wiki/ Electric_energy_consumption%23:̈text=Electric%20energy%20consumption%20 is%20the,2017%20it%20was%2021%2C372%20TWh. So the energy generation in the ISM year after year is ~30 times of the global electricity generation of the present time! It turns out that this energy generated by the ISM is quite comparable to a typical hurricane with a rain rate of, say, 15 mm/day over a radius of, say, 665 km, which gives rise to about 6.0 $\times$ 10^{14} watts. But the difference here is that hurricanes are episodic while the monsoons have been happening year after year!

2.2   Broad-Scale Features of the Asian Monsoon

The ISM circulation features in the lower troposphere (say, 850 hPa; Fig. 2.1a, b) comprise the following:

- Mascarene High with its counterclockwise flow centered around 20°S and 60°E.
- The Southeast trade winds of the Southern Indian Ocean that merge with the flow of the Mascarene High, which leads to:
- The cross-equatorial flow along the Eastern African Highlands with southerlies at ~10 ms^{-1}, which leads to:
- The Somali Jet or the Findlater Jet (after Findlater 1969) is a continuation of the cross-equatorial flow that accelerates as it exits from the east, off the Horn of Africa with episodic wind speeds reaching over 50 ms^{-1} and otherwise climatologically at ~15 ms^{-1}.

- The southwesterly flow over the Arabian Sea is the iconic lower tropospheric circulation of the ISM, which wraps around to form the monsoon trough over the Bay of Bengal and North-Central India.
- There is significant rainfall along the Western Ghats in the southwest corner of India and along the eastern foothills of the Himalayas.

In the upper troposphere (~200 hPa; Fig. 2.1a, c), the large anticyclone of the Tibetan High is clearly apparent. The TEJ with wind speeds upwards of 10 ms^{-1} is also clearly visible as it emanates from the outflow of the Tibetan High. These features have nearly a 180° longitudinal span and are identified as planetary scale features of the monsoon. The TEJ appears as part of the zonally symmetric upper-level easterlies (cf. Fig. 1.3).

The ISM by way of its off-equatorial location also has a bearing on the Hadley Circulation following the Kuo-Eliassen equation (cf. Sect. 1.2). For example, as the ISM intensifies, the meridional gradient of the diabatic heating is modulated in such a way that it could weaken the zonally symmetric Hadley Circulation. On the other hand, the exchange of the momentum flux at the interface of the tropical-subtropical latitudes could enhance or weaken the TEJ which will imply on the consequent strength of the Hadley Circulation and the ISM, by way of the Kuo-Eliassen equation (cf. 1.2). Either way, it is yet another example of the impact of the ISM on the general circulation.

The nearly symmetric opposite of the ISM is the IWM circulation. However, note the eastward shift of the domain in the winter relative to the summer of the Indian Monsoon between Fig. 2.2a, b and Fig. 2.1a, b, respectively. The IWM has the following defining features in the lower troposphere (Fig. 2.2a, b):

- The Siberian High-pressure system centered ~50°N and 125°E which serves as a counterpart of the Mascarene High of the ISM.
- The northeasterly flow serves as a counterpart of the southeasterly trades of the ISM.
- The wind surges of the northeasterly flow are observed along the western shores of the South China Sea with speeds nearing 15 ms^{-1}. These surges represent the southeastward advance of the cold Siberian air mass across the East Asian coast with rapid temperature drops associated with cold outbreak episodes.
- This north-northeasterly flow wraps itself around a monsoon trough that gradually migrates from north of the equator in December to south of the equator in January and February, which becomes then part of the Australian Summer Monsoon.

In the upper troposphere, at ~200 hPa (Fig. 2.2a, c), the West Pacific High of the IWM is a counterpart of the Tibetan High of the ISM. The STJ forms at the outer, northern flank of the West Pacific High, south of Japan. Although the Tibetan High forms over a region of high altitude (Tibetan Plateau, which serves as an elevated heat source in the summer), the West Pacific High forms over the region of very high rainfall and over the Western Pacific Warm Pool region. Again, the wide longitudinal

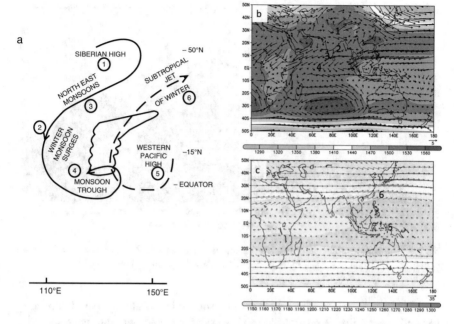

Fig. 2.2 (**a**) A schematic of the characteristic circulation features of the IWM and the corresponding climatological mean DJF (**b**) 850 hPa and (**c**) 200 hPa wind circulation and geopotential height from CFSR. The wind speed is in m/s and 850 hPa and 200 hPa heights are in meters and decameters, respectively. The numbers overlaid in panels (**b**) and (**c**) correspond to those in panel (1). (Panel (**a**) is reproduced from Krishnamurti et al. 2013)

expanse of the Tibetan High, the Siberian High, the West Pacific High, and the STJ clearly reflects the planetary scale features of the IWM.

2.3   The Evolution of the Indian Summer Monsoon

The ISM is very well known through the arrival of the monsoon rains in the southwest corner of India, in the coastal province of Kerala. For example, Ananthakrishnan and Soman (1988) note that on the day of the onset, the rain rate dramatically increases from around 5 mm/day before onset to nearly 20 mm/day by the day of the onset date and thereafter (Fig. 2.3). This onset of the ISM is closely monitored by the Indian Meteorological Department as well as the rest of the world since the agricultural yield of the commodity crops grown in India is critically dependent on the rains from the ISM. Given the very stable nature of the ISM with its existence from at least 7 to 9 million years ago when the Tibetan Plateau achieved its modern elevation and lateral extent (Clemens 2006), there is sufficient confidence that after the arrival of the summer monsoon rains in Kerala, the rest of the country

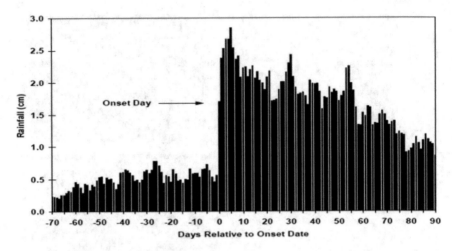

Fig. 2.3 The composite of daily rainfall climatology (cm) as a function of days relative to the onset of the monsoon in Kerala, India. (Reproduced from Krishnamurti et al. 2013)

would receive its share of the summer monsoon rains in due course of time (Fig. 2.4a). Similarly, the demise of the ISM is systematic and periodic, year after year (Fig. 2.4b). In Fig. 2.4a, one can note the early onset of the East Asian Monsoon, which is in early May, before the onset in Kerala. Similarly, the onset isochrones in the northwest part of India appear in late June, almost a month after the onset of the ISM rains in Kerala (Fig. 2.4a). The ISM withdrawal isochrones in Fig. 2.4b indicate that the rains begin to withdraw at the earliest from the northwest part of India and gradually recede in the southeast direction. The standard deviation in the onset isochrones is in the range of 10–20 days (Fig. 2.4a) and slightly higher for the withdrawal isochrones (~20–30 days; Fig. 2.4b). There are variants to this depiction of the onset and demise of the Asian Monsoon (e.g., Misra and DiNapoli 2014; Noska and Misra 2016; Bombardi et al. 2019), which, however, are still comparable to the isochrones shown in Fig. 2.4.

This evolution of the ISM rains and their withdrawal are coincident with the evolution of the circulation field both in the lower and in the upper troposphere (Figs. 2.1 and 2.2). One of the iconic features of the onset of ISM is the increase in the kinetic energy of the winds on the planetary scales (Fig. 2.5), which yet again reminds us of the global impact of the ISM. To construct Fig. 2.5, Krishnamurti et al. (2013) used Fourier decomposition on the zonal and the meridional winds at discrete vertical levels and latitude circles (over the belt of 0–30°N) and then vertically integrated them to compute the average kinetic energy ($\overline{\overline{K}}_n$) as

$$\overline{\overline{K}}_n = \frac{1}{\int_{P_{\text{top}}}^{P_{\text{surf}}} \int_{y_1}^{y_2} dydp} \int_{P_{\text{top}}}^{P_{\text{surf}}} \int_{y_1}^{y_2} \frac{u_n^2 + v_n^2}{2} dydp \qquad (2.1)$$

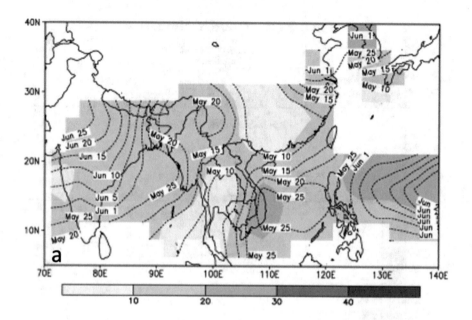

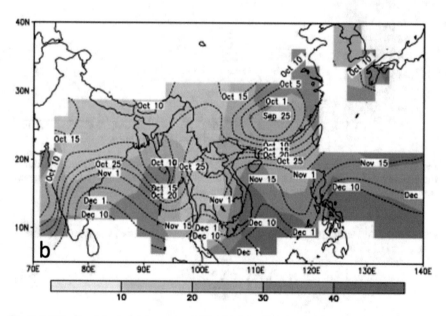

Fig. 2.4 The climatological (**a**) onset and (**b**) withdrawal isochrones (contours) and their corresponding standard deviation (shaded; in days) of the Asian Summer Monsoon based on the period (1979–1999). (Reproduced from Janowiak and Xie (2003); published (2003) by the American Meteorological Society)

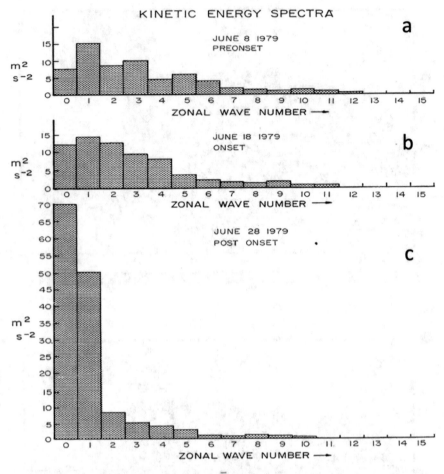

Fig. 2.5 The average kinetic energy spectra $((\overline{\overline{K}}_n)$ from Eq. 2.1; units m^2s^{-2}) as a function of zonal wave number computed over one season in 1979 for (**a**) pre-onset, (**b**) onset, and (**c**) post-onset time of the ISM. (Reproduced from Krishnamurti et al. 2013)

where the Fourier decomposition of the u and v components of the winds yields

$$u\left(\lambda\right) = \sum_{n=0}^{N} u_n e^{in\lambda} \tag{2.2a}$$

and

$$v\left(\lambda\right) = \sum_{n=0}^{N} v_n e^{in\lambda} \tag{2.2b}$$

where λ is the zonal wave number and u_n and v_n are the Fourier coefficients of nth wave number of the Fourier decomposition of the u and v components of the wind, respectively. p_{top} and p_{surf} are pressures at the top and the surface of the atmospheric column, and y_1 and y_2 are the latitude band over which the integral in Eq. 2.1 is computed. The average kinetic energy in Eq. 2.1, computed for a single year (1979), suggests an explosive growth at the time of the onset (Fig. 2.5b) from the pre-onset time (Fig. 2.5a), which is sustained in the post-onset time (Fig. 2.5c). This planetary scale signature of the onset of the ISM in the integrated kinetic energy (Eq. 2.1 and Fig. 2.5b) is symptomatic of the evolution of the cross-equatorial flow and the southwesterly flow in the lower troposphere and the development and sustenance of the TEJ in the upper troposphere of the ISM. In the boreal summer season, about half of the hemispheric exchange of mass flux occurs through the region of the Somali Jet, which is just 10% of the Earth's circumference (Findlater 1969). This yet again reminds us of the importance of the Asian Monsoon to the atmospheric general circulation.

2.4 The Tropical Indian Ocean

The Tropical Indian Ocean is an integral part of the Asian Monsoon that has several unique features that are unmatched in the other tropical oceans. For instance, the Tropical Indian Ocean is the only tropical ocean where the annual mean winds are westerly. Because of the strong seasonal cycle of the monsoon winds, the upper ocean of the Tropical Indian Ocean displays a rich variability across time and space scales, which interplay with the overlying atmosphere to produce a complex coupled ocean-atmosphere monsoon climate.

The schematic of the surface ocean currents in the Tropical Indian Ocean is shown in Fig. 2.6. During the summer monsoon, the SEC and the East African Coastal Current feed into the Somali Current, which bifurcates at ~4°N, flowing offshore eastward while the other branch recirculates across the equator as the Southern Gyre (Fig. 2.6a). Another couple of gyres (Great Whirl and Socotra Eddy) are formed further north, which all result in the deepening of the mixed layer and warming of the upper 100 m of the ocean from the resulting downwelling. Schott and Quadfasel (1982) indicated from observations that the appearance of the Great Whirl at the time of the onset of the summer monsoon was a response to the very strong anticyclonic wind stress curl, offshore from the Somali Coast. The Great Whirl gradually spins down, although it is still weakly discernible underneath at the time of the onset of the IWM circulation (Bruce et al. 1981).

In contrast, during the IWM, the confluence of the East African Coastal Current with the Somali Current at ~2°S-4°S leads to the eastward flowing South Equatorial Counter Current (Fig. 2.6b). Furthermore, the coastal currents including the West Indian Coastal Current and the East Indian Coastal Current undergo reversal from ISM (Fig. 2.6a) to the IWM (Fig. 2.6b) regime.

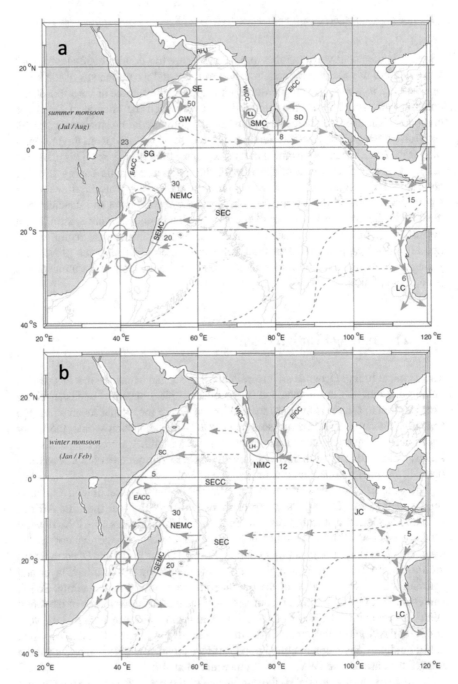

Fig. 2.6 Schematic of the surface ocean currents in the Tropical Indian Ocean during (**a**) ISM and (**b**) IWM with discharges indicated at some locations in red (in Sv $= 10^6$ m^{3}s^{-1}). The surface currents indicated in the figure are South Equatorial Current (SEC), South Equatorial Countercurrent (SECC), Northeast and Southeast Madagascar Current (NEMC and SEMC), East African Coastal Current (EACC), Somali Current (SC), Southern Gyre (SG) and Great Whirl (GW) and associated upwelling wedges, Socotra Eddy (SE), Ras al Hadd Jet (RHJ) and upwelling wedges off Oman, West Indian Coast Current (WICC), Laccadive High and Low (LH and LL), East Indian Coast Current (EICC), Southwest and Northeast Monsoon Current (SMC and NMC), South Java Current (JC), and Leeuwin Current (LC). (Reproduced from Schott and McCreary 2001)

Owing to the strong semi-annual surface eastward winds along the equator in the Indian Ocean, there is a strong appearance of eastward jets at the surface, which are called the equatorial or the Wyrtki Jets (Wyrtki 1973). The strong Wyrtki Jets (~100 cm/s) appear in the boreal spring and fall seasons. These jets carry warm upper-layer waters eastward, lower the sea level and the mixed layer depth in the Western Indian Ocean, and raise the sea level and deepen the mixed layer in the Eastern Equatorial Indian Ocean. Using a hierarchy of ocean models, Han et al. (1999) concluded that Wyrtki Jets are driven by the semi-annual component of the surface winds, accounting for nearly 81% of the amplitude of the jets. They also found that the freshwater influx during the boreal summer and fall seasons, which leads to the shoaling of the mixed layer depth along the Equatorial Indian Ocean, is also important in strengthening the Wyrtki Jet.

The annual mean equatorial undercurrent of the Indian Ocean is westward. An eastward equatorial undercurrent in the Indian Ocean is consistently observed in the boreal spring. This is unlike the other equatorial oceans where the equatorial undercurrent is typically eastward. Furthermore, the equatorial undercurrent in the Indian Ocean is transient because of the strong seasonally varying surface winds. This contrasts with the quasi-steady equatorial undercurrents in the Pacific and Atlantic Oceans under the prevailing easterly trade winds.

2.5   The Coupled Ocean-Atmosphere Phenomenon of the Monsoon

An important factor of the Northern Indian Ocean to consider is that it receives around 90 Wm^{-2} of net heat flux from the atmosphere during the spring season that far exceeds (by at least a factor of two or more) that is received by the Western Pacific Warm Pool (Webster 2006). And yet the Northern Indian Ocean is significantly cooler than the Western Pacific Warm Pool.

A simple back-of-the-envelope calculation for the warming rate of the upper 50 m of the North Indian Ocean in the boreal spring season, assuming that the heat flux received at the surface of the ocean is evenly distributed through the upper ocean layer (say, *mixed layer depth*), yields some interesting results. For example, for an average heat flux of 90 Wm^{-2} that impinges on the surface of the North Indian Ocean, it translates to a warming rate of ~13.4 Kyr^{-1} ($= \frac{dT}{dt} = \frac{1}{\rho_{SW} C_W} \frac{dF_{net}}{dz} = \frac{1}{1020 \times 3994} \times \frac{90}{50} \times 86400 \times 365$, where ρ_{SW}, C_{SW} are density and specific heat capacity of seawater. Likewise, Webster (2006) indicates such high but unrealistic warming rates over the Northern Indian Ocean for other seasons of the year and an annual mean heating rate of 7 Kyr^{-1} for an annual mean heat flux of 50 Wm^{-2}. It should be noted that the atmospheric heat flux over the Northern Indian Ocean displays a very strong seasonal cycle dictated by the downwelling shortwave flux peaking in boreal spring, while over the Western Pacific Warm Pool region, the seasonal cycle of the atmospheric heat flux is relatively weaker. These warming rates suggest that the heat flux received by the Northern Indian Ocean is nontrivial. Of course, these heating rates are not realized. It should be noted that the strong seasonal cycle of the SST and the upper North Indian Ocean cannot be achieved with such high warming rates. Furthermore, the diagnosed warming rate is far greater than the observed heating rate ($\sim$0.02 Kyr^{-1}; Gnanaseelan et al. 2017). Obviously, we are missing something important in the heat budget equation, which is explored in the following passages.

The ocean dynamics involving ocean currents are a significant component of the heat balance of the Northern Indian Ocean (Godfrey 1996). In other words, the mass flux of import of cold and export of warm water from the upper North Indian Ocean and/or the thickening of the mixed layer is necessary to allow for the large atmospheric flux to impinge on the surface of the ocean without raising the SST substantially. The regulation of SST by clouds following the mechanism of Ramanathan and Collins (1991) or that by surface wind evaporation feedback Wallace (1992) is not significant as convection and evaporation are weak during the spring and early summer months over the Northern Indian Ocean when the atmospheric heat flux is at its annual peak.

The importance of the meridional ocean heat transport in regulating the seasonal cycle is clearly illustrated in Fig. 2.7. Loschnigg and Webster (2000) using an ocean model that displayed reasonable fidelity in simulating the Tropical Indian Ocean when forced with observed atmospheric forcing showed that the meridional heat flux, M, along a line of constant latitude, in the upper ocean, given by Eq. (2.3) is a significant component of the seasonal cycle of the heat budget of the Tropical Indian Ocean.

$$M = \rho_W C_W \int_x \int_z TV dx dz \qquad (2.3)$$

and

$$S = \rho_W C_W \int_y \int_x \int_z \frac{\partial T}{\partial t} dy dx dz \qquad (2.4)$$

Simply, then ignoring the lateral transports (e.g., through the Indonesian Throughflow)

$$S \cong M + F_{atm} \qquad (2.5)$$

where T, V, S, and F_{atm} are ocean temperature, meridional ocean current component, heat storage in the upper ocean, and total atmospheric heat flux, respectively. Loschnigg and Webster (2000) zonally integrated Eq. 2.3 between longitudes of 35°E and 115°E and used 29°S to 25°N as meridional boundaries in their study.

In Fig. 2.7a, the net atmospheric heat flux shows a semi-annual variation with dual seasonal peak in the spring and fall seasons and a dual minimum in the summer and winter seasons. The dual minimum in the net atmospheric heat flux is a result of the reduction in downwelling shortwave flux from increased monsoon

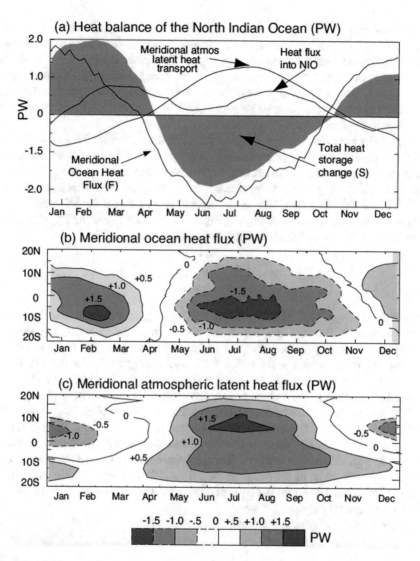

Fig. 2.7 (**a**) The meridional ocean heat transport, the ocean heat storage, the meridional atmospheric heat transport, and the net atmospheric heat flux into the Northern Indian Ocean. The latitude time cross section of the zonally averaged between 35°E and 115°E climatological (**b**) meridional heat transport and (**c**) latent heat transport. All units are in Petawatts (PW = 10^{15} W). (Reproduced from Webster 2006)

clouds in the summer and because of the lower solar declination angle in the winter. Furthermore, the monsoon winds are strong in the summer and in the winter leading to strong evaporation. Figure 2.7a suggests that the oceanic meridional heat flux is for the most part opposite in sign to the atmospheric heat flux. So we observe a strong southward heat flux of M in late March through October of the year and northward flux of M from November through the end of March of the following year (Fig. 2.7a). Similarly, the atmospheric heat flux is positive from mid-February through October and becomes negative from November through January of the following year (Fig. 2.7a). The heat storage in the upper North Indian Ocean reaches a minimum and is below the annual mean from mid-April through mid-October and reaches a maximum around February and is above the annual mean from mid-October through mid-April (Fig. 2.7a).

The latitudinal cross section of M in Fig. 2.7b indicates that the year is divided clearly between the time of southward flux (spring and summer) and northward flux (late fall and winter). The strongest meridional flux occurs around 10°S in all seasons, which is near the zone of maximum meridional SST gradient (Webster 2006; Fig. 2.7b). Webster (2006) observed that the meridional flux of latent heat in the overlying atmosphere (L in Eq. 2.6; Fig. 2.7c) is nearly equal and opposite to the meridional heat flux in the upper north Indian Ocean (Fig. 2.7b).

$$L = \frac{Lv}{g} \int_x \int_{p_{top}}^{p_{bot}} qv \, dx \, dp \tag{2.6}$$

where L_v, q, p_{top}, and p_{bot} are latent heat of vaporization, specific humidity, and pressure at the top and bottom of the atmospheric column, respectively. In other words, Fig. 2.7b, c is very powerfully conveying the coupled ocean-atmosphere phenomenon of the Indian Monsoon by suggesting that the amount of heat transported by the Indian Ocean from the summer to the winter hemisphere is equal and opposite to the amount of heat transported by the overlying atmosphere. It is to be appreciated that this coupling is accomplished through the wind-driven ocean heat transport and atmospheric circulation that is driven by surface heat flux and buoyancy gradients. This led Loschnigg and Webster (2000) to the schematic shown in Fig. 2.8, which suggests that the wind-driven Ekman Ocean transport, which is to the right of the wind in the NH and left of the wind in the SH, is driving the cross-equatorial ocean heat transport. The expression for the Ekman Ocean transport is provided below.

In the boundary layer, the horizontal momentum equations are

$$fv = \frac{1}{\rho}\frac{\partial p}{\partial x} + F_x \tag{2.7}$$

$$fu = -\frac{1}{\rho}\frac{\partial p}{\partial x} + F_y \tag{2.8}$$

where f, u, v, p, ρ, F_x, and F_y are Coriolis force, zonal and meridional winds, pressure, density, and zonal and meridional components of friction, respectively. Equations 2.7 and 2.8 can be written in vector form:

$$f\vec{V} = f\left(\vec{V}_g + \vec{V}_{ag}\right) = \hat{k} \times \frac{1}{\rho}\nabla P + \vec{F} \tag{2.9}$$

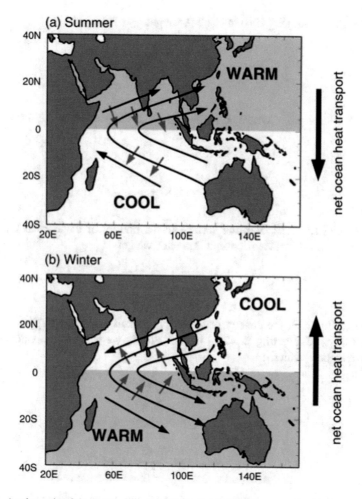

Fig. 2.8 A schematic of the seasonal cycle of the Indian Monsoon system for (**a**) summer (JJAS) and (**b**) winter (DJF). The curved arrows denote the atmospheric winds driven by the large-scale thermal contrast denoted by 'warm' and 'cool'. The gray arrows are the Ekman transport vectors forced by the winds. The large vertical arrow to the right of each panel denotes the direction of the net upper ocean heat transport. (Reproduced from Loschnigg and Webster (2000); © American Meteorological Society. Used with permission)

where $\overrightarrow{V}_g$ and $\overrightarrow{V}_{ag}$ are geostrophic and ageostrophic components of the winds. But from Geostrophic balance, we know that

$$f\overrightarrow{V}_g = \hat{k} \times \frac{1}{\rho}\nabla P \qquad (2.10)$$

So by substituting Eq. 2.10 in Eq. 2.9, we get

$$f \overrightarrow{V}_{ag} = \overrightarrow{F} = \hat{k} \times \frac{1}{\rho} \frac{\partial \overrightarrow{\tau}}{\partial z} \tag{2.11}$$

But $\overrightarrow{F} = \hat{k} \times \frac{1}{\rho} \frac{\partial \overrightarrow{\tau}}{\partial z}$, where $\overrightarrow{\tau}$ is the horizontal wind stress on the surface of the ocean.

Therefore,

$$f \overrightarrow{V}_{ag} = \hat{k} \times \frac{1}{\rho} \frac{\partial \overrightarrow{\tau}}{\partial z} \tag{2.12}$$

Integrating Eq. 2.12 from the surface to the bottom of the Ekman layer (Z_e), where the effect of surface friction disappears, we get

$$f \overrightarrow{V}_{Ekman} = \hat{k} \times \frac{\overrightarrow{\tau}}{\rho_o} \tag{2.13}$$

Here in Eq. 2.13, the density of water, ρ, is taken out of the integral and held constant to ρ_o. Converting Eq. 2.13 in scalar form, we get the components of the Ekman transport, which are

$$u_{Ekman} = \frac{\tau_y}{f\rho_o} \tag{2.14}$$

and

$$v_{Ekman} = -\frac{\tau_x}{f\rho_o} \tag{2.15}$$

where τ_x and τ_y are zonal and meridional wind stress, respectively. In the upper ocean, a large fraction of the ocean heat transport is accomplished by the Ekman Ocean transport (Eq. 2.13) between the latitudes of 4°N–10°N and by the geostrophic component north of 10°N (Loschnigg and Webster 2000).

It may be noted that Ekman theory breaks down as one approaches the equator. Sahami (2003), however, suggests a Rossby duct hypothesis for the cross-equatorial heat transport in the Indian Ocean (Fig. 2.9). Sahami (2003) notes that pairs of Rossby waves straddling the equator form a wave train between the Ekman transport that begins around 6°S and 6°N (Fig. 2.9). The Rossby wave structures that straddle the equator (where the Ekman transport becomes infinitesimally small) are the dominant equatorial waves forced by the cross-equatorial winds of the monsoon flow. These equatorial Rossby waves allow the transport across the equator through the ducts displayed in Fig. 2.9. A composite of the meridional ocean heat transport computed as a function of lead/lag around the date of the onset and demise of the ISM rainfall shows the continuity of the heat transport across the equator (Fig.

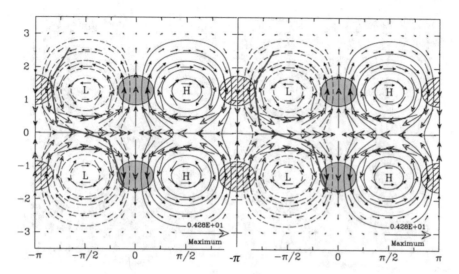

Fig. 2.9 The horizontal structure of zonally propagating equatorial Rossby wave (for $n = 1$) overlaid with red streamlines indicating the Rossby duct mechanism. (Adapted from Kiladis et al. 2009)

2.10; Noska and Misra 2016). Figure 2.10a suggests that about 120 days before the onset, the southward heat transport is confined to latitudes south of 12°S. But as the ISM evolves, the southward upper oceanic heat transport advances further north and dominates across 25°S to 20°N post onset of the ISM. Similarly, as the ISM retreats, the southward heat transport also retreats southward (Fig. 2.10b). The Wyrtki Jets (both in the spring and the fall seasons) exhibit significant intraseasonal and interannual variations. These perturbations (or anomalies) of the jets after arriving at the eastern end of the equatorial Indian Ocean Basin are reflected as Rossby waves that propagate poleward in either hemisphere as eastern boundary waves. One set of these waves radiates northward into the Bay of Bengal and the eastern Arabian Sea (Perigaud and Delecluse 1992; Basu et al. 2000). The other set of southward-propagating waves affects the Indonesian boundary and throughflow regime (Sprintall et al. 2000). Therefore, the Rossby duct hypothesis proposed in Sahami (2003) for the cross-equatorial heat transport in the Indian Ocean is quite relevant in the context of the variations of the Wyrtki Jets.

2.6 The Tibetan Plateau

One of the defining aspects of the monsoon is the differential land-ocean thermal contrast that leads to a large-scale wind response (Fig. 2.8). It is argued that seasonality of rainfall alone does not construe a monsoon-like regime. For example, the trade-wind regime could have a very strong seasonal cycle in precipitation,

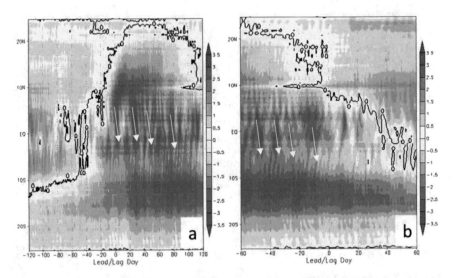

Fig. 2.10 The climatological zonally averaged meridional ocean heat transport (0.1 PW) computed as lead/lag function of the (**a**) onset date and (**b**) demise date of the IMR from CFSR. The positive and negative values on the x-axis correspond to the meridional ocean heat transport leading and lagging the (**a**) onset and (**b**) demise dates, respectively. The shaded positive and negative values denote northward and southward meridional ocean heat transport, respectively. The negative and positive values on the x-axis refer to the meridional heat transport leading and lagging the (**a**) onset and (**b**) demise date of the Indian summer monsoon rains, respectively. The white arrows in the two panels emphasize the diagonal alignment of the ocean heat transport across the equator, which is similar to the Rossby duct mechanism following Sahami (2003) and also illustrated in Fig. 2.9. (Adapted from Noska and Misra 2016)

almost monsoonal in appearance with strong ocean-land-atmosphere interactions. But the dynamics of the trade wind regimes and the monsoonal wind regimes are not the same.

The Tibetan Plateau is a massive orographic feature roughly covering a domain of 25°N–45°N and 70°E–105°E with a mean elevation of about 4000 m above sea level. The Tibetan Plateau has mechanical and thermodynamical impacts on the atmospheric circulation. In the former, the Plateau diverts the circulation over and around it while in the latter by its high elevation serves as an elevated heat source, providing a huge thermal contrast to elicit a wind response. The early studies focused on the mechanical forcing of this topography in splitting the STJ (e.g., Bolin 1950). Some of these early works identified the evolution of the Asian Monsoon based on the splitting of the STJ (e.g., Murakami 1951, 1958; Matsumoto 1990). For example, Matsumoto (1990) noted that the withdrawal of the ISM coincided with an abrupt temperature drop over the Eurasian continent in late September and a sudden southward shift of the STJ around the Tibetan Plateau.

A detailed heat budget analysis of the Tibetan Plateau conducted by Yanai and Li (1994) revealed that in the pre-onset period of the ISM, strong sensible heat flux (which can reach up to 100 Wm^{-2}) from the large ground-air temperature difference is the primary source of diabatic heating. This diabatic heating from sensible heat flux in the boundary layer of the atmosphere is balanced by cold horizontal advection from the prevailing westerlies over the Plateau during the pre-monsoon period. However, after the onset of the (Southeast Asian) Monsoon rains in early May, diabatic heating from condensational heating, especially in the eastern part of the Plateau, becomes significant. After the onset of the monsoon rains, the condensational heating over the Plateau is nearly balanced by the adiabatic cooling due to ascent. Yanai et al. (1992) indicate throughout the summer (from pre-onset to withdrawal of the Asian Monsoon), the Tibetan Plateau serves as a heat source (largely contributed by sensible heat flux) but the Plateau becomes a heat sink in the winter. Yanai and Li (1994) also describe a robust diurnal variation in the atmospheric boundary layer over the Plateau with the diurnal range of the ground surface temperature in the order of 30–40 °C, with the diurnal peak at local noon (0600 UTC at 90°E). Similarly, the wind speed in the surface layer also displays a strong diurnal variation reaching a diurnal minimum in the early morning and reaching a maximum at 1400 and 1800 local time (Yanai and Li 1994).

It is the sensible heat flux over the Tibetan Plateau in the spring season that leads to a large-scale reversal of the temperature gradient between land and ocean (Fig. 2.11). Yanai et al. (1992) noted a reversal of the meridional temperature gradient between 5°N and 25°N, which is generally negative before onset and reverses to a positive value after the ISM is established. Noska and Misra (2016) composited the temperature gradient computed between the latitudes of 5°N and 25°N at 300 hPa and plotted it as a function of days lagging and leading the date of onset of the ISM (Fig. 2.10a). Noska and Misra (2016) show that the reversal of this temperature gradient (traced by the zero contour line in Fig. 2.11a) over the Tibetan Plateau longitudes (from around 105°E to about 70°E) leads the onset of the ISM rains. This lead time between the reversal of the temperature gradient and the onset of the ISM rains reduces from 18 days around 105°E to about 0 days by 70°E (Fig. 2.11a). Similarly, the reversal of the temperature gradient between 5°N and 25°N shifts from positive to negative values gradually from the western to the eastern longitudes of the Tibetan Plateau with the demise date of the ISM rains leading the reversal by a few days as one moves further east from 70°E (Fig. 2.11b).

This view of the Tibetan Plateau as an elevated heat source driving the Asian Monsoon is now being challenged (Chou et al. 2001; Prive and Plumb 2007; Boos and Kuang 2010). The alternative theory is that the Tibetan Plateau serves like a thermal insulator preventing dry and cold extra-tropical air from ventilating the moist and warm air over continental India. Boos and Kuang (2010) showed that the diabatic heating over Tibetan Plateau locally enhances rainfall at its southern edge but the overall monsoon circulation is unaffected by its removal provided a narrow orography representing the Himalayas is preserved. However, Park et al. (2009) from their series of hierarchical numerical modeling study with and without the Tibetan Plateau topography indicate that the Tibetan Plateau serves as a source for mechanically driven downstream moisture convergence of the background zonal westerlies impinging on the southern part of the Plateau to give rise to the relatively early onset of the monsoon rains over the Bay of Bengal and South China Sea

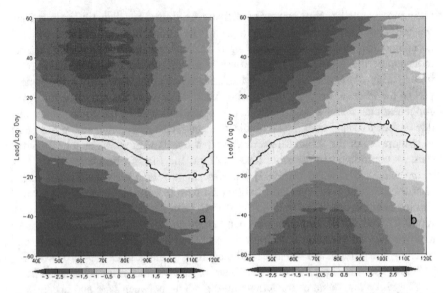

Fig. 2.11 The climatological meridional temperature gradient computed from Climate Forecast System Reanalysis as the difference in temperature at 300 hPa between 25°N and 5°N as a lead/lag function of the (**a**) onset date and (**b**) demise date of the IMR. Positive and negative shaded values denote southward and northward meridional gradients, respectively. The negative and positive values on the y-axis refer to the meridional temperature gradient leading and lagging the (**a**) onset and (**b**) demise date of the IMR, respectively. (Adapted from Noska and Misra 2016)

relative to the ISM. They also further claim that the effect of the Tibetan Plateau is greatest in May. In the mature phase of the monsoon in July and August, the strength of the monsoon is unaffected by either the presence or the absence of the Tibetan Plateau (Park et al. 2012).

2.7   The Asian Monsoon Variations

2.7.1   The Monsoon-ENSO Teleconnection

A very common graph that typifies the monsoon-ENSO teleconnection is illustrated in Fig. 2.12. This figure shows the lead/lag correlation between the June–September seasonal rainfall anomaly of the All-India averaged rainfall with monthly mean SST over the Niño3 and Niño3.4 region over the eastern equatorial Pacific Ocean, computed over a relatively long period from 1871 to 1998 (Krishnamurthy and Kinter 2003). A striking feature of this relationship between the summer monsoon rainfall over India and SST variability of the Niño3 region is that the strongest correlation of −0.6 occurs from October to January following the JJAS monsoon season (Fig. 2.12). Figure 2.12 also shows that the simultaneous correlation (at zero

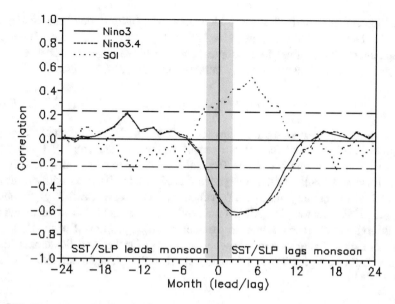

Fig. 2.12 The observed lead/lag correlation of All-India averaged JJAS seasonal anomaly of rainfall with monthly SSTA averaged over Niño3 (solid), Niño3.4 (dashed) region, and the SOI (dotted). The shaded area represents the JJAS the season of the ISM rainfall; long dashed lines represent the 99% confidence interval according to the t-test. These correlations were computed for 1871–1998. (Reproduced from Krishnamurthy and Kinter 2002)

lag) is also significant but slightly weaker at −0.5. There is almost no significant correlation with Niño3 SSTA preceding the monsoon season (or preceding the month of June). We find a similar relationship of the ISM rainfall with Niño3.4 SSTA as the Niño3 region (Fig. 2.12). A similar association of the ISM rainfall is also observed with the Southern Oscillation Index (SOI) but with the sign opposite to that with the Niño3 or Niño3.4 SSTA. Figure 2.12 serves as a good template to test the hypothesis that the Indian Monsoon may influence the evolution of ENSO. After all, lag/lead correlations do not necessarily imply cause and effect but serve as a good basis to propose a hypothesis to be tested.

One of the difficulties of understanding the interaction between the ISM and ENSO is that they are an integral part of the same coupled ocean-land-atmosphere climate system. Therefore, separating their influence is complex. Furthermore, the Indian Monsoon variability is well known to be strongly influenced by internal (chaotic) variability. But Kirtman and Shukla (2000) investigated the mutual influence of the ISM and ENSO in a tactful manner. They used the Cane-Zebiak coupled model confined to the Tropical Pacific (Zebiak and Cane 1987) to simulate ENSO with and without the influence of the variability of the ISM. The variability of the ISM in the coupled model was parameterized in the form of zonal wind stress anomalies (meridional wind stress anomalies are comparatively small over the Equatorial Pacific and are neglected in this study) over the Tropical Pacific in the

June–September period. This zonal wind stress variability during June–September over the Tropical Pacific was linearized to conform to two processes: (i) coupled air-sea interaction confined to the tropical Pacific and (ii) large-scale changes in the Walker Circulation that are remotely influenced by the ISM. Therefore,

$$\tau_{total} = \tau_{Pacific} + \tau_{monsoon} \qquad (2.16)$$

where τ_{total}, $\tau_{Pacific}$, and $\tau_{monsoon}$ are total zonal wind stress anomaly over the Tropical Pacific, zonal wind stress anomaly internal to the Tropical Pacific, and zonal wind stress anomaly over the Tropical Pacific due to monsoon variability.

Kirtman and Shukla (2000) integrated an AGCM for 50 years with prescribed climatological monthly mean SST (which avoids ENSO variability) to obtain $\tau_{monsoon}$. This climatological AGCM run produced a varying ISM (owing to internal variability), which enabled them to parameterize $\tau_{monsoon}$ (zonal wind stress over Tropical Pacific Ocean due to monsoon variability) for the June–September in the form of a linear regression as a function of All-India averaged summer monsoon rainfall, IMR as

$$\tau_{monsoon} = \alpha(x, y) \times IMR \qquad (2.17)$$

where $\alpha(x, y)$ are the linear regression coefficients. So in this case, Kirtman and Shukla (2000) were able to derive the zonal wind stress anomalies over the Tropical Pacific as a function of IMR, which was independent of the ENSO variability. It is, however, well known that a strong or a weak monsoon is associated with stronger or weaker Walker Circulation over the Tropical Pacific that also includes stronger or weaker trade wind easterlies over the Tropical Pacific (Nigam 1994), respectively. The Regression Eq. (2.17) was able to confirm this relationship.

Kirtman and Shukla (2000) integrated the coupled ocean-atmosphere Cane-Zebiak model over the Tropical Pacific for 300 years separately with and without the parameterized monsoon zonal wind stress anomalies (following Eq. 2.17). They compared this pair of integrations to conclude the following:

- Prescribing a weak ISM (namely, using $\tau_{monsoon}$ by applying IMR $= -2 \frac{mm}{day}$ in Eq. 2.17) on an existing El Niño event in the Tropical Pacific leads to overall enhancement and lengthening of the El Niño event. The $\tau_{monsoon}$ has the effect of enhancing the westerly wind stress anomaly in the Niño4/Niño3.4 region.
- Prescribing a strong ISM (namely, using $\tau_{monsoon}$ by applying IMR $= +2 \frac{mm}{day}$ in Eq. 2.17) on an existing El Niño event in the Tropical Pacific leads to a significant weakening of the El Niño event with $\tau_{monsoon}$ producing easterly wind stress anomalies in the Niño4/Niño3.4 region.
- Prescribing a weak ISM on an existing La Niña event in the Tropical Pacific weakened the La Niña event, although the impact is not as large as in the case of the El Niño event. Furthermore, as was the case with the warm event, the maximum response of the SST and thermocline anomalies in the Equatorial Pacific was 2–4 months after the monsoon season.

- Prescribing a strong ISM on an existing La Niña event in the Tropical Pacific leads to further strengthening of the La Niña event, although again they find that the impact is not as large as in the case of an existing El Niño event. However, there is an indication that the strong monsoon forcing on a La Niña event leads to the earlier development of the following warm event.
- Zonal wind stress anomalies over the Equatorial Pacific associated with weak monsoon events can trigger the earlier development of El Niño events.

Kirtman and Shukla (2000) conducted further experiments with the Cane-Zebiak model wherein they parameterized the IMR as a function of Niño3 SSTA (Nino3A):

$$IMR = \beta \times Nino3A \qquad (2.18)$$

where $\beta \equiv$ linear regression coefficient $= \frac{0.7mm}{day^o C}$. They used Eq. 2.18 in Eq. 2.17 to generate $\tau_{monsoon}$. They also ran experiments where IMR was randomized (but the standard deviation was similar to that obtained from Eq. 2.18) so that it was not a function of the Niño3 SSTA and $\tau_{monsoon}$ was nonzero either. They showed from these experiments that parameterized IMR (Eq. 2.18) broadened the Niño3 SST anomaly spectrum, suggesting greater irregularity of the ENSO variations. The randomized ISM variability completely broke down the lead/lag monsoon-ENSO relationship (e.g., correlation in Fig. 2.12 was hovering around the zero line), while this lead/lag relationship between ISM and ENSO was preserved in an experiment where 35% of the IMR was determined from Niño3 SSTA and the remaining 65% was random, suggesting that the synergy between the variations of the ISM and ENSO is subtle amidst internal or chaotic variability.

2.7.2   The Tropospheric Biennial Oscillation

The TBO is one of the characteristic features of the Asian-Australian Monsoon rainfall (Meehl 1994; Webster et al. 1998; Webster 2006). The TBO refers to the spectral peak in the Asian-Australian Monsoon in the 2–3-year period range with the anomalous wet monsoon in one summer followed by an anomalous dry monsoon in the following summer. The TBO is pervasive in spatial extent extending from Indonesia to the Indian Monsoon region. Incidentally, the spectrum of the Niño3.4 SSTA also shows a secondary biennial (~2 year) peak. Meehl (1994) suggested that TBO occurs as a natural variability of the coupled ocean-atmosphere-land monsoon system that is sustained by the memory residing in the ocean mixed layer and snow over the Tibetan Plateau (Fig. 2.13).

To understand the schematic showing the sequence of events related to the TBO, one must be cognizant of the persistent SST anomaly structures in the Equatorial Indian and Equatorial Pacific Oceans (Fig. 2.14). In the Equatorial Pacific Ocean, there is a strong tendency for the SSTA to persist from the boreal summer (JJA) to winter (DJF) seasons as shown in Fig. 2.14a, c. Similarly, the Equatorial Indian

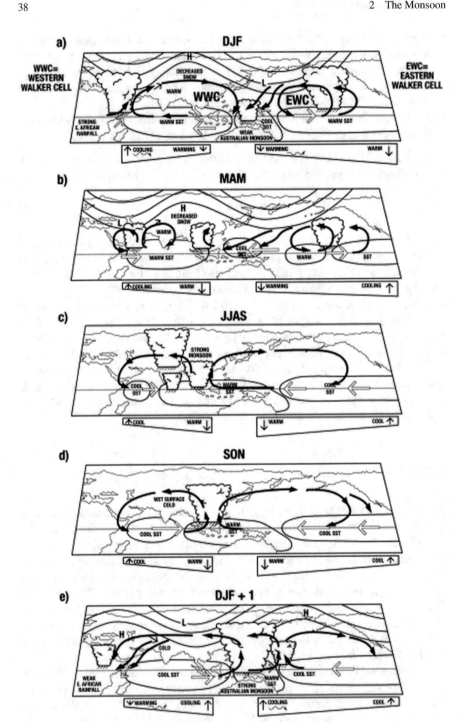

Ocean displays a tendency to persist with SSTA from the end of the boreal summer season to the end of the following spring season (Fig. 2.14b, c). There is, however, a clear persistence minimum of the SSTA in the Equatorial Pacific in the MAM season (Fig. 2.14b, c) that is referred as to the 'predictability barrier' (Webster and Yang 1992). It is so referred because climate models have extremely poor seasonal prediction skills in predicting the SSTA in the Equatorial Pacific during the boreal spring season when climatologically the zonal SST gradients are the weakest in the year. The persistence characteristics of the Equatorial Indian and Equatorial Pacific Oceans shown in Fig. 2.14 support the biennial components of the Asian-Australian Monsoon complex.

Now, returning to Fig. 2.13, in the winter before the strong Asian Monsoon (DJF), the Central and the Western Indian Ocean are warm while the Eastern Indian Ocean and the Indonesian Sea are cool with a prevailing warm ENSO-like event over the Central Pacific, which is also associated with weak Australian Summer Monsoon (Fig. 2.13a). The anomalous convection over Central Pacific is associated with an atmospheric Rossby wave response over Asia that produces a ridge at 500 hPa and warms the land surface temperature (Fig. 2.13a). This ridging leads to less snow over Eurasia (Figs. 2.13a, b). Therefore, in the following spring season (MAM), the Tibetan Plateau warms significantly as there is less of the downwelling shortwave flux used to melt the snow (Fig. 2.13b). The land-ocean contrast becomes strong enough to yield a strong Asian Monsoon (Fig. 2.13c). In the meanwhile, the strong convection over East Africa in boreal winter (Fig. 2.13a) and spring (Fig. 2.13b) causes westerly wind stress in the Western Indian Ocean, which sets off eastward-propagating downwelling equatorial Kelvin wave, that would deepen the thermocline in the Western Indian Ocean (Fig. 2.13b). On the other hand, the surface easterly anomalies in the far Western Pacific and the Eastern Indian Ocean with weak Australian Monsoon (Fig. 2.13a) persist in the following spring season (Fig. 2.13b), which generate upwelling Kelvin wave in the Equatorial Pacific. These Kelvin waves act to raise the thermocline in the Central and in the Eastern Pacific, thus transitioning the SSTA from warmer anomalies in MAM (Fig. 2.13b) to cooler anomalies in JJAS (Fig. 2.13c). It may be noted that both the western and eastern Walker Cells diminish from DJF (Fig. 2.13a) to MAM (Fig. 2.13b).

Fig. 2.13 A schematic explaining the sequence of events of a Tropospheric Biennial Oscillation starting from (**a**) the winter before the strong Asian Summer Monsoon, (**b**) the spring before the strong Asian Summer Monsoon, (**c**) during the strong Asian Summer Monsoon season, (**d**) in the fall following the strong Asian Summer Monsoon, and (**e**) the strong Australian Monsoon in the boreal winter after the preceding strong Asian Summer Monsoon. Large, unfilled arrows indicate surface wind anomalies. The wedge-shaped outline below each panel represents thermocline orientation, with the small arrows in those wedges indicating the anomalous movement of the thermocline. Wavy arrows within the wedges indicate Kelvin waves, which are upwelling if the arrow is pointing up or downwelling if the arrows at the end are pointing downward. The solid black arrows represent the eastern and western Walker Cells. (Reproduced from Meehl and Arblaster (2002); © American Meteorological Society. Used with permission)

(a) Persistence of JJA SST to DJF

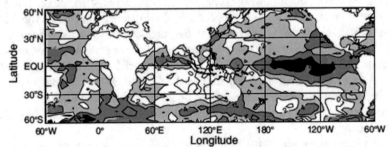

(b) Persistence of DJF SST to JJA

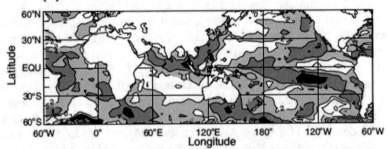

(c) SST persistence along equator

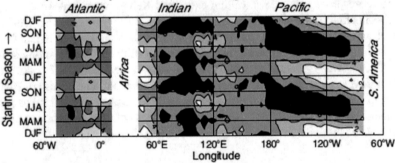

(d) SST persistence along equator (no ENSO)

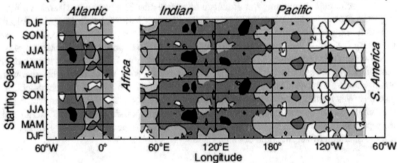

During JJAS, strong convection of the Asian Monsoon is associated with surface westerly wind anomalies with shallower thermocline and cooler SSTA in the Western Indian Ocean and deeper thermocline and warmer SSTA in the Eastern Indian Ocean (Fig. 2.13c). Heavier rain over the Asian Monsoon cools the land surface, and stronger eastern Walker Circulation results in stronger easterly trades over the Pacific that cool the SST in the Central and Eastern Pacific Ocean (Fig. 2.13c).

In the SON season following the strong Asian Summer Monsoon, the seasonal cycle migrates convection to the southeast that encounters the warm SST over the Eastern Indian Ocean and the Indonesian Sea, which is set up by the dynamical response to westerly winds in the Equatorial Indian Ocean in JJAS (Fig. 2.13d). Finally, in the DJF season following the strong Asian Summer Monsoon, the Australian Summer Monsoon reaches peak strength with surface westerly wind anomalies appearing just south of the equator from the Southeastern Tropical Indian Ocean as part of the strong western Walker Circulation (Fig. 2.13e). In the meanwhile, the surface westerly anomalies begin to appear in the far Western Equatorial Pacific Ocean that set off downwelling Equatorial Kelvin waves that begin to deepen the thermocline in the Eastern Pacific Ocean and set up the transition for the next chain of events leading to the opposite phase of the TBO, culminating with weak ISM in the following JJAS and followed by a weak Australian Monsoon in the subsequent DJF season.

Kim and Lau (2001) using a coupled model suggest that the biennial tendency of the ENSO evolution is tied to the coupling of ENSO to monsoon wind forcing over the Western Pacific Ocean. Therefore, the implication here is that without the monsoon wind forcing, ENSO is likely to be less biennial. It is also observed that there are epochs when either TBO or ENSO components are stronger than others (Torrence and Webster 1999). For example, there was little biennial variability in either the IMR or Niño3 SSTA between 1920 and 1950. Meehl and Arblaster (2002) suggest that when the TBO transition conditions are weak, then the system is likely to be less biennial with more spectral power shifting to lower frequencies. Anomalous Tropical Pacific and Indian Ocean SSTs are the dominant transition conditions in the TBO with anomalous meridional temperature gradients over Asia and the convection of the Australasian Monsoon system.

Fig. 2.14 The six-month persistence of SSTA depicted as correlations of SSTA from (**a**) June–August to the following December–February and (**b**) December–February to the following June–August. c) The seasonal cycle of the six-month persistence of SSTA depicted as correlations of SSTA at each ocean point along the equator and (**d**) correlations of (**c**) repeated after removing the Niño3 SST variations from each grid point using a linear regression for monthly Niño3 SSTA on to each ocean point along the equator. Only significant correlations are shaded at 0.2 (lightest), 0.4, and 0.6 (darkest) at 95% confidence interval. These correlations are computed from 1871 to 1996. (Reproduced from Webster 2006)

2.7.3   The Indian Ocean Dipole

The IOD mode is one of the most widely recognized basin scale variations in the Indian Ocean (Saji et al. 1999). It is defined by an index that is the difference in the area averaged SST over the Western Indian Ocean (50°E–70°E and 10°S and 10°N) and Eastern Indian Ocean (90°E–110°E and 0°–10°S). The positive phase of the IOD corresponds to warmer than average SSTs in the Western Indian Ocean and cooler than average in the Eastern Indian Ocean. Similarly, the negative phase of the IOD refers to cooler than average SSTs in the Western Indian Ocean and warmer than average in the Eastern Indian Ocean. The evidence of the dipole appears in early to mid-summer and peaks in the fall/early winter (OND) period before rapidly decaying in the boreal spring season. The two phases of the IOD are not symmetric: (i) the amplitude of the positive phase is stronger than the negative phase of the IOD, and (ii) in the negative phase, the SSTA, especially in the Western Indian Ocean, is much weaker than in the positive phase.

The interest in the IOD was raised considerably when in JJAS 1997, the ISM recorded a near-normal rainfall despite one of the strongest El Niño events in the Equatorial Pacific (SSTA in JJAS 1997 over Niño3.4 was ~3 °C). Krishnamurthy and Kirtman (2009), from spectral decomposition of the ISM rainfall, indicated that the rainfall component forced by the strong positive IOD mode of 1997 counteracted the remote El Niño forcing to arrive at a near-normal ISM rainfall. Several other studies have confirmed this finding. The association of the positive phase of the IOD with surplus ISM rainfall is also characterized by the negative precipitation anomalies over Eastern Indian Ocean and associated positive rainfall anomalies over the monsoon trough in the Bay of Bengal. Annamalai et al. (2003) associate this rainfall pattern with a local meridional overturning circulation with evidence of southerly surface winds originating off Sumatra and converging in the Bay of Bengal.

The IOD is also characterized by variations in the Wyrtki Jet. For example, in the 1993–1994 IOD event, the Wyrtki Jet in the boreal fall of 1993 had an estimated transport of 35 Sv, whereas in the boreal spring, the Wyrtki Jet had a maximum transport of 5 Sv. The weakening of the Wyrtki Jet lowered the SST and shoaled the thermocline in the Eastern Equatorial Indian Ocean (Meyers 1996). Furthermore, Potemra et al. (1997) from their ocean modeling study indicated that the Indonesian Throughflow was higher by about 5 Sv than the mean seasonal cycle of the model simulation in the spring of 1994 owing to the strong pressure gradient between the Pacific and Eastern Indian Ocean.

There are, however, a few points of contention with the IOD. Some studies debate the usage of 'dipole' in IOD because in several years of strong IOD, the SSTA in the Western Indian Ocean is very weak. So essentially in such instances, the 'dipole' is inexistent and is defined by the SSTA in the Eastern Indian Ocean only. In some instances of IOD, the SSTA in the Eastern Indian Ocean is abruptly diminished in the summer and fall months, and in such cases, the IOD is largely defined by the SSTA in the Western Indian Ocean for much of the year. This abrupt diminishment of the IOD in boreal spring is a response to the strong monsoonal wind forcing in the summer. Some studies claim that IOD is a manifestation of ENSO in the Tropical Indian Ocean. Others claim that IOD is intrinsic to the Tropical Indian Ocean and is triggered by coupled ocean-atmosphere instability that is essentially self-sustaining (Webster et al. 1999).

2.8   The Other Regional Monsoons

Besides the Asian-Australian Monsoon complex, the community broadly also recognizes the North American, the South American, and the African Monsoons (Fig. 2.15). Each of these regions is unique in its ways but more broadly shares the monsoonal features of a strong seasonal cycle of precipitation (Fig. 2.15), which is characterized by strong gradients of temperature, pressure, and moisture. Additionally, Misra (2008), through a set of modeling studies, shows that the summer rainfall in these regions is uniquely characterized by the neighboring air-sea interactions, which render the terrestrial rainfall to have local auto-decorrelation time in excess of 3 days through robust land-atmosphere interaction. In contrast, the uncoupled, atmosphere-only integrations display a significantly reduced auto-decorrelation time of the rainfall (~1 day), suggesting that the daily rainfall is less persistent during the rainy season in the absence of the air-sea interaction. But many of these regional monsoons also differ significantly from the Asian Monsoon. For example, the analogy to the schematic of the ISM circulation in Fig. 2.1 to describe the South American Monsoon circulation is made after the annual mean winds are removed (Zhou and Lau 1998). Similarly, the North American Monsoon like the South American Monsoon does not have a distinct winter counterpart as the Asian Monsoon.

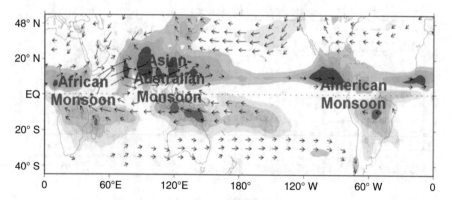

Fig. 2.15 The domains of the regional monsoons indicated by the difference of the 850 hPa winds and precipitation between the JJA and DJF seasons. (Reproduced from Zhou et al. 2016)

Trenberth and Stepaniak (2004) argue that in the tropics where the Coriolis force is weak, the meridional movement of energy largely occurs through the large-scale overturning circulation of the atmosphere, like the ITCZ, Hadley Circulation, and Monsoon Circulation. As a result of maintaining the global energy balance, they suggest that the moisture in the atmosphere is forced toward ascending regions of the air, where they rise, cool, and condense, resulting in strong latent heating, generating rains in the monsoon and the ITCZ regions that drive the upward branch of their meridional overturning cells. In the subtropics, the descending branch of these overturning cells subsides and warms, suppressing cloud formation and allowing the compensatory radiative cooling to balance the energy. Therefore, the regional monsoons arise as a requirement to maintain the global energy balance and are collectively with the Asian Monsoon, termed as the global monsoon. Wang and Ding (2008) claim that the global monsoon represents the dominant seasonal variation of precipitation and circulation in the global tropics.

Considering the varied definitions of the regional monsoons, like those based on the seasonal reversal of winds and precipitation (e.g., Ramage 1971) or temperature (Yanai et al. 1992) or vertically integrated moisture transport (Fasullo and Webster 2003), Trenberth et al. (2000) suggest that the vertical cross sections of the divergent wind circulations could reveal the areas of regional monsoons more comprehensively. For example, the definition of the monsoons in Ramage (1971) does not

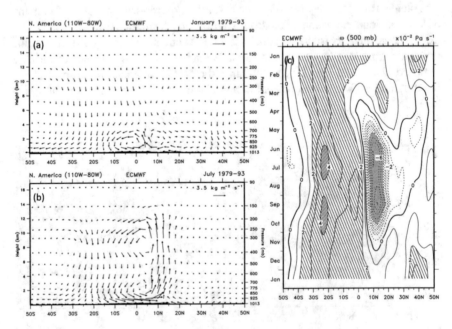

Fig. 2.16 The zonally averaged (110°W–80°W) climatological (1979–1993) divergent wind circulation (kg m^2 s^{-1}) from ECMWF reanalysis for (**a**) January and (**b**) July. (**c**) The seasonal cycle of the zonally averaged (110°W–80°W) vertical motion (ω; Pa s^{-1}) with stippling indicating upward and hatching indicating subsidence. (Reproduced from Trenberth et al. 2006)

include the American Monsoons. Trenberth et al. (2000) show through regional zonal mean cross sections of the divergent wind circulations in five monsoonal regions including Africa (between 10°E and 40°E), Australia-Asia (60°E–180°), North America (110°W–80°W), South America (80°W–40°W), and West Africa (30°W–10°E) a persistent overturning circulation during the rainy season that is typified by maximum vertical motion around 400 hPa, divergence that is strongest at around 150 hPa, and convergence in the lower troposphere with a maximum at around 925 hPa. An illustration of their argument is provided in Fig. 2.16 for the North American Monsoon. This figure shows the meridional overturning circulation centered around 15°N in the summer occurs over the depth of the troposphere (~12–14 km deep) with subsidence across the equator. Trenberth et al. (2006) also show a shallow (~<4 km) tropospheric overturning circulation that is persistent both in the summer and winter for the North American Monsoon (Fig. 2.16). In fact, they allude to the importance of these shallow overturning circulations to the overall Monsoon Circulation in the context of the African Monsoons and the South American Monsoons. Furthermore, they suggest that the appearance of the upward motion (ω) at 500 hPa in the boreal summer season attests to the seasonality of this circulation and consequently indicates a monsoonal circulation, although the winter counterpart, unlike the Asian-Australian Monsoon, is nonexistent (Fig. 2.16).

Chapter 3
The Role of the Diabatic Heating in the Tropical Atmosphere

Abstract The chapter places the importance of atmospheric convection in tropical climate and its discovery from early field studies. Early concepts of atmospheric convection are discussed alongside new concepts developed in the recent decade. The equatorial wave theory is introduced in this chapter and concluded with a discussion of the concept of organization of atmospheric convection.

Keywords Diabatic heating · Moist static energy · Dry static energy · Mesoscale convective system · Stratiform rain · Convective rain · Entrainment · Convectively available potential energy · Clouds · Conditional instability of the second kind · Quasi-equilibrium · Cumulus parameterization · Equilibrium level · Entropy · Radiative-convective equilibrium · Blackbody · Tropical waves · Shallow water equation · Coriolis · Schrodinger equation · Rossby wave · Gravity wave · Kelvin wave · Moisture mode · Weak temperature gradient · Organization of convection · Enthalpy

3.1 The Ubiquity of Convection in the Tropics

One of the defining features of the tropics is the rather ubiquitous presence of deep convection that either appears organized (e.g., ITCZ) or scattered at a given instant of time. Early weather forecasts based on dry barotropic models had a very limited skill in the tropics that was extended by including moist physics, after the realization of the significant importance of deep convection in the thermodynamics and the dynamics of the circulation system (Riehl and Malkus 1958). The large-scale thermodynamic and moisture budget studies from observational fields in the tropics revealed that the vertical motion in deep convection (or cloud mass flux) is far larger than the large-scale vertical motion (or mean vertical flux of the large scale), thus causing compensating subsidence between the clouds (Yanai et al. 1973). To complete the heat and the moisture budgets, Yanai et al. (1973) concluded that the warming and the drying of the large-scale atmosphere by subsidence (forced by the upward cloud mass flux) are counteracted by the detrainment of condensed water and its evaporation from the plethora of shallow cumulus clouds in the lower

© Springer Nature Switzerland AG 2023
V. Misra, *An Introduction to Large-Scale Tropical Meteorology*, Springer Atmospheric Sciences, https://doi.org/10.1007/978-3-031-12887-5_3

troposphere. These conclusions have broadly been accepted ever since as a guiding principle to understand the synergy between deep convection and the large-scale heat and moisture variations in the tropics.

Let us elaborate on this large-scale budget further to illustrate the importance of moist physics in understanding the tropical climate. The heat budget equation of the atmosphere can be written in flux form with pressure (p) as the vertical coordinate:

$$q_1 \equiv \overbrace{\frac{\partial \overline{s}}{\partial t} + \nabla.\overline{s\,\overrightarrow{V}} + \frac{\partial \overline{s\omega}}{\partial p}}^{\text{LHS}} = \overbrace{q_R + L\,(c - e) - \nabla.\overline{s'\overrightarrow{v}'} - \frac{\partial \overline{s'\omega'}}{\partial p}}^{\text{RHS}} \tag{3.1}$$

where the overbar denotes spatial average to represent the 'large scale' as in Yanai et al. (1973) and in other studies, it has been used to denote temporal average (e.g., monthly mean) to represent the large scale (e.g., Nigam et al. 2000), s is the dry static energy$=c_pT + gz$, q_R is the net (shortwave + longwave) radiative heating rate, $L(c - e)$ is the net condensational heating (where c and e refer to condensation and evaporation of condensed water, respectively), $\nabla.\overline{s'v'}$ is the horizontal diffusion of dry static energy, and $\frac{\partial \overline{s'\omega'}}{\partial p}$ represents the vertical diffusion of sensible heat flux. In Eq. 3.1, all terms in the LHS represent the large scale and are relatively easily calculated from the analysis. The terms in the RHS of Eq. 3.1 represent terms involving moist physics that must be diagnosed from some form of a cloud and radiation model. Yanai et al. (1973) called the estimation of the heat budget from the LHS of Eq. 3.1 famously as q_1 (or $\frac{1}{g}\int_{p_s}^{P_{top}} q_1 dp \equiv Q1$) or apparent heat source of the large-scale motion system consisting of radiative heating, condensational heating, and convergence of vertical eddy transport of sensible heat flux. It may be noted that q_1 is apparent because the sub-grid scale diabatic heating (or cooling) is indirectly estimated from the large-scale fields. Yanai et al. (1973) concluded that the large-scale heating of the environment is primarily from the compensating, adiabatic large-scale subsidence since the cloud mass flux exceeds the mean vertical mass flux from large-scale convergence.

Similarly, the moisture budget equation can be written in flux form with p as the vertical coordinate:

$$q_2 \equiv -L\overbrace{\left(\frac{\partial \overline{q}}{\partial t} + \nabla.\overline{q\,\overrightarrow{V}} + \frac{\partial \overline{q\omega}}{\partial p}\right)}^{\text{LHS}} = \overbrace{L\,(c - e) + L\nabla.\overline{q'\overrightarrow{v}'} + L\frac{\partial}{\partial p}\overline{q'\omega'}}^{\text{RHS}} \tag{3.2}$$

where q is the specific humidity, $\nabla.\overline{q'v'}$ is the horizontal diffusion of moisture, and $\frac{\partial}{\partial p}\overline{q'\omega'}$ is the vertical diffusion of moisture. Once again, the LHS of Eq. 3.2 can be estimated from the 'large-scale' analysis, while the RHS requires estimates from a model. Yanai et al. (1973) called the estimation of the moisture budget from the large scale famously as q_2 (or $\frac{1}{g}\int_{p_s}^{P_{top}} q_2 dp \equiv Q2$) or apparent moisture sink due to net condensation and vertical divergence of eddy transport of moisture (or diffusion of latent heat flux). Once again, q_2 is referred to as an apparent moisture sink because it

Fig. 3.1 The profile of compensating subsidence warming (dot-dashed line), evaporation of detrained cloud condensate (dotted line), their sum (solid line), and net $q1$ (dashed line) for convection in phase III of GATE easterly wave case. (Reproduced from Gregory and Miller 1989)

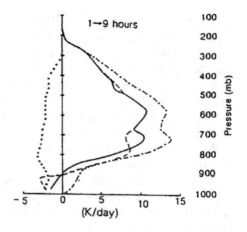

is estimating the moisture consumed by the sub-grid scale condensational processes indirectly from the large-scale fields. Yanai et al. (1973) concluded that the drying of the environment is primarily from the large-scale subsidence, which is counteracted to an extent that evaporates in the relatively dry environment by the large amounts of water vapor and cloud condensate detrained from the clouds.

Several studies have computed q_1 and q_2 from observational field campaign data and reanalysis and have shown that compensatory environmental subsidence is an important balancing term for vertical mass flux in deep convective clouds. For example, Gregory and Miller (1989) used a two-dimensional cloud resolving model with a prescribed environmental (or large-scale) lapse rate and specific humidity profile and showed that the compensating subsidence closely follows q_1 and slightly exceeds it at all levels, which is balanced by the evaporation term (Fig. 3.1). In Fig. 3.1, Gregory and Miller (1989) computed q_1 from the large-scale data collected from the field campaign in GATE (a field campaign conducted over the Northern Tropical Atlantic Ocean off of West Africa), and evaporation of the detrained cloud condensate and vertical cloud mass flux are estimated from a cloud model. But the compensating environmental subsidence is equivalent to the vertical cloud mass flux (with the opposite sign). These results have also been replicated from data collected in other tropical regions as well (e.g., Yanai et al. 1973; Johnson et al. 1999).

The vertical distribution of latent heating ($L(c\text{-}e)$ in Eq. 3.1) is also significant from its implication on the atmospheric and even oceanic circulation. The vertical structure of the latent heating also signifies the type of precipitation feature that is producing it. For example, a bottom-heavy profile is usually associated with the isolated convective system, which is characteristic of having only convective precipitation. In contrast, a mature mesoscale convective system (precipitation features with an area greater than 2000 km²) is likely to show a top-heavy profile of latent heating owing to the significant amount of stratiform precipitation. Johnson et al. (1999) indicated the trimodal characteristics of tropical convection. Prior to this study, the tropics were known to carry shallow trade wind cumulus clouds and deep cumulonimbus clouds to give the bimodal distribution of clouds and associated latent heating profile. However, Johnson et al. (1999) discovered from the radar data collected during the TOGA-COARE and GATE field campaigns that the tropics have three cloud types: shallow cumulus, congestus, and cumulonimbus with radar echo tops for these three cloud types near three prominent stable layers corresponding to the trade wind stable layer at ~2 km, the freezing line (0 °C) at ~5 km, and the tropopause at ~15 km, respectively. These stable layers inhibit further cloud growth and promote cloud detrainment. Johnson et al. (1999) further indicate that congestus and cumulonimbus clouds contribute to the total convective rainfall while the shallow cumulus clouds contribute to moistening and preconditioning the atmosphere for deep convection.

By combining cloud resolving simulations and discerning the properties of cloud and precipitation detected by TRMM precipitation radar and microwave imager observations, Liu et al. (2015) made some very interesting observations of latent heating profiles in the tropics:

- Over both land and ocean, mesoscale convective systems contribute most of the latent heating with their contribution nearing 100% in the upper troposphere (>7–9 km above the surface).
- Over the land, convective latent heating dominates over stratiform latent heating with the combined latent heat peak around 4–7 km (Fig. 3.2a–c).

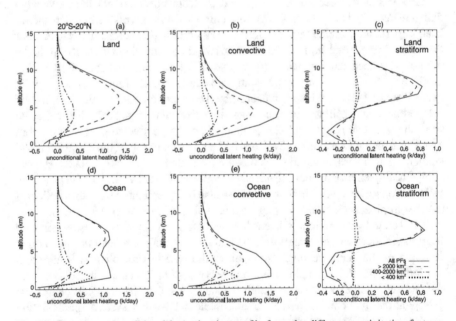

Fig. 3.2 The mean unconditional latent heating profile from the different precipitation features over tropical (20°S–20°N) (**a–c**) land and (**d–f**) ocean for (**a, d**) total = convective + stratiform latent heating, (**b, e**) convective latent heating only, and (**c, f**) stratiform latent heating only. (Reproduced from Liu et al. (2015); © American Meteorological Society. Used with permission)

- Over the ocean, convective and stratiform latent heating magnitudes are comparable, with the combined latent heat profile showing a dual peak. The lower peak is contributed by shallow precipitation features and the upper peak from mesoscale convective systems, with the dominant contribution of stratiform latent heating above 7 km altitude (Fig. 3.2d–f).
- The mesoscale convective systems have larger stratiform fractions over oceans and relatively smaller fractions over continents relative to convective fraction. Monsoonal mesoscale convective systems have properties intermediate between those of continental and oceanic mesoscale convective systems.
- There is significant heterogeneity in the latent heat profiles of continental regions compared to the oceanic regions. This is largely to do with the difference in size and intensity of precipitation features. For example, in tropical Africa, the precipitation features with an area >10,000 km^2 contribute significantly to latent heat compared to that over Amazon. Similarly, precipitation features in tropical Africa with radar echo top >12 km contribute significantly to latent heat compared to that over Amazon.
- Although all three modes of convection (shallow, congestus, and cumulonimbus) are detected over tropical oceans, the latent heating contribution from congestus is relatively small and varies from one ocean basin to another.

In reviewing tropical field programs over 50 years period, Zipser (2003) suggested that the concept of the 'hot tower' for cumulonimbus clouds in the tropical oceans and land areas is somewhat unfounded. This concept of 'hot tower' refers to the undiluted core of cloud mass flux in the cumulonimbus clouds that extend from the cloud base to the top of the cloud. It is now well known from several decades of observations over tropical oceans involving aircraft penetrations into cumulonimbus clouds that actual liquid water content in tropical clouds rarely exceeds 0.4 of the *adiabatic cloud water content* because of dilution from entrainment and mixing of environment air beside the fallout of the condensate from the clouds (Zipser 2003). Zipser (2003) explains that despite the dilution of these updraft cores by entrainment, oceanic cumulonimbus clouds can reach the tropical tropopause from the combined effects of buoyancy from the freezing of the condensate and shifting of the condensation to sublimation latent heating rate. In situations where condensate forming in a rising parcel subsequently freezes, the additional latent heat of fusion following the ice adiabat reduces the moist adiabatic lapse rate of the rising parcel, making it further buoyant than if the parcel was following the moist adiabat. Zipser (2003) suggests that to account for the large differences in the vigor of the convective clouds between tropical land and ocean, one needs to consider for a slight reduction in entrainment rates in the updrafts of land convection, which can substantially increase their vertical velocity, reducing the likelihood of falling condensate and increased supercooled liquid water content. In such circumstances, the cloud or the rainwater freezes instantly at temperatures between −4 and −15 °C, with the additional latent heat of fusion reducing the moist adiabatic lapse rate and making the parcel more buoyant. Therefore, as these moist air parcels reach higher heights owing to their higher acquired buoyancy, where the temperature could be less than −15 °C, then water vapor sublimates to ice, releasing the latent heat of sublimation (= latent heat of condensation + latent heat of fusion) instead of latent heat of condensation, making tropical land convection invariably more vigorous than tropical oceanic convection.

But the cloud resolving modeling study of Seeley and Romps (2016) clearly indicates that under radiative-convective equilibrium conditions, the cloud buoyancy is not significantly different between model simulations with and without ice microphysical processes. Their main argument for the relative insensitivity of cloud buoyancy (determined by the difference between the temperature of the cloud parcel and the environment) to ice microphysical processes is that the environmental temperature profile is equally influenced by the developing clouds. It may be mentioned that the radiative-convective equilibrium is a good approximation to the tropical atmosphere with its 'C'-shaped relative humidity profile (Romps 2014) and trimodal cloud fraction profiles (Dessler et al. 2006). This is true because the fast gravity waves enforce moist-convective lapse rates even where there is little local convective heating in the tropics. Seeley and Romps (2016) show that the latent heat of fusion released by deposition and freezing increases the temperature of both the clouds (any grid cell with nonprecipitating condensed water mass fraction greater than 10^{-5} Kg/Kg and vertical velocity >1 m/s) and their environment (domain mean at 200 m grid spacing over a 36 km × 36 km domain and time mean profile over 25 days; Fig. 3.3). Figure 3.3 shows that the warming of the environment temperature is nearly identical to the warming of the cloud temperature due to ice microphysical processes between the altitudes of 500 m and 11 km, where the clouds are positively buoyant in the mean. The breakdown of the sources of cloud buoyancy (b) is given by the following equation:

Fig. 3.3 The absolute temperature difference of the environment (black line) and cloud (red line; see text for the description of differentiating cloud from the environment) between the cloud resolving model simulation with and without ice. (Reproduced from Seeley and Romps 2016)

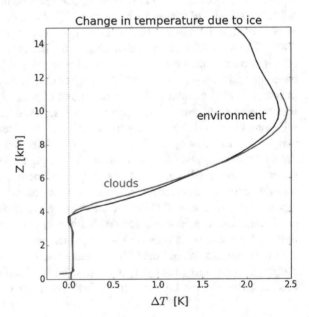

$$
b \cong g
\begin{bmatrix}
\overbrace{\dfrac{\Delta T}{T}}^{\text{Term 1}} + \overbrace{\left(\dfrac{R_v}{R_d} - 1 \right) \Delta q_v}^{\text{Term 2}} - \overbrace{\Delta q_{\text{con}}}^{\text{Term 3}}
\end{bmatrix}
\tag{3.3}
$$

where Terms 1, 2, and 3 are buoyancy due to differences in temperature, moisture, and condensate loading between undiluted air parcels (cloud) and the environment, respectively. Figure 3.4a shows that all these terms are comparable between the ice and no-ice cloud model runs, and in fact, in Fig. 3.4b, it is shown that above the melting level, the updraft velocities are slightly lower in the simulation with ice in comparison to the simulation without ice. If the release of latent heat of fusion above the melting level is not the reason for the top-heavy tropical undiluted buoyancy profile, then what is it? Again, Seeley and Romps (2016) by way of elegant argument suggest that the top-heavy buoyancy profile in the tropics is a result of the cloud-environment temperature difference. They indicate that the smallness of the saturation specific humidity in the upper troposphere forces the difference in the moist static energy between the undiluted parcel and the environment as a strong function of sensible heat ($c_p \Delta T$) and a weaker function of latent enthalpy ($L \Delta q_v$). So if the effect of entrainment is to adjust the temperature profile by an amount Δh^*, where $h^* = c_p T + L_v q_v^* + gz$, then the temperature difference between the cloud and environment at a given height is given by $\frac{\Delta h^*}{\frac{\partial h^*}{\partial T}}$, where $\frac{\partial h^*}{\partial T} = c_p + L_v \frac{\partial q_v^*}{\partial T}$. In the lower

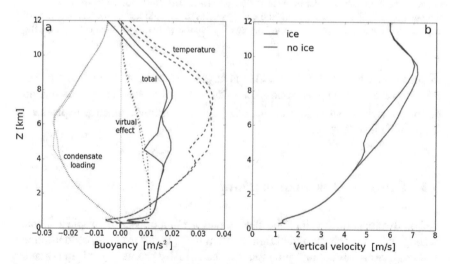

Fig. 3.4 (a) The mean buoyancy of cloud updrafts from the simulation of a cloud resolving model with ice (blue) and without ice (red). The total buoyancy of clouds (solid lines; LHS of Eq. 3.3), buoyancy due to temperature difference (Term 1 of Eq. 3.3; dashed line), virtual effect (Term 2 of Eq. 3.3; dash-dotted line), and condensate loading (Term 3 of Eq. 3.3; dotted line). (b) The mean vertical velocity of cloud updrafts with (blue line) and without (red line) ice simulation. (Reproduced from Seeley and Romps 2016)

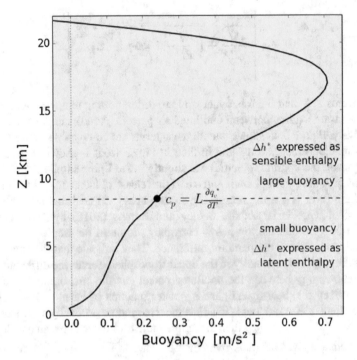

Fig. 3.5 An illustration of buoyancy of an adiabatically lifted near-surface parcel from a radiative-convective equilibrium simulation over an SST of 310 K. The black dot marks the level between the layer where the saturated moist static energy excess of an undiluted parcel (Δh^*) is primarily expressed as latent enthalpy ($L\Delta q_v$) and the layer where it is primarily sensible enthalpy ($c_p\Delta T$)

troposphere, where $L_v \frac{\partial q_v^*}{\partial T} > c_p$, Δh^* is largely expressed as latent enthalpy (Fig. 3.5). However, in the upper troposphere, where the parcel-environment temperature difference is the greatest because $c_p \gg L_v \frac{\partial q_v^*}{\partial T}$, Δh^* is predominantly expressed by the sensible heat (Fig. 3.5).

3.2 CISK vs. Quasi-equilibrium

Before dwelling further into the details and the importance of tropical diabatic heating (which includes, in addition to latent heat release from the condensational process, heating or cooling from radiative and diffusive processes), it is important to clarify an important and commonly prevailing misconception about diabatic heating in the tropical atmosphere. A commonly misstated statement is that 'the large-scale tropical circulations are driven by latent heat release in deep convection'. Such statements, although they may appeal to intuition, are perpetuated by theories of Conditional Instability of the Second Kind (CISK) proposed by Charney and

Eliassen (1964). The CISK theory in its simplest form suggests that the large-scale moisture convergence supplies the moisture for atmospheric convection and the latent heat release of the condensational process within the clouds causes a horizontal pressure gradient force, which drives the large-scale moisture convergence. Emanuel et al. (1994) succinctly argue that this concept of CISK is inherently wrong. CISK assumes that the latent heating produced in atmospheric convection leads to the production of kinetic energy (convergence is after all an advective process maintained by kinetic energy). But for this assumption to be true, heating and temperature variations must be positively correlated (as the covariance of these two variables generates available potential energy following Lorenz (1955)). Emanuel et al. (1994) indicate, however, that convection damps available potential energy (implying covariance of heating and temperature is negative) in the absence of surface evaporation.

The seeds of the concept of latent heat release in deep convection as an energy source for large-scale tropical circulations were sowed earlier in Riehl and Malkus (1958) that followed a field study called the 'Thunderstorm Project' in Florida and Ohio. Emanuel et al. (1994) indicate that this idea of deep convection driving large-scale circulation stems from the implied covariation of heating and temperature when the diabatic heat source is applied (externally), which in general produces temperature perturbations that are positively correlated with heating. However, in the real atmosphere, where clouds are internal to (or interact with) the environment and affect other sources of diabatic (e.g., radiation and diffusion) and adiabatic heat sources, this covariation of heating and temperature cannot be taken for granted. Arakawa (2003) suggests a more comprehensive view of atmospheric convection, as one that couples the radiative-dynamical-hydrological-oceanographic processes through redistribution of latent heating, sensible heating and momentum in the atmosphere and influencing absorption, emission, and reflection of radiation. We will elaborate further upon these concepts in the following paragraphs.

Arakawa (1969) originally came up with this concept of quasi-equilibrium to provide a closure for a cumulus parameterization scheme that he proposed. This concept of quasi-equilibrium had a profound impact on our understanding of the large-scale flow and the deep convection. The quasi-equilibrium theory suggests that the rate of production of available energy by large-scale processes is equal to its consumption and destruction by the convective clouds. Alternatively, it can be stated that the short timescale, chaotic convection is in statistical equilibrium with its environment, which is evolving much more slowly with near equality between the rate of production of energy by the large scale and its consumption and destruction by convection.

It may be noted that this available potential energy is nothing but CAPE, which is proportional to the area contained between the moist adiabatic ascent and the environment lapse rate between the *level of free convection* and the *equilibrium level* (or level of neutral buoyancy). As a consequence of quasi-equilibrium, the virtual temperature profile of the convecting atmosphere is approximately a sole function of the *subcloud layer entropy* (or subcloud layer *equivalent potential temperature*). This is because CAPE in the convecting atmosphere while not zero is approximately invariant and convection adjusts the virtual temperature profile to the moist adiabat tied to the subcloud layer entropy.

Here, the notion of deep convection is an ensemble of clouds that detrain at distinct cloud top heights, with the towering cumulus cloud feeding off (or entraining the environment air) in which the numerous but distinct ensemble of shallow cumulus clouds detrains at lower altitudes. Since these clouds develop and decay at much shorter timescales compared to the timescales over which the large-scale circulation is generating the available potential energy, Arakawa (1969) came up with this concept of quasi-equilibrium wherein the ensemble of clouds is in a form of statistical equilibrium that entails available potential energy being almost quasi-invariant with time. One must be extremely cautious not to interpret this notion as though CAPE is invariant with time. There is ample evidence to show that CAPE in a given location can vary from a few hundred J/Kg to zero J/Kg in a matter of days (Stevens et al. 1997). The invariance of CAPE implied in the quasi-equilibrium theory suggests that changes in CAPE are small compared to those that would occur if convection were somehow suppressed while the large scale continues to build CAPE with time. The concept of the quasi-equilibrium theory has an observational basis, wherein the rate of change of CAPE by large-scale processes alone (for a particular rate of entrainment), computed by analysis of Marshall Islands (in the Central Pacific Ocean) sounding data, is much larger compared to the observed rate of change of CAPE that is nearly zero and invariant (Fig. 3.6; Arakawa and Schubert 1974). Therefore, by the quasi-equilibrium theory, since the available potential energy is nearly invariant, there cannot be a significant conversion of this energy to the kinetic energy of the large-scale disturbances as CISK demands.

In contrast, physical parameterizations based on CISK (e.g., Kuo convection scheme; Kuo 1965, 1974) posit a statistical equilibrium of the rate of supply of moisture by the large scale and its rate of consumption by convection. Raymond and Emanuel (1993) suggest that most statistical equilibrium theories are rooted in causality. For example, turbulence is caused by instability in the large scale. Raymond and Emanuel (1993) argue that the statistical equilibrium in Kuo convection scheme fundamentally violates the causality because convection is not caused by large-scale supply of moisture.

Emanuel et al. (1994) indicate that the subcloud boundary layer equivalent temperature (or entropy) is controlled by radiative cooling, surface sensible and latent heat fluxes, turbulent entrainment at the top of the subcloud layer, and the flux of low entropy air from the middle troposphere by convective unsaturated downdrafts driven by evaporation of precipitation. Therefore, subcloud layer entropy is found to increase with an increase in SST, an increase in surface wind speed, an increase in entropy just above the subcloud (or boundary) layer, and an increase in entropy of the downdraft. It should be noted, however, that downdrafts in convection reduce subcloud layer entropy. Since the typical tropical environment profile of entropy shows a minimum in the middle troposphere, the reduction of the subcloud layer entropy is dependent on the relative magnitude of the downdraft mass flux and the entropy deficit in the lower and middle troposphere.

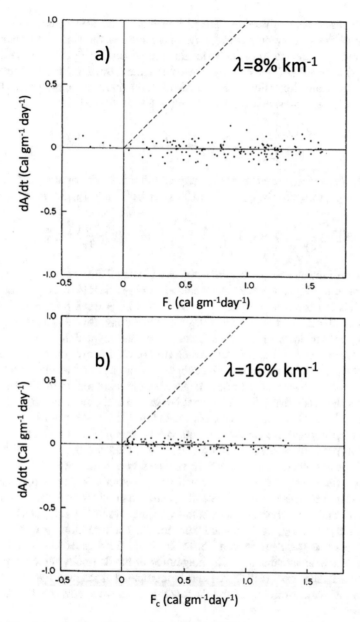

Fig. 3.6 The observational verification of the quasi-equilibrium assumption for cloud type with entrainment rate (λ) equal to (**a**) 8% km^{-1} and (**b**) 16% km^{-1} from sounding data collected from the Marshall Islands (in the tropical Central Pacific Ocean provided by Yanai et al. 1973). The ordinate is the observed rate of change of available potential energy (or cloud work function A), and the abscissa is the available potential energy due to large-scale processes only. (Reproduced from Arakawa and Schubert (1974); Published (1974) by the American Meteorological Society)

Using the quasi-equilibrium theory, Emanuel et al. (1994) interpret steady-state circulations in the convective atmosphere such as the Hadley Circulation as that driven by horizontal gradients (in the case of Hadley Circulation meridional gradients) of subcloud layer entropy. They derive the following criterion to maintain steady large-scale circulation in the convecting atmosphere by applying the thermal wind relationship (by justifying that it is applied to sufficiently large scale):

$$1 + \frac{1}{f} \nabla . \frac{1}{f} (T_s - T_T) \nabla \theta_{eb} \geq 0 \tag{3.4}$$

where T_s, T_T, and θ_{eb} are surface temperature, tropopause temperature, and subcloud layer entropy, respectively. For the zonally symmetric case, Eq. 3.4 reduces to

$$1 + \frac{1}{f^2} \frac{\partial}{\partial y} \left((T_s - T_T) \frac{\partial \theta_{eb}}{\partial y} \right) - \frac{\beta}{f^3} \left((T_s - T_T) \frac{\partial \theta_{eb}}{\partial y} \right) \geq 0 \tag{3.5}$$

The use of the quasi-equilibrium theory to understand the steady-state circulation is illustrated in Fig. 3.7a, where the subcloud layer gradient is weakly violating Eq. 3.5. In this instance, the horizontal gradient of SST is weak (Fig. 3.7a), and the horizontal advection in the subcloud layer and the convective downdrafts over the warm water both act to reduce the subcloud layer entropy and drive its distribution to meet the criterion of Eq. 3.5. As a result, the convection over the colder water (to the right in Fig. 3.7a) continues, albeit slightly weak. In other words, the adiabatic warming from the subsidence of the large-scale circulation over the colder waters is smaller than radiatively forced descent between clouds, and thus, the large-scale circulation is unable to suppress the convection over the colder waters.

In the case when the criterion in Eq. 3.5 is strongly violated, i.e., when the horizontal gradient of SST is strong (Fig. 3.7b), then the large-scale circulation becomes strong enough to completely suppress convection over the cold water, forming a trade wind inversion. Alternatively, the adiabatic warming from large-scale descent balances the radiative cooling between the clouds over the cold waters. In this instance, quasi-equilibrium stops working over the cold waters, and the subcloud layer entropy is decoupled from the temperature aloft. The temperature profile over the warm water and above the trade wind inversion over the cold water is nearly moist adiabatic. The magnitude of the thermally direct, large-scale circulation in this case (Fig. 3.7b) can be computed from requiring the balance between adiabatic warming of the descending air and radiative cooling over the cooler waters.

3.3   Radiative-Convective Equilibrium

The radiative-convective equilibrium refers to a statistical equilibrium between the radiative processes and convective processes. This equilibrium process is often used to understand the long-term average of the temperature profile. This theory implies

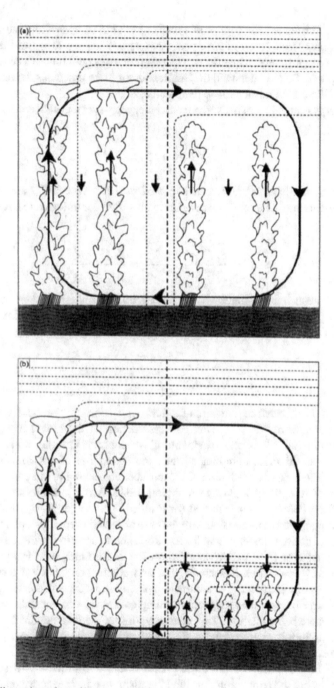

Fig. 3.7 Illustration of steady-state circulation in a convecting atmosphere for a case when the gradient in SST temperature is (**a**) weak and (**b**) strong. The SST is warmer to the left and cooler to the right. The dashed lines are contours of saturated entropy (with higher values to the left). (Reproduced from Emanuel et al. 1994)

that the equilibrium temperature profile results from the compensatory radiative and convective processes. The radiative processes essentially try to warm the surface and cool the troposphere, while the convective processes cool the surface and warm the troposphere. So to understand the implication of these processes, let us consider the case of the global mean surface temperature.

The energy emitted from the surface of a blackbody emitter (with emissivity $= 1$) is

$$F = \sigma T_s^4 \tag{3.6}$$

where $\sigma \equiv$ Stefan-Boltzmann constant $= 5.67 \times 10^{-8}$ $Wm^{-2}K^{-4}$ and T_s is the surface temperature. The total solar radiation absorbed per unit area (SRA) is given by

$$SRA = \frac{S_o (1 - \alpha) \pi R^2}{4\pi R^2} = \frac{S_o (1 - \alpha)}{4} \tag{3.7}$$

where $S_o \equiv$ mean solar insolation at the top of the atmosphere or solar constant $\approx$ 1388 Wm^{-2}, α is planetary albedo ≈ 0.3, and R is the radius of Earth. Substituting SRA from Eq. 3.7 for F in Eq. 3.6 and solving for T_s, we get

$$T_s = \left(\frac{S_o (1 - \alpha)}{4\sigma} \right)^{0.25} = 255.8K \tag{3.8}$$

This global mean surface temperature falls way short of the observed global mean temperature of ~288 K. This underestimation of the global mean surface temperature in Eq. 3.8 is understandable if we concede that the climate system of the Earth is far more complicated than the absorption of solar radiation and emission of the longwave radiation at the surface. For example, there is significant absorption of re-emitted longwave radiation (~340 Wm^{-2}). Therefore, if this is added to SRA before it is substituted for F in Eq. 3.6, we get $T_s = 318.4$ K. This surface temperature is more in line with the case of radiative equilibrium. Deploying a complete radiative transfer model for the atmosphere, one can attain a temperature profile that adheres to radiative equilibrium (an equilibrium state in the absence of non-radiative processes like convection and conduction). Such a temperature profile is very unrealistic for our climate system with near surface temperatures being very warm, upper air temperatures being cold with very unstable lapse rates, and troposphere being shallow. This generalization of the temperature profile from radiative equilibrium is true even in the presence of the major variable gases like water vapor.

We can now realize that the atmospheric convective process is a significant process to restore the temperature profile to realism. In non-convective regions (like the semi-arid regions), convection manifests in diffusive (or conduction) process and or advective transport of mass and heat energy. From the parcel theory, we know that for dry air parcels:

$\Gamma < \Gamma_d$} the atmosphere is stable to upward displacements
$\Gamma = \Gamma_d$} the atmosphere is neutral to upward displacements
$\Gamma > \Gamma_d$} the atmosphere is absolutely unstable to upward displacements

where Γ is the lapse rate of the parcel and Γ_d is the dry adiabatic lapse rate (= 9.8 °C/km). Pure radiative equilibrium yields lapse rates that are absolutely unstable ($\Gamma > \Gamma_d$).

So in dry environments, radiative-dry convective equilibrium prevails, whereby the dry convection removes the instability, and the lapse rate is rendered neutral to upward displacements. The temperature profile from this adjustment is far more realistic but still too warm near the surface and too cool in the upper air. In moist environment:

$\Gamma < \Gamma_m$} the atmosphere is absolutely stable to upward displacements
$\Gamma_m < \Gamma < \Gamma_d$} the atmosphere is conditionally unstable to upward displacements
$\Gamma = \Gamma_m$} the atmosphere is neutral to upward displacements
$\Gamma > \Gamma_d$} the atmosphere is absolutely unstable to upward displacements

where Γ_m is the saturated lapse rate. Unlike Γ_d, which is a constant, Γ_m depends on the moisture content and accounts for the latent heating and freezing from phase changes of water, which renders it less than Γ_d.

The lapse rate from radiative-dry convective equilibrium is greater than Γ_m. So convective processes come into play to adjust the temperature profile to a moist adiabatic lapse rate, resulting in cooler near surface and warmer upper air temperatures, verifying far more with observations. The radiative-convective equilibrium operates relatively quite well to understand the long-term mean temperature profile. However, the tropical atmosphere variations will not happen if the thermodynamic profile of the atmosphere was always moist neutral. The seasonal, diurnal, interannual, and other long and short time variations of insolation result in convective processes to constantly act to restore the radiative-convective equilibrium in a time mean sense.

3.4 Matsuno-Gill Atmosphere

In the late 1950s, it was proposed that the equatorial region of the planet may have some 'trapped motions' owing to the rotation of the planet (Yoshida 1959; Bretherton 1964). In the seminal paper of Matsuno (1966), the shallow water equations were solved on an equatorial beta plane linearized about a basic state at rest to show that the solutions produced inertia-gravity waves, Kelvin waves, Rossby waves, and mixed Rossby-gravity (also called Yanai) waves. Soon after the mixed Rossby-gravity waves were observed, Kelvin waves were also observed in the equatorial stratosphere (Yanai and Maruyama 1966) and in the equatorial oceans (Wunsch and Gill 1976). These waves were dry or free modes, which are uncoupled to convection or any other forcing. However, their initial source was assumed to be moist convection in the troposphere since the vertical structure of these waves showed upward energy dispersion, consistent with forcing from below (Holton 1972, 1973). In the tropical troposphere, these waves are forced by convection, and in turn, these waves organize convection (e.g., the MJO), hence termed convectively coupled equatorial waves (Kiladis et al. 2009).

The shallow water equations are preferred for deriving the theory of these equatorial waves because they are relatively simple, and the horizontal scales of these waves are much larger than their vertical scales. The shallow water equations govern vertically independent motions of a rotating fluid which is incompressible and homogeneous. The shallow water model has essentially one vertical level so that it cannot directly include factors that vary with height. In cases where the mean state is relatively simple, the vertical variations can be separated from the horizontal variations by several sets of shallow water equations that describe the mean state. As the solutions of the shallow water model discussed below will be based on a linearized basic state, any linear combination of the solutions will also serve as a solution. The governing equations of the shallow water model for inviscid flow are the horizontal momentum equations (3.9 and 3.10) and the mass continuity equation (3.11):

$$
\overbrace{\frac{\partial u}{\partial t}}^{\text{Term 1}} + u\frac{\partial u}{\partial x} + v\frac{\partial u}{\partial y} - fv = -\frac{\partial \phi}{\partial x} \tag{3.9}
$$

$$
\overbrace{\frac{\partial v}{\partial t}}^{\text{Term 1}} + u\frac{\partial v}{\partial x} + v\frac{\partial v}{\partial y} + fu = -\frac{\partial \phi}{\partial y} \tag{3.10}
$$

$$
\frac{\partial u}{\partial x} + \frac{\partial v}{\partial y} + \frac{\partial w}{\partial z} = 0 \tag{3.11}
$$

Here, $\phi \equiv$ geopotential height $= gh$, where h is the height. Since we are interested in steady-state solutions, Term 1 in Eqs. 3.9 and 3.10 disappears. Vertically integrating the mass continuity equation from top to bottom of the fluid depth, we get

$$
w_{\text{top}} - w_{\text{bot}} = -h\left(\frac{\partial u}{\partial x} + \frac{\partial v}{\partial y}\right) \tag{3.12}
$$

Assuming a fixed bottom (which implies $w_{\text{bot}} = 0$) and free surface (which implies $w_{\text{top}} \neq 0$), we can rewrite Eq. 3.12 as

$$
w_{\text{top}} = \frac{\partial h}{\partial t} = -h\left(\frac{\partial u}{\partial x} + \frac{\partial v}{\partial y}\right) \tag{3.13}
$$

We then split the state variables of Eqs. 3.9, 3.10, and 3.13 as the sum of the basic state (represented by overbar) and perturbation about this state (represented by prime). This is followed by linearizing Eqs. 3.9, 3.10, and 3.13, about the atmospheric basic state, which is at rest (which implies $\bar{u} = \bar{v} = \nabla\bar{\phi} = 0$; horizontal pressure gradients of the mean state are absent). Linearization assumes that forced motion is sufficiently weak and can be treated by linear dynamics. We then get

$$\frac{\partial u'}{\partial t} - fv' = -g\frac{\partial h'}{\partial x} \tag{3.14}$$

$$\frac{\partial v'}{\partial t} + fu' = -g\frac{\partial h'}{\partial y} \tag{3.15}$$

$$\frac{\partial h'}{\partial t} + H\left(\frac{\partial u'}{\partial x} + \frac{\partial v'}{\partial y}\right) = 0 \tag{3.16}$$

where in Eq. 3.16, H is the mean depth of the fluid. We now apply β plane approximation. This approximation introduces only a 14% error up to 30° from the equator (Gill 1982). This approximation as it is applied near the equator entails the following::

Making a small angle assumption in the tropics, we can say $\sin\lambda \approx \lambda$. Therefore,

$$f = 2\Omega\sin\lambda \approx 2\Omega\lambda$$

Similarly, making the small angle assumption in the tropics, we can say $\cos\lambda = 1$. Therefore,

$$\beta = \frac{\partial f}{\partial y} = \frac{2\Omega\cos\lambda}{R} \approx \frac{2\Omega}{R}$$

where R is the radius of the Earth. In essence, the β plane approximation amounts to setting f to linearly vary in space as

$$f = f_o + \beta y \tag{3.17}$$

where f_o is the Coriolis parameter at a reference latitude λ_o and y is the north-south displacement about this reference latitude. For the special case of a β plane centered about the equator, Eq. 3.17 reduces to

$$f = \beta y \tag{3.18}$$

Equation 3.18 is called the equatorial β plane approximation. We apply this approximation to avoid the complicated nonlinear terms that would otherwise appear in the momentum equations and make the solutions of Eqs. 3.14, 3.15 and

3.16 much more complicated. So after the β plane approximation is applied to Eqs. 3.14 and 3.15, we get

$$\frac{\partial u'}{\partial t} - \beta y v' = -g \frac{\partial h'}{\partial x} \tag{3.19}$$

$$\frac{\partial v'}{\partial t} + \beta y u' = -g \frac{\partial h'}{\partial y} \tag{3.20}$$

Applying a waveform of solution that takes the form:

$$u'(x, y, t) = U(y)e^{i(kx - \eta t)} \tag{3.21}$$

$$v'(x, y, t) = V(y)e^{i(kx - \eta t)} \tag{3.22}$$

$$h'(x, y, t) = \overline{H}(y)e^{i(kx - \eta t)} \tag{3.23}$$

where $U(y)$, $V(y)$, and $\overline{H}(y)$ are the amplitudes of each of the variables due to the wave that varies in the y direction, k is the zonal wave number, and η is the frequency of the waveform. Substituting Eqs. 3.21, 3.22, and 3.23 in Eqs. 3.19, 3.20, and 3.16, respectively, we get

$$-i\eta U - \beta y V = -ikg\overline{H} \tag{3.24}$$

$$-i\eta V + \beta y U = -g \frac{\partial \overline{H}}{\partial y} \tag{3.25}$$

$$-i\eta \overline{H} + H \left(ikU + \frac{\partial U}{\partial y} \right) = 0 \tag{3.26}$$

From Eq. 3.24, we get

$$U = \frac{\left(ikg\overline{H} - \beta y V \right)}{i\eta} \tag{3.27}$$

Substituting Eq. 3.27 in Eqs. 3.25 and 3.26, we get

$$\left(\beta^2 y^2 - \eta^2 \right) V - ikg\beta y \overline{H} - i\eta g \frac{\partial \overline{H}}{\partial y} = 0 \tag{3.28}$$

$$\left(\eta^2 - gHk^2\right)\overline{H} + i\eta H \left(\frac{\partial \eta}{\partial y} - \frac{k\beta y V}{\eta}\right) = 0 \tag{3.29}$$

From Eq. 3.29, we get

$$\overline{H} = i\eta H \left(\frac{\left(\frac{k\beta y V}{\eta} - \frac{\partial \eta}{\partial y}\right)}{\left(\eta^2 - gHk^2\right)}\right) \tag{3.30}$$

Substituting Eq. 3.30 in Eq. 3.28, we get

$$\frac{\partial^2 V}{\partial y^2} + \left(\frac{\eta^2}{gH} - \frac{\beta^2 y^2}{gH} - k^2 - \frac{\beta k}{\eta}\right) V = 0 \tag{3.31}$$

We have the functional form for U in Eq. 3.27, $\overline{H}$ in Eq. 3.30, and now we require a functional form for V from Eq. 3.31 to complete the specification of the structure of the wave. We will consider two asymptotic solutions to Eq. 3.31.

Asymptotic Solution 1 Consider the non-divergent limit for Eq. 3.31. This implies that $H \to \infty$ from Eq. 3.16. Therefore, Eq. 3.31 reduces in this case to

$$\frac{\partial^2 V}{\partial y^2} = \left(k^2 + \frac{\beta k}{\eta}\right) V \tag{3.32}$$

Equation 3.32 will have wave solutions of the following form:

$$V = Ae^{iny}, \text{ which implies } \frac{\partial^2 V}{\partial y^2} = -n^2 V \tag{3.33}$$

where n is the meridional wave number. Substituting Eq. 3.33 in Eq. 3.32, we get

$$\eta = \frac{-\beta k}{\left(k^2 + n^2\right)} \tag{3.34}$$

Equation 3.34 is the dispersion relationship for Rossby waves. It may be noted that this dispersion relationship is for a basic state at rest. Incidentally, the dispersion relationship in Eq. 3.34 is identical to that for Rossby waves in mid-latitude flow. Therefore, equatorial Rossby waves are also potential vorticity conserving waves.

Asymptotic Solution 2 Consider the non-rotating limit for Eq. 3.31, which implies $\beta = 0$. Then Eq. 3.31 reduces to

$$\frac{\partial^2 V}{\partial y^2} = -\left(\frac{\eta^2}{gH} - k^2\right) V \tag{3.35}$$

Again, using the solution of the form in Eq. 3.33 in Eq. 3.35, we get

$$\eta = \pm\sqrt{gH\left(k^2 + n^2\right)} \equiv \eta_{\text{pure gravity}} \tag{3.36}$$

Equation 3.36 describes the dispersion relationship of two-dimensional external gravity waves, one propagating eastward and the other westward. They are external as opposed to internal gravity waves because we have assumed a free surface at the top unlike finding the solution for a density stratified fluid, like the ocean. They are pure gravity waves that are not influenced by Earth's rotation.

From these two asymptotic solutions, we observe that the shallow water equations support divergent, irrotational gravity wave (with buoyancy as the restoring force) and non-divergent, rotational Rossby wave (with potential vorticity as the restoring force). To seek the general form of the solution to Eq. 3.31, we require that

$$V \to 0 \text{ as } y \to \pm\infty \tag{3.37}$$

to satisfy the requirement of the equatorial β plane approximation (i.e., the waves are bounded within the acceptable meridional limits of the equatorial β plane approximation). Matsuno (1966) showed that the solution to Eq. 3.31 that satisfies the condition stated in Eq. 3.37 occurs only if

$$\frac{\sqrt{gH}}{\beta}\left(\frac{\eta^2}{gH} - k^2 - \frac{\beta k}{\eta}\right) = 2n + 1, n = 0, 1, 2, 3\ldots \tag{3.38}$$

The LHS of Eq. 3.38 must be equal to an odd integer for the solution of Eq. 3.31 to be viable under the equatorial β plane approximation. Substituting Eq. 3.38 in Eq. 3.31, Matsuno (1966) obtained a form of the Schrodinger equation that provides the structural form of the equatorial waves. The Schrodinger equation for an oscillator, therefore, is

$$\frac{\partial^2 V}{\partial y^2} + (2n + 1)\,V = 0 \tag{3.39}$$

The solution of the Schrodinger equation takes the following form:

$$V(y) = Ae^{\left[\left(\frac{-y^2\sqrt{gH}}{2\beta}\right)P_n\left(\frac{(gH)^{\frac{1}{4}}y}{\sqrt{\beta}}\right)\right]} \tag{3.40}$$

The solution of this Schrodinger equation (3.40) involves a Hermite polynomial, P of order n. This order n is a form of the meridional mode and appears as a nodal point that can be identified by the point where the meridional wind goes to zero in the north-south direction.

Alternatively, Eq. 3.38 can be rewritten as

$$\frac{\eta^2}{gH} - k^2 - \frac{\beta k}{\eta} = (2n+1)\frac{\beta}{\sqrt{gH}} \tag{3.41}$$

Equation 3.41 is a cubic equation in η. Therefore, there are three values of η when n and k are specified. For $n \geq 1$, the three values of η are unique for the whole range of k. Two of these are the (eastward and westward propagating) inertia-gravity waves, and the remaining one is the Rossby wave. The inertia-gravity waves acknowledge the presence of Earth's rotation as opposed to the pure gravity waves of Eq. 3.36. Figure 3.8 illustrates the three frequencies for arbitrary values of k and for some definite meridional mode n, on a linear scale. Kelvin wave is not captured as one of the three frequencies, and that will be considered as a special case of this dispersion relationship in Eq. 3.39.

To extract the Rossby wave solution, we assume low frequency (since we know Rossby waves operate on synoptic scales, often called planetary scale waves). This assumption allows us to ignore $\frac{\eta^2}{gH}$ in Eq. 3.38, and rearranging the equation, we get

$$\eta_{\text{Rossby Wave}} = \frac{-\beta k}{k^2 + (2n+1)\frac{\beta}{\sqrt{gH}}} \tag{3.42}$$

Equation 3.42 provides the frequency of the equatorial Rossby wave, and its phase speed ($c_{\text{Rossby Wave}}$) is given by

$$c_{\text{Rossby Wave}} = \frac{\eta_{\text{Rossby Wave}}}{k} = \frac{-\beta}{k^2 + (2n+1)\frac{\beta}{\sqrt{gH}}} \tag{3.43}$$

The propagation direction of the equatorial Rossby wave is westward (because η and k have opposite signs; Fig. 3.8). The phase speed in Eq. 3.43 is a function of wave number, and therefore, Rossby wave is dispersive. Using appropriate values of n in Eq. 3.40, we can solve for V and obtain η from Eq. 3.42. Then using this value of V in Eq. 3.30, we can solve for $\overline{H}$ and then use Eq. 3.27 to solve for U. Thereupon, we can compute u', v', h' from Eqs. 3.21, 3.22, and 3.23 to obtain the spatial structures for the equatorial Rossby wave. For $n = 1$, the equatorial Rossby wave structure shows a pair of low/high pressure systems straddling the equator (Fig. 3.9e).

To extract the high-frequency inertia-gravity waves, we neglect $\frac{\beta k}{\eta}$ in Eq. 3.38 because this term becomes smaller as η increases compared to the other terms. Again, rearranging Eq. 3.38, we get

$$\eta_{\text{Inertia-Gravity}} = \pm\left((2n+1)\beta\sqrt{gH} + k^2 gH\right)^{\frac{1}{2}} \tag{3.44}$$

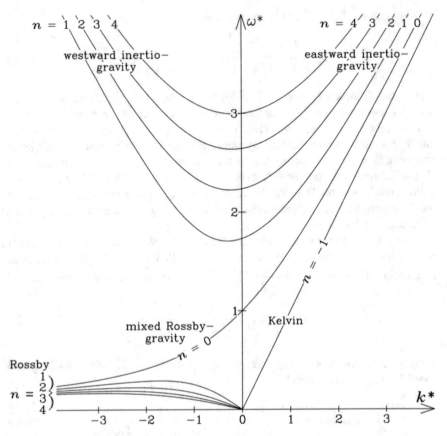

Fig. 3.8 The solution of Eq. 3.40 showing non-dimensionalized frequencies ($\omega^* = \frac{\eta}{(\beta\sqrt{gH})^{1/2}}$) along the ordinate for arbitrary values of the non-dimensionalized zonal wave number ($k^* = k\left(\frac{\sqrt{gH}}{\beta}\right)^{1/2}$) along the abscissa. It may be noted that solutions with the same and opposite signs of frequency and wave number are eastward- and westward-propagating solutions, respectively. (Reproduced from Kiladis et al. 2009)

$$c_{\text{Inertia–Gravity}} = \frac{\pm\left((2n+1)\,\beta\sqrt{gH} + k^2 gH\right)^{\frac{1}{2}}}{k} \tag{3.45}$$

In Eqs. 3.44 and 3.45, positive and negative signs for frequency and wave number relate to eastward- and westward-propagating inertia-gravity waves, respectively. For large k (i.e., small wavelength) and small n, Eqs. 3.44 and 3.45 become

$$\eta_{\text{pure gravity}} = \pm k\sqrt{gH} \tag{3.46}$$

$$c_{\text{pure gravity}} = \pm\sqrt{gH} \tag{3.47}$$

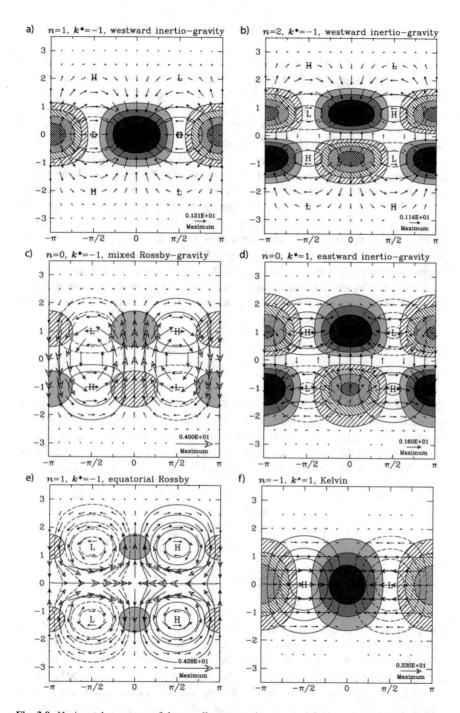

Fig. 3.9 Horizontal structures of the zonally propagating equatorial waves that arise as solutions to the shallow water equations on an equatorial β plane. k^* is a nondimensional zonal wave number, and the fields have been non-dimensionalized by taking units of frequency and wave number as noted in the caption of Fig. 3.8. The equator is represented by 0 in the ordinate. Divergence is denoted by hatching and convergence by shading at a 0.6-unit intervals. The contours represent geopotential with a contour interval of 0.5 units, and dashed lines represent negative geopotential. The scale of the maximum wind vector in each panel is specified in the bottom right corner. (Reproduced from Kiladis et al. 2009)

Equations 3.46 and 3.47 provide the frequency and phase speed of pure gravity waves ($\beta = 0$). Unlike the phase speed of Rossby waves (Eq. 3.43) or the phase speed of inertia-gravity waves (Eq. 3.45), the phase speed of the pure gravity wave is independent of the wave number. Therefore, pure gravity waves are non-dispersive. From Fig. 3.8, it can also be seen that all inertia-gravity waves propagating either eastward or westward approach their phase asymptotically toward that of the pure gravity waves (e.g., Kelvin wave; Equation 3.47). Again, the spatial structure of the inertia-gravity waves can be arrived in the same way as the equatorial Rossby wave structure was obtained. The structures of the inertia-gravity waves are shown in Fig. 3.9a, b. For $n = 1$, the westward-propagating inertia-gravity wave shows a very symmetrical structure about the equator with divergence/convergence maximized at the equator in phase with the meridional wind and in quadrature with the zonal winds and the height field (Fig. 3.9a). Similarly, the spatial structure of the eastward-propagating inertia-gravity wave for $n = 1$ can be visualized from Fig. 3.9a multiplied by -1. At $n = 2$, these wave structures of the inertia-gravity waves develop a node at the equator and a node on either side of the equator (Fig. 3.9b).

Setting $n = 0$ in Eq. 3.41 yields

$$\eta_{n=0} = k\sqrt{gH}\left(\frac{1}{2} \pm \frac{1}{2}\left(1 + \frac{4\beta}{k^2\sqrt{gH}}\right)^{1/2}\right) \tag{3.48}$$

The positive root of Eq. 3.48 is an eastward moving inertia-gravity wave. However, the negative root of Eq. 3.48 corresponds to a westward moving inertia-gravity wave for small wave number (or long waves $k \to 0$) and a westward-propagating Rossby wave for zonal scales characteristic of synoptic scale. As a result, this negative root of Eq. 3.48 represents the frequency of a mixed Rossby-gravity wave (see Fig. 3.8). Therefore,

$$\eta_{\text{Mixed Rossby-Gravity}} = \frac{k\sqrt{gH}}{2}\left(1 - \left(1 + \frac{4\beta}{k^2\sqrt{gH}}\right)^{1/2}\right) \tag{3.49}$$

$$c_{\text{Mixed Rossby-Gravity}} = \frac{\sqrt{gH}}{2}\left(1 - \left(1 + \frac{4\beta}{k^2\sqrt{gH}}\right)^{1/2}\right) \tag{3.50}$$

Once again, the spatial structure of the mixed Rossby-gravity waves can be obtained as in the case of Rossby waves and inertia-gravity waves. The spatial structures of the equatorial Rossby component of the westward-propagating mixed Rossby-gravity wave at zonal scales characteristic of synoptic scale disturbances appear as a lobe of high and low pressures, symmetric about the equator (Fig. 3.9c). The stronger inertia-gravity wave component of the eastward-propagating mixed Rossby-gravity wave for long waves (or small wave numbers) also appears as a pair of high and low pressures straddling the equator but with higher intensity (Fig. 3.9d).

As mentioned earlier, equatorial Kelvin waves are not part of the solution of Eq. 3.41. But these waves are observed very distinctly both in the equatorial oceans and the atmosphere. To extract the structure of the equatorial Kelvin waves, we must revisit Eqs. 3.16, 3.19, and 3.20. These waves have no meridional component of motion. Therefore, $v' = 0$ and Eqs. 3.19, 3.20, and 3.16 reduce to

$$\frac{\partial u'}{\partial t} = -g\frac{\partial h'}{\partial x} \tag{3.51}$$

$$\beta y u' = -g\frac{\partial h'}{\partial y} \tag{3.52}$$

$$\frac{\partial h'}{\partial t} + H\left(\frac{\partial u'}{\partial x}\right) = 0 \tag{3.53}$$

Applying the wave solutions of Eqs. 3.21, 3.22, and 3.23 in Eqs. 3.51, 3.52, and 3.53, respectively, we get

$$\frac{\eta}{k} = \frac{g\overline{H}}{U} \tag{3.54}$$

$$\beta y U = -g\frac{\partial \overline{H}}{\partial y} \tag{3.55}$$

$$-i\eta\overline{H} + iHkU = 0 \tag{3.56}$$

Substituting Eq. 3.54 in Eq. 3.56, we get the frequency of the equatorial Kelvin wave:

$$\eta_{\text{KelvinWave}} = +k\sqrt{gH} \tag{3.57}$$

Therefore,

$$c_{\text{KelvinWave}} = +\sqrt{gH} \tag{3.58}$$

Differentiating Eq. 3.54 with respect to y, we get

$$g\frac{\partial \overline{H}}{\partial y} = \frac{\eta_{\text{KelvinWave}}}{k}\frac{\partial U}{\partial y} \tag{3.59}$$

Substituting Eq. 3.59 in Eq. 3.55, we get

$$\frac{1}{U}\frac{\partial U}{\partial y} = -\frac{k\beta y}{\eta_{\text{KelvinWave}}} = -\frac{\beta y}{c_{\text{KelvinWave}}} \tag{3.60}$$

Integrating Eq. 3.60, we get

$$U\left(y\right) = U_o e^{-\left(\frac{\beta y^2}{2c_{\text{KelvinWave}}}\right)} \tag{3.61}$$

U_o is a constant and is the amplitude of the zonal wind perturbation associated with Kelvin Wave. The exponent in Eq. 3.61 must be negative so that $U(y)$ in Eq. 3.61 decays away from the equator. This is the reason why only the positive sign of the phase speed of the Kelvin wave in Eqs. 3.57 and 3.58 is retained. Therefore, the equatorial Kelvin wave is eastward propagating. Setting the exponent of Eq. 3.61 to 1, we have

$$\frac{\beta y^2}{2c} = 1 \tag{3.62}$$

or

$$y = \sqrt{\frac{2c_{\text{KelvinWave}}}{\beta}} \tag{3.63}$$

Equation 3.63 provides the decay width of the equatorial Kelvin wave or also called the Rossby radius of deformation, which denotes the length scale where the rotational forces become as important as the buoyancy force. Setting $n = -1$ in Eq. 3.39 recovers the frequency of the eastward-propagating inertia-gravity wave (Eq. 3.44). But since we are dealing with a pure gravity wave, Eq. 3.57 follows as the frequency of the equatorial Kelvin wave. Therefore, in Fig. 3.8, the eastward-propagating equatorial Kelvin wave appears for $n = -1$. Revisiting Fig. 3.8, we can see that the gap between the low frequency of the equatorial Rossby waves and the high frequency is filled by the Yanai (or mixed Rossby-gravity waves) and the equatorial Kelvin waves. The Kelvin wave structure when $n = -1$ has predominantly zonal wind near the equator with latitudinal symmetry of the height and wind fields about the equator and the in-phase relationship between height and wind (Fig. 3.9f).

    The phase speed of all these equatorial waves in Eqs. 3.43, 3.45, 3.47, 3.50, and 3.58 is dependent on H, the depth of the shallow water fluid. This depth, H, determines the phase speed and the scale of the waves. In the context of the real atmosphere, this depth is called the equivalent depth, which cannot be perceived as a physical depth but as a parameter of the equatorial wave theory. The equivalent depth is the depth of the shallow water wave that gives the observed phase speed in the real atmosphere. Haertel and Kiladis (2004) found that the deep convective heating profile with a single maximum in the mid-troposphere (e.g., Fig. 3.2b, e) projected very strongly on gravity wave modes with a phase speed of 49 m/s (or

$H = 245$ m) while the latent heating profile with opposite signs in the upper and lower troposphere (e.g., Fig. 3.2c, f) projected on gravity wave modes with a phase speed of 23 m/s (or $H = 54$ m). It may be noted that atmospheric motions with a barotropic structure (which has a constant profile in the troposphere) display large equivalent depths. The vertical modes associated with medium equivalent modes have a single phase reversal in the mid-troposphere that represents baroclinic motions. In the equatorial β plane, the free waves with large equivalent depths have large latitudinal extents, and such atmospheric motions propagate to high latitudes with a barotropic vertical structures (such as the teleconnection patterns emanating from the tropics). On the other hand, free waves of medium equivalent depths excited, say, in the mid-latitudes tend to propagate equatorward. Such tropical motions have a baroclinic structure, exhibiting the midlatitude influence on the tropics.

The theoretical framework of the equatorial waves is also firmly grounded by observations. Wheeler and Kiladis (1999) showed the projection of the satellite-derived brightness temperature on zonal wave number-frequency space using Fourier analysis (Fig. 3.13a, b). The dominant signals in these figures are identified with all the equatorial waves. But Sun and Wu (2020) make a very important claim that these convectively coupled equatorial waves are local and zonally asymmetric in contrast to Figs. 3.10a, b that suggest that these waves circumnavigate. They indicate that because the atmosphere is dissipative with timescales of 5–10 days, slowly propagating waves like the mixed Rossby-gravity wave with a group velocity of about 5 ms^{-1} can be observed within a few thousand kilometers of the convective region. Therefore, they suggest that there are preferred wave activity regions in the tropics rather than tropical waves over the whole tropical belt. Furthermore, Fig. 3.10 suggests power to these waves at Wavenumber 0 that is not physical. Finally, the power spectrum shown in Fig. 3.10 is a ratio of the raw power to the power of a smoothed red noise background spectrum. This ratio plays a crucial role in determining the appearance of the waves, as illustrated in Fig. 3.10. But the (amount of) smoothing, which is optimal in Fig. 3.10, lacks a clear objective criterion and is subjective.

Gill (1980) solved the shallow water equations with friction and prescribed heating to arrive at some very illuminating solutions on the tropical circulations. Gill (1980) used the frictional damping to take the form of Rayleigh damping (i.e., linear drag is proportional to wind speed) and thermal damping to take the form of Newtonian cooling (i.e., the heating rate is proportional to temperature perturbation). In this model, the vertical velocity is proportional to the heating rate that is imposed. Therefore, the prescribed heating enters the shallow water model through the mass continuity equation. The thermal and frictional damping rates are also assumed to have the same timescale. Gill (1980) neglected high-frequency inertia-gravity waves and short Rossby waves in their solution of the shallow water equations. The solution of the shallow water equations for heating that is imposed symmetrically and asymmetrically to the equator is shown in Figs. 3.11 and 3.12, respectively.

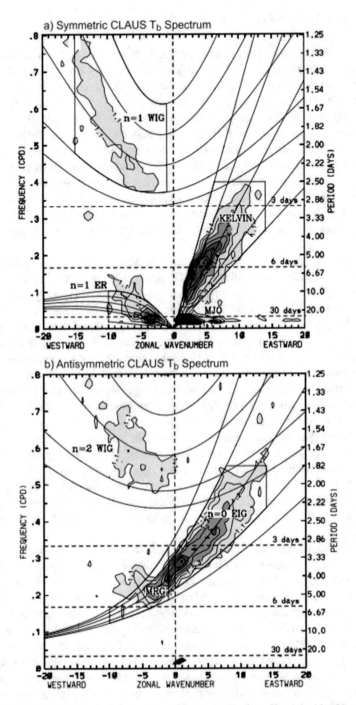

Fig. 3.10 The Fourier analysis showing the power spectrum of the Cloud Archive User Service (CLAUS) brightness temperature in wave number-frequency for (**a**) symmetric and (**b**) asymmetric components for July 1983 to June 2005 summed between 15°N and 15°S. The shaded values are plotted as a ratio of the raw power and the power in a smoothed red noise background spectrum. Shading shows signal (or ratio) is significant at greater than 95% confidence interval. Dispersion curves for Kelvin wave at $n = -1$, equatorial Rossby wave at $n = 1$, westward-propagating inertia-gravity waves at $n = 1$ and 2, eastward-propagating inertia-gravity wave at $n = 0$, and mixed Rossby-gravity waves are plotted for equivalent depths of 8, 12, 25, 50, and 90 m. (Reproduced from Kiladis et al. 2009)

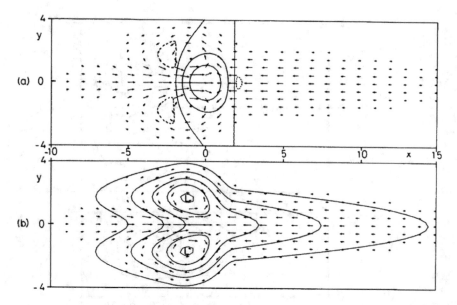

Fig. 3.11 Solutions from the shallow water model for imposed symmetric heating about the equator. (**a**) Contours of vertical velocity (positive is solid contour) with wind field for the lower troposphere. The vertical velocity is proportional to the imposed heating and acquires the same shape as the imposed heating function. (**b**) Contours of perturbation pressure field (negative is solid contour) with wind vectors for the lower troposphere. (Reproduced from Gill 1980)

The solution to the symmetric heating has two parts (Fig. 3.11b). One part comprises the equatorial Rossby wave, and the other part is the equatorial Kelvin wave (Fig. 3.11b). The positive contours of the upward vertical motion circumscribe the heating function (Fig. 3.11a). The contours of pressure perturbation, which is negative everywhere, suggest an equatorial trough to the east of the forcing region, with relatively higher pressure on the west of the forcing region (Fig. 3.11b). There are two lobes of low pressure straddling the equator to the west of the forcing (Fig. 3.11b). The easterlies prevail to the east of the forcing region and westerlies to the west of the forcing region (Fig. 3.11a, b). If the forcing region (where heating is applied) is taken to represent the Maritime Continent region, then this symmetric solution represents the Walker circulation over the Pacific with easterly trades flowing toward the forcing region and vertical motion rising over the forcing region. These solutions are very reminiscent of the equatorial Kelvin wave, which is damped out as it progresses eastward (in the Gill model), and Rossby wave (westward of the forcing region) with an $n = 1$ structure. The elongated structure of the equatorial trough relative to the rather muted structure of the Rossby wave, west of the forcing region, is a result of the different phase speeds of the Kelvin and the Rossby waves. Using Eq. 3.43, the phase speed of the equatorial Rossby wave at $n = 1$ and for long wave approximation ($k \to 0$, which implies $k^2 \sim 0$) suggests that the phase speed of the equatorial Rossby wave is about three times slower than that of the equatorial

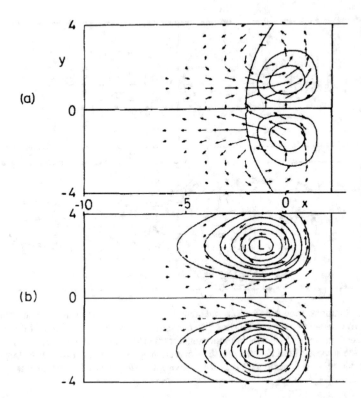

Fig. 3.12 Solutions from the shallow water model for imposed asymmetric heating about the equator. (**a**) Contours of vertical velocity are proportional to the imposed heating rate. The upward vertical motion is in the northern hemisphere ($0 < y < 4$), and the contours in the heating hemisphere acquire the same shape as the heating function. Subsidence dominates in the opposite hemisphere ($-4 < y < 0$). (**b**) Contours of perturbation pressure, which is negative in the northern hemisphere (low pressure) and positive in the southern hemisphere (high pressure)

Kelvin wave. This gives the equatorial Kelvin wave a spatial decay rate of three times more than the Rossby wave.

The solution of asymmetric heating also has two components (Fig. 3.12). The first component comprises the mixed Rossby-gravity wave, and the second component comprises the Rossby wave. The positive contour of the vertical velocity to the north of the equator circumscribes the prescribed heating pattern, while the dashed contours to the south of the equator suggest subsidence (Fig. 3.12a). The pressure contours and the cyclonic flow to the north of the equator in Fig. 3.12b suggest a low-pressure center and a high-pressure center with the anticyclonic flow to the south. The cross-equatorial flow is down the pressure gradient. Of note in Fig. 3.12b is the absence of the equatorial Kelvin wave response and the resemblance to the Rossby wave response for $n = 2$. The mixed Rossby-gravity wave is confined to the forcing region, and long mixed Rossby-gravity waves do not propagate. The asymmetric solution in Fig. 3.12 essentially describes the Hadley Circulation with

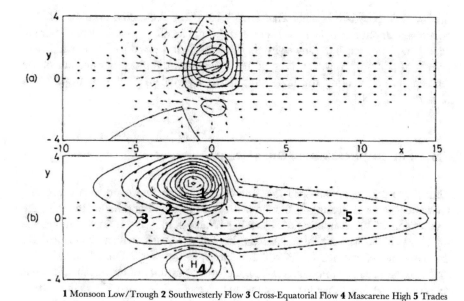

1 Monsoon Low/Trough 2 Southwesterly Flow 3 Cross-Equatorial Flow 4 Mascarene High 5 Trades

Fig. 3.13 The solution of the shallow water equation from the addition of the solution from the symmetric (Fig. 3.11) and asymmetric (Fig. 3.12) heating. (**a**) Contours of vertical velocity with a contour interval of 0.3 indicate the dominance of heating north of the equator along with the lower tropospheric wind vectors. The flow to the east of the forcing region is similar to Fig. 3.11, entirely contributed by the symmetric part of the heating. (**b**) Contours of perturbation pressure with a contour interval of 0.3. The features of the flow are identified in the index at the bottom of the panel. (Reproduced from Gill 1980)

the ascending part of the circulation in the heating hemisphere and the descent in the opposite hemisphere.

The linear addition of the two (symmetric and asymmetric; Fig. 3.13) for the total solution resembles much of the lower troposphere circulation patterns for boreal summer with easterlies in the Pacific, westerlies in the Indian Ocean, the anticyclonic circulation of the Mascarene High in the south, the cross-equatorial flow, the southwesterly flow, and the Indian Monsoon trough, with the Indian Monsoon rainfall providing the asymmetrical part of the latent heating and the ITCZ over the Indian and the Pacific Ocean offering the symmetric component of the prescribed heating (Fig. 3.13b). The beauty of this solution as presented in Gill (1980) is that it appears as a linearly separable solution of the symmetric and asymmetric components of heating in addition to formally providing the theory of the importance of the diabatic heating to the tropical atmospheric circulation.

These elegant Gill solutions are, however, not bereft of any criticisms. One criticism is that the damping coefficients (for both Newtonian and Rayleigh damping) employed in the Gill solutions are considerably larger than can be justified by the physical processes occurring in the free troposphere (Wu et al. 2000a). Another criticism is that the Gill solution for the response of the wind in the lower troposphere is somewhat artificial because convective heating that has significant amplitude in the mid-troposphere would ideally, given the large damping coefficients, not reach the lower troposphere to elicit a response in the wind field (Wu et al. 2000b). However, it could be argued that turbulent entrainment of the momentum could bring the heat-induced wind signal to the lower troposphere (Chiang et al. 2001).

3.5   The Implication on the Oceans

Thus far, we have emphasized the atmosphere, which seems obvious given that the condensational heating in the clouds resides in the atmosphere. However, the implication of the condensational heating is also felt in the oceans through the influence of the latent heat release on the surface winds, which are critical for the dynamics of the oceans. In addition, precipitation also affects the freshwater flux into the oceans, which then affects the surface salinity of the oceans. This has relatively less impact than the surface winds, as precipitation just affects the surface layer of the ocean and is somewhat localized to the regions where rainfall occurs. Clouds, however, also affect the net heat flux into the ocean, where the types of clouds (e.g., high, low, or mid-level) have different implications on the net heat flux (cf. Sect. 5.4). Such cloud types are dependent among other factors, on the vigor, and the extent of atmospheric convection. Be it as may be, we will focus our attention on the implication of the condensational heating on the surface wind in this section. Surface winds affect the surface and even the deep ocean currents, locally and remotely.

As noted in the previous section, knowing the spatial distribution of latent heating will produce a wind response as a linear solution to the linearized shallow water equations. Another theory for surface wind follows from Lindzen and Nigam (1987), which determines the surface wind based on SST. The Lindzen and Nigam (1987) model posits that the surface winds are driven by pressure gradients imposed by the turbulent fluxes in the boundary layer with lower pressure over warm SST and high pressure over cool SST. Although convection does not appear in this argument for surface winds, the SST gradients are, however, to some extent associated with the presence and the absence of deep convection and the associated clouds. Chiang et al. (2001), however, indicate that the free-tropospheric pressure gradient (associated with deep convection) tends to dominate the departure of the zonal component of the surface winds from its zonal mean while the SST gradients have a dominance on the departure of the meridional component of the winds from its zonal mean. Observational studies also attest to this conclusion from Chiang et al. (2001), which finds that the time mean meridional pressure gradient imposed on the atmospheric boundary layer by the SST dominates on the meridional component of the surface wind (Raymond et al. 2004). However, Raymond et al. (2004) find that at sub-seasonal and synoptic scales, the pressure gradient imposed by the free troposphere dominates on the meridional component of the surface wind.

Nigam and Chung (2000) suggest that the vertical distribution of the latent heating affects the surface winds. Their argument follows from the thermodynamic equation, which can be written as

$$\frac{\partial T}{\partial t} + \vec{V} . \nabla T - \frac{T}{\theta} \left(\frac{\partial \theta}{\partial p} \right) \omega = \dot{Q} \tag{3.64}$$

where $\dot{Q}$ is the diabatic heating rate (inclusive of radiative, turbulent heat flux divergence and condensational heating). Since in the tropical convective systems the dominant thermodynamic balance is between diabatic (condensational) heating and adiabatic cooling (from the vertical ascent in deep convection), it follows from Eq. 3.64 that

$$-\omega \approx \frac{\dot{Q}}{\sigma} \tag{3.65}$$

where $\sigma = \frac{T}{\theta} \left(\frac{\partial \theta}{\partial p} \right)$. By using the mass continuity equation and substituting Eq. 3.65 for ω in it, we get

$$\nabla . \vec{V} = -\frac{\partial \omega}{\partial p} \approx \frac{\partial \left(\frac{\dot{Q}}{\sigma} \right)}{\partial p} \tag{3.66}$$

Equation 3.66 shows that divergence in the tropics is related to the vertical gradient of the diabatic heating rate. This equation suggests that a bottom-heavy or a top-heavy latent heating profile will have a corresponding implication on the divergence to be either bottom heavy or top heavy, respectively. In fact, Nigam and Chung (2000) implicated the bias of a bottom-heavy latent heating profile for the strong near-equatorial easterly winds in the Pacific in a climate model simulation. It is generally thought that the adiabatic component of ω $\left(\omega_{adia} = -\left(\frac{\partial T}{\partial t} + \vec{V} . \nabla T \right) \right)$ is generally much smaller given that thermal gradients in the tropics are very weak and the local time rate of change of temperature is also small.

3.6 The Weak Temperature Gradient Balance

It is well known that the horizontal temperature and density gradients in the tropics are weak. This is primarily because in the tropics, f is small and the Rossby radius of deformation is large. Consequently, the internal gravity waves rapidly redistribute the heating anomalies over a large area (throughout the tropics) to maintain weak horizontal temperature gradients. Therefore, it is a common practice to neglect horizontal temperature variations in idealized modeling studies of the tropical climate (e.g., Neelin and Held 1987, Sobel and Bretherton 2000). This leads us to an approximate balance, indicated by Eq. 3.62, wherein adiabatic cooling balances the diabatic heating, which is called the WTG balance or approximation (Sobel et al. 2001).

In the parcel theory, the subcloud layer entropy (θ_e) and the environment lapse rate dictate the buoyancy of the air parcels. Sobel et al. (2001) and other studies before recognized, however, that moisture in the free troposphere (above the boundary layer) also dictates convection as the convective clouds entrain environment air and affect the buoyancy of the rising air parcels. Additionally, it is observed that the tropical boundary layer exhibits a quasi-steady moist static energy leading to the concept of boundary layer quasi-equilibrium (Emanuel 1995; Raymond 1995). In examining convection in the Western Pacific Warm Pool, Raymond (1995) indicated that convection is initiated when the boundary layer equivalent potential temperature exceeds a threshold value, which is determined by conditions above the cloud base (or the tropospheric virtual temperature profile). The study finds that convection is regulated by the balance between the tendency of the surface fluxes to increase and convective downdrafts to decrease the boundary layer equivalent temperature. This boundary layer quasi-equilibrium balance in the precipitating regions is achieved by the import of low entropy air through the convective downdrafts in the boundary layer that is balanced by the gain from the surface fluxes. Raymond (1995) further finds that clear-air entrainment from above and radiative cooling of the boundary layer plays a secondary role in the boundary layer equivalent potential temperature budget. Therefore, the free troposphere moisture can affect convection through this modulation of the subcloud layer entropy.

Adames and Maloney (2021) using a 2D (zonal and pressure plane) of the vertically truncated shallow water equations very nicely illustrate how the atmosphere adjusts to the WTG balance in response to anomalous heating/cooling (Fig. 3.14). As shown in Fig. 3.14a, when the monopole anomalous heating is turned on, an overturning circulation is initiated with the location of the vertical upward motion fixed at the heat source and flanked by descent. Thereafter, in about 30 min (Fig. 3.14b), the gravity waves propagate away from the region of heating and warm the troposphere through adiabatic subsidence. In the meanwhile, the WTG balance is nearly achieved with adiabatic cooling almost balancing the diabatic warming (Fig. 3.14b). After 1.5 h (Fig. 3.14c), the gravity waves have propagated far enough, and the atmosphere is in complete WTG balance. The difference in the adjustment to monopole (Fig. 3.14a–c) and dipole (Fig. 3.14d–f) heating anomalies is that in the former, the domain mean temperature changes while in the latter, the domain mean temperature anomalies are eliminated. It should be, however, recognized that adjustment to the WTG atmosphere in the 3D atmosphere with rotation is far more complex than what is illustrated in Fig. 3.14.

In the example shown in Fig. 3.14, the timescale to adjust to WTG balance $(\tau_{WTG}) = \frac{\text{speed of gravity waves}}{\text{Region's horizontal length scale}} = \frac{50\text{m/s}}{300\text{km}} = 1.6$ h. However, the timescale for achieving WTG balance can range from minutes for mesoscale convection to days in planetary scale circulations (Adames et al. 2019). Furthermore, it should be noted that for the first baroclinic modes (with a single maximum in the vertical profile of

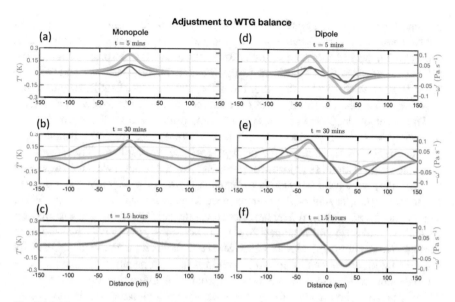

Fig. 3.14 The anomalous tropospheric temperature response diagnosed from a 2D vertically truncated shallow water model following Nicholls et al. (1991) for (**a–c**) monopole heat source and (**d–f**) dipole heat source at (a, d) time = 5 min after the heating/cooling is turned on, (**b, e**) time = 30 min, and (**c, f**) time = 90 min after the heating/cooling is turned on. The blue line, red line, and thick pink line show the mid-tropospheric temperature anomalies, the vertical velocity anomalies, and the vertical velocity that would balance the heat source, respectively. (Reproduced from Adames and Maloney 2021)

ω through the depth of the troposphere), the gravity wave speed is higher than for higher baroclinic modes. The gravity wave speed decreases with decreasing vertical wavelength.

The time to adjust to the WTG balance therefore is found to be critical for the coupling of the large scale to convection. If the adjustment is fast and the boundary layer moist static energy remains nearly fixed, then changes in the free troposphere moisture become critical for the genesis and organization of convection (Adames and Maloney 2021). This leads to moisture modes.

3.7 Moisture Modes

Yu and Neelin (1994) discovered the moisture modes as wave solutions in which the free troposphere water vapor dictates the dynamics of convection. In these waves, precipitation anomalies co-locate with positive precipitable water anomalies. This feature is observed in MJOs (Raymond and Fuchs 2009; Sobel and Maloney 2013) and is often identified as moisture mode (Adames and Maloney 2021). Detailed moisture budget studies reveal that both horizontal and vertical advection and the remaining q_2 terms (e.g., condensation, evaporation, diffusion processes) also play critical roles at the various stages of the propagating moisture modes. For example, vertical advection of moisture is found to be important for the propagation of the MJO from the Maritime Continent to the Western Pacific (Adames and Wallace 2015; Feng et al. 2015). On the other hand, the surface latent heat fluxes are found to be important when MJO amplifies over the Indian Ocean (de Szoeke et al. 2015).

The onset of deep convection in the tropics produces two distinct responses: (i) gravity waves are produced to adjust the atmosphere to WTG balance and (ii) occurrence of the compensatory drying of the troposphere. Timescales are associated with both these adjustment processes: for the former, we have τ_{WTG}, and for the latter, we have a convective adjustment (τ_D). For precipitating cases when $\tau_D \ll \tau_{WTG}$ (i.e., drying of the column occurs far more rapidly before the WTG balance can be achieved), it leads to 'stratiform instability' or 'moisture-stratiform instability' (Adames and Maloney 2021). In such cases, the gravity waves fail to homogenize the temperature anomalies. Such anomalies then modulate CAPE/CIN and thereby modulate convection in the regions they propagate into. The moisture anomalies in such instabilities are, however, smaller than the moisture modes.

In instances when $\tau_D \gg \tau_{WTG}$, the temperature anomalies are homogenized rapidly, and the waves are governed by the moisture; that is, the waves are moisture modes. In the case when $\tau_D \sim \tau_{WTG}$, it gives rise to the so-called mixed systems that exhibit intermediate behavior between Matsuno-Gill equatorial wave solutions and moisture modes. The continuum of the waves bracketing these cases is illustrated in Fig. 3.15. In this figure, the top vertex of the triangle is occupied by dry gravity waves where $\tau_D \ll \tau_{WTG}$, and the bottom right vertex is occupied by moisture modes ($\tau_D \gg \tau_{WTG}$). In between these vertices, the space is occupied by the mixed moisture gravity waves (Fig. 3.15). Theoretical considerations suggest that the moisture modes and Rossby waves (both are low frequency waves) also lie along a continuum, which Adames and Maloney (2021) show as forming the bottom of the triangle in Fig. 3.14. Some studies have shown that the structures of some of the moisture modes resemble Rossby waves in the tropics (e.g., Adames and Ming 2018; Diaz and Boos 2019).

3.8   Organization of Convection

'Organization of convection' has become a maxim in the current literature of atmospheric science, with satellite imageries providing a continuous stream of events showing convective cells arranged in specific geometries (e.g., TCs, squall lines, mesoscale convective systems, etc.). In this section, we will try to define the organization of convection more objectively and try and understand why such organization could be occurring.

The organization of convection in the context of the formation of TCs takes a more obvious tone, given that the cyclonic circulation develops in association with the arrangement of the convection in spiral rainbands from initially disorganized

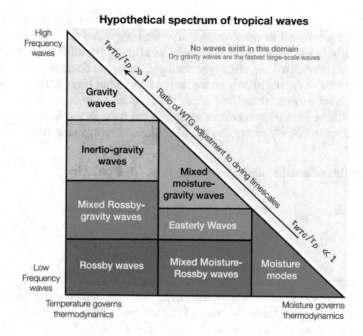

Fig. 3.15 A schematic distribution of the spectrum of tropical waves. All moist waves (which become increasingly moist from left to right) are considerably slower than the dry gravity waves that occupy the top vertex of the triangle. τ_{WTG} and τ_D refer to the timescales of WTG balance adjustment and convective drying adjustment, respectively. (Reproduced from Adames and Maloney 2021)

cells of convection. Krishnamurti et al. (1994) conducted energy budget diagnostics of an Atlantic Hurricane formation to objectively define the organization of convection. They simulated Hurricane Frederic of 1979 over a 72-h period when it matured from a tropical easterly wave with wind speeds of 6 ms^{-1} on 0000 UTC August 30 to a hurricane at 37 ms^{-1} on 0000 UTC September 2 in a global NWP model. The model simulated tropical storm force winds at the 24th hour of the forecast, delayed by about 16 h in relation to the observed estimates (Krishnamurti et al. 1994). Krishnamurti et al. (1994) used the regional domain bounded by 0.5°N to 20.5°N and 29.5°W to 49.5°W to conduct the regional energetics for Hurricane Frederic. An obvious drawback of this methodology over a regional domain is the formulation of the boundary flux terms that appear for completeness. Krishnamurti et al. (1994) indicate that the domain must be so chosen that it is large enough to make the boundary flux terms insignificant and the internal processes within the domain dominate.

The time history of the regional energetics of the hurricane formation showed that in the first 36 h of the integration, zonal available potential energy (P_Z) and zonal kinetic energy (K_Z) increased monotonically and thereafter vacillated (Fig. 3.16). The eddy available potential energy (P_E) increased monotonically through the 72-h integration of the model, while the eddy kinetic energy (K_E) showed a monotonic increase up to the first 54 h and then displayed an explosive growth as the hurricane intensified to its maximum windspeed toward the end of the 72-h integration (Fig. 3.16). K_Z displayed the largest magnitude among P_Z, P_E, and K_E energy terms over the history of the model integration (Fig. 3.16). The energy transformations between these four terms were explained in terms of two energy generation terms and four important energy conversion terms following Lorenz (1967) and schematically shown in Fig. 3.17. The two energy generation terms are as follows:

1. Generation of zonal available potential energy (G_Z), given by

$$G_z = -\frac{R_d}{c_p} \int_{p_T}^{p_s} \frac{\overline{\theta^* Q^*}}{p\frac{\partial \overline{\theta}}{\partial p}} \rho dx dy dz \qquad (3.67)$$

G_z represents the covariance of heating and temperature due to the local meridional overturning (Hadley) cell. The overbars, square brackets, prime, and asterisk mean domain average, zonal mean, deviations from the zonal mean, and deviation of the zonal average from the area average, respectively. Q^* is the heating rate per unit mass ($\equiv c_p \left(\frac{dT}{dt}\right) - \alpha \omega$). The other symbols ($u$, v, p, , ω) take the usual meaning. ϕ and a are the latitude and radius of Earth, respectively. The monotonic increase of G_Z over the 72-h integration period (Fig. 3.18e) is related to the local Hadley Circulation in the Eastern Atlantic Ocean, evident from the prevailing ITCZ in the satellite imagery.

2. Generation of eddy available potential energy (G_E) is given by

$$G_E = -\frac{R_d}{c_p} \int_{p_T}^{p_s} \frac{\overline{\theta' Q'}}{p\frac{\partial \overline{\theta}}{\partial p}} \rho dx dy dz \qquad (3.68)$$

G_E represents the covariance of heating and temperature of the local zonally asymmetric component of the meridional overturning circulation (or the local Hadley Cell). This term showed an initial monotonic growth in the first 24 h related to the asymmetry in the ITCZ, but the explosive growth in the last 12 h of the integration is related to the zonal asymmetry from the organized convection of the hurricane (Fig. 3.18f).

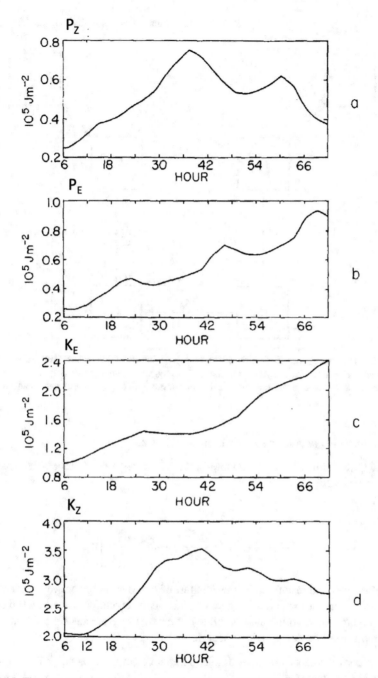

Fig. 3.16 The time evolution of the area (0.5°N–20.5°N and 29.5°W–49.5°W) averaged (**a**) zonal available potential energy (P_Z), (**b**) eddy available potential energy (P_E), (**c**) eddy kinetic energy (K_E), and (**d**) zonal kinetic energy (K_Z) with units of 10^5 Jm^{-2} from a 72-h global model simulation of Hurricane Frederic of 1979 from 0600 UTC August 30, 1979, through 0000 UTC September 2, 1979. (Reproduced from Krishnamurti et al. (1994); © American Meteorological Society. Used with permission)

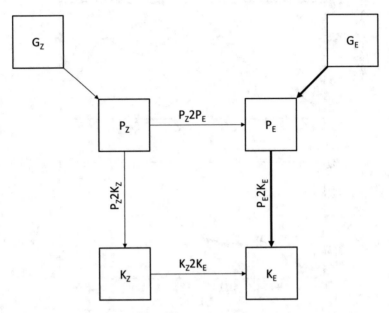

Fig. 3.17 The energy transformation terms in the evolution of Hurricane Frederic (1979) as indicated in Krishnamurti et al. (1994). The arrows suggest the energy transformation direction, and the thicker arrows dictate the organization of convection as suggested in Krishnamurti et al. (1994)

The four energy conversion terms are as follows:

1. The conversion of zonal available potential energy to eddy available potential energy (P_Z2P_E) is given by

$$P_Z2P_E = -\int_{P_T}^{P_s} \left[\frac{1}{\sigma}\overline{v'T'}\frac{\partial \overline{T^*}}{a\partial\phi} + \frac{1}{\sigma}\overline{\omega'T'}\frac{\partial \overline{T^*}}{\partial p} \right] dp \qquad (3.69)$$

P_Z2P_E is found usually to be the smallest of the four energy conversion terms in the time history of the energy budget of a hurricane evolution because the thermal field is quasi-zonal with weak meridional temperature gradients (Fig. 3.18a). This term represents the downgradient heat flux.

2. The conversion of zonal available potential energy to zonal kinetic energy (P_Z2K_Z) is given by

$$P_Z2K_Z = -\frac{1}{g}\int_{P_T}^{P_s} \frac{R_d}{p}\overline{\omega^*T^*}dp \qquad (3.70)$$

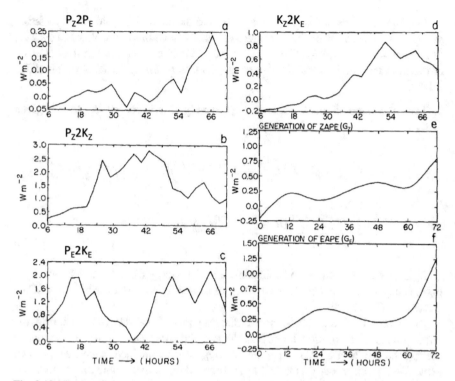

Fig. 3.18 Time evolution of the energy transformations (Wm-2) from (**a**) zonal available potential energy to eddy available potential energy, (**b**) zonal available potential energy to zonal kinetic energy, (**c**) eddy available potential energy to eddy kinetic energy, and (**d**) zonal kinetic energy to eddy kinetic energy and generation of (**e**) zonal available potential and (**f**) eddy available potential energy. These transformations have been area averaged over a domain of 0.5°N–20.5°N and 29.5°W–49.5°W from the 72-h simulation of Hurricane Frederic of 1979 from a global atmospheric model. (Reproduced from Krishnamurti et al. 1994; © American Meteorological Society. Used with permission)

The local meridional overturning (Hadley) cell, which represents the ascent of warm air and the descent of cold air along the latitude circle, contributes to P_Z2K_Z. Krishnamurti et al. (1994) found this energy conversion term to steeply increase after 18 h of integration and then vacillated (Fig. 3.18b). The rise of this term coincided with the development of the closed cyclonic flow of the TC in the easterly wave.

3. The conversion of eddy available potential energy to eddy kinetic energy (P_E2K_E) is given by

$$P_E2K_E = -\frac{1}{g}\int_{p_T}^{p_s} \frac{R_d}{p}\overline{\omega'T'}dp \qquad (3.71)$$

P_E2K_E represents the ascent of warm air and the descent of cold air on the scale of AEW, which usually begins to increase as the wave begins to develop as a cyclone (Krishnamurti et al. 1994). The first pulse of the increase occurred in the first 18 h of the forecast and the second pulse as the TC was maturing into a hurricane (Fig. 3.18c).

4. The conversion of zonal kinetic energy to eddy kinetic energy (K_Z2K_E) is given by

$$K_Z2K_E = \frac{1}{g}\int_{PT}^{Ps} \cos\phi \overline{u'v'}\frac{\partial\left(\frac{[u]}{\cos\phi}\right)}{a\partial\phi}dp + \frac{1}{g}\int_{PT}^{Ps}\overline{v'^2}\frac{\partial[v]}{a\partial\phi}dp + \frac{1}{g}\int_{PT}^{Ps}\frac{\tan\phi}{a}u'^2\,[v]\,dp$$

$$+\frac{1}{g}\int_{PT}^{Ps}\overline{\omega'u'}\frac{\partial[u]}{\partial p}dp + \frac{1}{g}\int_{PT}^{Ps}\overline{\omega'v'}\frac{\partial[v]}{\partial p}dp$$

$$(3.72)$$

K_Z2K_E represents the barotropic conversion of energy from zonal flow acceleration to eddy kinetic energy. This term begins to steeply rise after the first 30 h of the integration (Fig. 3.18d).

Krishnamurti et al. (1994) noted that the growth of the AEW (during the initial stages of their model integration) and the eventual maturity to hurricane (during the last 36 h of their 72-h model integration) coincided when there was an explosive growth of eddy kinetic energy from eddy available potential energy (cf. Eq. 3.71). That is, the covariance of heating (or vertical velocity) and temperature was maximized on the scale of the regional domain, which happened when the convective cells acquired a certain quasi-circular geometry, as the hurricane evolved from the AEW. Therefore, by examining the time evolution of the regional energy budget terms in the case of the formation of Hurricane Frederic, Krishnamurti et al. (1994) concluded that the organization of convection alludes to the generation of the eddy available potential energy and its conversion to eddy kinetic energy on the scale of the disturbance. These two energy transformation terms are highlighted in Fig. 3.17. Such diagnostics were also conducted to define the organization of convection in the context of monsoon depressions (Krishnamurti et al. 2005) and MJO (Krishnamurti et al. 2016).

Self-aggregation of convection, wherein there is evidence of spontaneous organization of convection into one or several long-lasting clusters surrounded by large areas of dry air that occurs under spatially homogenous boundary conditions and forcing is another form of organization of convection (Moncrieff 2004; Wing 2019). There are many processes like wind shear and SST gradients that can organize convection. But Wing et al. (2021) suggests that self-aggregation represents a fundamental instability of the radiative-convective equilibrium, which is distinct from these other mechanisms of organization of convection. A growing consensus in the field is that feedback involving longwave radiation and water vapor, or clouds, is essential for initiating and maintaining self-aggregation (Wing et al. 2017). A discerning feature of self-aggregation of convection is the drying of the non-convective regions and an increase in the moisture gradient of the regional domain (i.e., drier parts of the domain get drier and wetter parts get moister). Since the domain is dominated by the fractional area of drier areas than the convective area, the domain mean results in significant drying from self-aggregation (Fig. 3.19). Figure 3.19 shows from observations and idealized modeling studies that there is decrease in free tropospheric relative humidity with a peak reduction in the middle troposphere.

Self-aggregation, by drying the mean state, makes the system more efficient in tropospheric radiative cooling and decreases energy gain by the surface, reducing high and mid-level cloud cover, and warming the domain mean tropospheric temperature (Wing 2019). For example, Wing and Emanuel (2014) find that domain mean OLR increases by about 10–30 Wm^{-2} from self-aggregation of convection. Similarly, the domain mean temperature of the free troposphere increases with self-aggregation because the free troposphere in the local region around convection becomes moister (Wing 2019). The domain mean temperature is dictated by the local region where convection occurs (as noted in Fig. 3.14). The parcels in the convective region of the domain rise along a warmer moist adiabat, and given the moister, free troposphere around the convection, it reduces the influence of entrainment and drives the troposphere toward the warmer moist adiabat.

There is some evidence to suggest that low clouds increase and high and mid-level clouds decrease from self-aggregation (Cronin and Wing 2017). Due to the

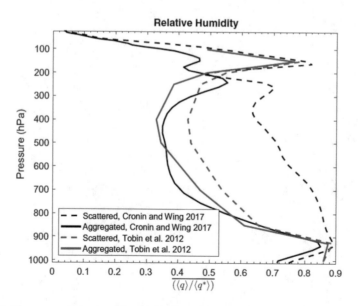

Fig. 3.19 The mean relative humidity profiles for scattered (dashed) and self-aggregated (solid) convection. The black lines show domain-mean profiles from numerical model simulations following Cronin and Wing (2017), while the red lines are from satellite observations of scattered and aggregated convection following Tobin et al. (2012). (Reproduced from Wing 2019)

opposing responses of high and low clouds to self-aggregation, the reflected short-wave radiation changes little, and the net radiative flux into the top of the atmosphere is reduced (Wing and Cronin 2016). Furthermore, mean precipitation is found to increase from self-aggregation by 23% due to higher precipitation efficiency with higher enthalpy fluxes and larger atmospheric radiative cooling (Cronin and Wing 2017). There is inconclusive evidence of temperature influencing self-aggregation. However, it is observed that the strength and scale of self-aggregation do depend on temperature, although the precise nature of this functionality is elusive (Wing 2019).

Chapter 4
Intertropical Convergence Zone

Abstract The chapter introduces theory for the latitudinal location of the Inter-Tropical Convergence Zone (ITCZ) followed by a description of the ITCZs in the three tropical ocean basins. The chapter is concluded with a discussion of how ITCZs can engender tropical cyclone genesis.

Keywords Evaporation · Lapse rate · Boundary layer · Inertia-gravity wave · Brunt Vaisala frequency · Atlantic Meridional Overturning Circulation · Thermocline · Convergence zone · Tropical cyclones · African easterly jet · Barotropic instability · Baroclinic instability · Potential vorticity · Shear · Trough · Radiation · Sea surface temperature · African easterly wave · South Pacific Convergence Zone · South Atlantic Convergence Zone

4.1 The Latitudinal Location of the ITCZ

The ITCZ is a prominent component of the atmospheric general circulation that is recognized by the latitudinal maximum in precipitation (Fig. 4.1a, b). The ITCZ is a manifestation of the zonally symmetric circulation (or the Hadley Circulation; cf. Chap. 1) and plays a critical role in the interhemispheric energy and mass balance while also influencing the variations of climate on many temporal scales. Therefore, the importance of the ITCZ cannot be over-emphasized. The ITCZ can be identified at any given instant of time, for example, in the blended satellite image from GOES-17 in Fig. 4.2. The Eastern Pacific ITCZ in Fig. 4.2a appears as a cluster of clouds along a tropical synoptic wave. The ITCZ fluctuates between a more well-defined ITCZ (Fig. 4.2b) and a broken-up state as shown in Fig. 4.2a, typically on a timescale of around 2 weeks. However, for the purposes of illustration, Fig. 4.2 shows two distinct examples of a broken-up and a well-formed ITCZ separated over a span of 9 months.

Although the ITCZ is ubiquitous in its longitudinal distribution across the tropics, there is a significant asymmetry both zonally and hemispherically. In this chapter, we will be discussing the zonally symmetric (or zonally averaged) ITCZ and some

V. Misra, *An Introduction to Large-Scale Tropical Meteorology*, Springer Atmospheric Sciences, https://doi.org/10.1007/978-3-031-12887-5_4

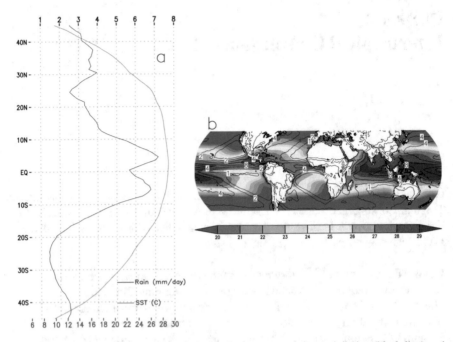

Fig. 4.1 (a) The zonal mean climatological annual mean oceanic precipitation (black line) and SST (red line) and (b) the corresponding spatial distribution of the climatological annual mean SST (shaded; °C; from OISSTv2) and precipitation (contoured; mm/day; from TRMM3B43). The SST climatology was computed over the period of 1990–2010 and rainfall climatology over the period of 1998–2015

of the regional features of the oceanic ITCZ. The continental ITCZ is far more complicated and in some instances is part of the regional monsoons.

A primary, hemispherically asymmetric feature of the climate system is that the maximum climatological annual mean and the corresponding zonal mean precipitation locate just north of the equator (~5°N; Fig. 4.1a, b). The ITCZ, as shown in Fig. 4.1a, b, does not necessarily align with the corresponding latitudinal peak in SST. Although the latitudinal peak in zonal mean SST and precipitation are very close to each other, they are not coincident (Fig. 4.1b). This suggests that warm SST is not a necessary condition for ITCZ to locate but nonetheless may have some significant role to play in locating the ITCZ. The discrepancy between the latitudinal peaks of SST and precipitation gets further exacerbated when we examine within regional ocean basins and in the various seasons rather than just the annual mean. For example, over the Indian Ocean, the latitude of the warmest SSTs and the latitude of maximum precipitation are farther apart than over the Pacific or the Atlantic Ocean (Fig. 4.1b).

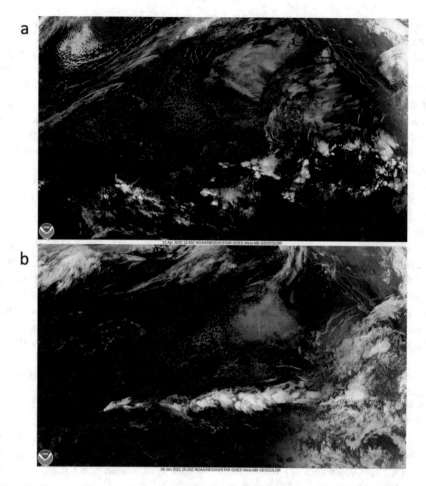

Fig. 4.2 A blended GOES-17 true color daytime and multispectral infrared at night, satellite imagery for an example of (**a**) broken-up state of ITCZ valid at 1330 UTC of April 22, 2020 and (**b**) well-formed ITCZ valid at 1530 UTC of January 6, 2020. (Reproduced from https://www. star.nesdis.noaa.gov/GOES/sector.php?sat=G17&sector=tpw)

4.1.1   The Role of the Coriolis Force

One of the earliest theories proposed on ITCZ was that of Charney (1971), which was based on CISK (Ooyama 1964). The CISK theory essentially suggests that boundary layer moisture convergence results in ascent, consequent condensation, and latent heat release, which causes a pressure drop and a horizontal pressure gradient that further enhances the convergence leading to a positive feedback loop. Originally, CISK was proposed as a theory for tropical cyclogenesis (Ooyama 1964). Charney (1971) adapted the CISK mechanism for the zonally symmetric circulation of the ITCZ. However, CISK is favored where the Coriolis parameter is large. This is best illustrated by the schematics shown in Fig. 4.3. In the instance when $f = 0$, then the boundary layer winds converge head-on at the center (Fig. 4.3a). In contrast, when $f \neq 0$, then the boundary layer winds converge in a spiral path (Fig. 4.3b) leading to higher wind speed (as a result of the creation of the tangential wind in a spiral path). In the case when $f \neq 0$, when the converging boundary layer winds acquire higher wind speed, it results in more evaporation from the ocean surface and, as a result, far more moisture convergence, fueling stronger convection than in the case when $f = 0$. Therefore, if CISK were the sole mechanism, then ITCZ would tend to locate where f is large, like over the poles. However, because the moisture content in the boundary layer of the equatorial atmosphere is much higher than in the polar regions, Charney (1971) argued that the ITCZ located itself closer to the equator. Charney's theory of the ITCZ was indeed quite influential as many other studies up to the mid-1990s continued to use and advance this theory for ITCZ (e.g., Hess et al. 1993; Waliser and Somerville 1994).

The CISK theory for ITCZ, however, appeared to be in conflict with some experimental results conducted with an AGCM. For example, Sumi (1992) obtained an ITCZ near the equator in an aqua-planet AGCM integration in which continental regions were replaced with ocean surface with prescribed uniform SST and pre-scribed radiative cooling in such a way that the horizontally averaged temperature was restored to a prescribed lapse rate. In other words, despite removing the limiting factor of lack of sufficient moisture at the poles by prescribing globally uniform warm SST in the aqua-planet globe with prescribed radiative cooling, Sumi (1992) obtained ITCZ in the equatorial region, contrary to the CISK theory. Subsequently, several other similar aqua-planet AGCM integrations with uniform SST have shown that ITCZ locates itself in the equatorial region (Kirtman and Schneider 2000; Chao 2000).

Chao and Chen (2004) indicate that the Coriolis parameter has another important effect on convection besides its influence on the boundary layer winds discussed earlier. This other effect of f relates to the vertical stratification effect. This effect of f is best illustrated by the following equations.

The divergence (D) equation with just the forcing terms containing f is given by

$$\frac{\partial D}{\partial t} = f\xi + \dots \tag{4.1}$$

Similarly, the vorticity (ξ) equation with just the forcing terms containing f is given by

$$\frac{\partial \xi}{\partial t} = -fD + \dots \tag{4.2}$$

The other terms on the RHS of Eqs. 4.1 and 4.2, which are not mentioned, are not important for this discussion. Substituting for ξ from Eq. 4.1 into Eq. 4.2 under an f-plane setting, we get

Fig. 4.3 Schematic of the converging boundary layer winds when (**a**) $f = 0$ and (**b**) $f \neq 0$. (Reproduced from Chao and Chen 2004)

f=0

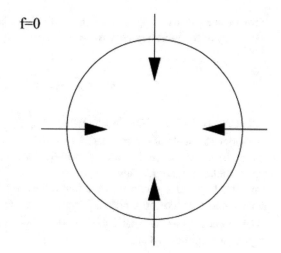

f≠0

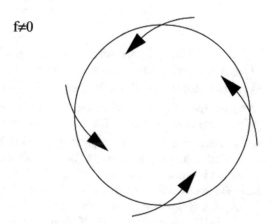

$$\frac{\partial^2 D}{\partial t^2} = -f^2 D + \ldots \tag{4.3}$$

Equation 4.3 is a form of an equation that governs a simple spring. Therefore, just as a spring resists compression and stretching, f in Eq. 4.3 serves as a resistance to convergence and divergence. In other words, f, through its vertical stratification effect, is resisting vertical motion in the center of the convection and compensating downward motion in the surrounding areas. Thus, if this vertical stratification effect of f were acting alone, then the preferred location of the ITCZ would be at the equator.

To reconcile these two conflicting effects of f on convection and therefore provide a pathway for the determination of the location of the ITCZ, Chao and Chen (2004) focused their attention on the squared frequency (ω^2) of the inertia gravity waves (given in Eq. 4.4). It so happens that convection occurs when the squared frequency of the inertia gravity wave (ω^2) turns negative (cf. Equation 8.4.23 in Gill 1982).

$$\omega^2 = f^2 + \alpha^2 N^2 + |F| \tag{4.4}$$

where $N^2 \equiv$ Brunt − Vaisala Frequency $= \left(\frac{g\partial\theta}{\theta\partial z}\right)$, α is the ratio of the vertical to the horizontal scale of the convective cell, and $|F|$ is the stabilization due to friction. It may be noted that in the case of an individual convective cloud, α could be large and f could be ignored. However, in the case of the ITCZ, they are cloud clusters embedded within a synoptic wave that allows for large horizontal scales (or α is small), and f cannot be ignored. Therefore, in an aqua-planet AGCM integration, ITCZ would appear at a latitude (ϕ) when $\frac{\partial\omega^2}{\partial\phi} = 0$. This implies that ITCZ will appear at a latitude, where

$$\frac{\partial f^2}{\partial \phi} = -\frac{\partial \alpha^2 N^2}{\partial \phi} + |F| \tag{4.5}$$

The two effects of f are summarized in Eq. 4.5. The gradient, $\frac{\partial f^2}{\partial \phi}$ in the LHS of Eq. 4.5, represents the vertical stratification effect of f, which makes the equator as the attractor for the ITCZ. Similarly, $-\frac{\partial \alpha^2 N^2}{\partial \phi}$ represents the latitudinal gradient of f-modified surface fluxes on the stability of the column (N^2), which makes the poles as the attractor for the ITCZ. It is, however, an impossible task to obtain an analytic expression for ($\frac{\partial \alpha^2 N^2}{\partial \phi}$) given that the ITCZ is a cluster of convective clouds embedded in a synoptic wave. In the case of the ITCZ, α can acquire a spectrum of values because of the range of cloud systems that ITCZ comprises from the individual convective system (of at most a few kilometers) to that of cloud clusters as part of a synoptic wave of several thousands of kilometers. Therefore, Chao and Chen (2004) conducted a series of idealized aqua-planet AGCM integrations with uniform SST to isolate the various impacts on the ITCZ formation. The analysis of their results is summarized in the following points:

- Specifying uniform surface fluxes and radiative cooling rate produced a weak ITCZ (diagnosed by a corresponding latitudinal peak in zonal mean precipitation) at the equator in the presence of f varying with latitude; when additionally f was set to a constant value, then even the weak ITCZ disappeared from the equator. These experiments suggest the significance of the vertical stratification effect of f in locating the ITCZ close to the equator.
- The ITCZ became very prominent (or strong) at the equator when surface fluxes were made interactive in addition to varying f with latitude while still keeping a uniform radiative cooling rate. This experiment suggests that the effect of f on the boundary layer winds and the consequent interaction between convection and surface fluxes cause vigorous convection but are not sufficient to overcome the

equatorward forcing due to inertial stability or the vertical stratification effect of f.

- Additionally, when the prescribed radiative cooling is replaced with interactive radiation in an experiment with interactive surface fluxes and f varying with latitude, the very prominent ITCZ moved slightly away from the equator. This experiment showed that the radiative cooling in cloud free regions of the subtropical latitudes enforces a stronger compensatory subsidence, which consequently forces the ascent in the ITCZ to be stronger. For this latter point, the ITCZ has to move away from the equator in order for the effect of f on boundary layer winds to be effective. In another additional experiment, when Chao and Chen (2004) prescribed surface winds while keeping the surface fluxes and radiation interactive and f varying with latitude, the prominent ITCZ became slightly weaker and was pushed back to the equator, suggesting that the interaction between convection and surface fluxes is reduced when wind-evaporation feedback is cut off.

4.1.2   The Role of the Oceans

Several studies have emphasized the role of the oceans in the hemispherically asymmetric location of the climatological annual mean ITCZ. This is best illustrated by the schematic of the energy budget shown in Fig. 4.4. It is argued that the asymmetry in the interhemispheric oceanic heat transport requires the ITCZ to be located north of the equator. From Fig. 4.4, it is clear that 0.4 PW (1 Petawatt $= 10^{15}$ W) of energy across to the NH from the SH is carried out by the northward ocean heat transport. It so happens that the Atlantic Meridional Overturning Circulation (AMOC) accounts for 0.4 PW of ocean heat transport across the equator in the Atlantic Ocean (Marshall et al. 2014). This northward oceanic heat transport is largely compensated by the southward atmospheric heat transport (Fig. 4.4), which is ~0.2 PW. This southward atmospheric heat transport is achieved by shifting the ITCZ north of the equator, because ITCZ represents the ascending cell of the zonally symmetric circulation (or the Hadley Cell; cf. Chap. 1). Although the hemispheric net TOA radiative forcing is symmetric about the equator, with the SH receiving $+0.2$ PW and NH losing -0.2 PW. Therefore, the net total (of ocean and atmosphere) heat transport is small and northward at $+0.2$ PW. As a result, the atmosphere and the ocean of the NH are slightly warmer (by ~2 °C) compared to that of the SH (Fig. 4.5). This relative warmth of the NH supports the estimate of NH emitting more OLR than SH and therefore a net northward energy transport across the equator in the coupled climate system.

Therefore, the climatological annual mean location of the ITCZ north of the equator is argued to be consistent and indeed a consequence of the NH atmosphere being heated more strongly than in the SH as a consequence of the asymmetric, northward oceanic heat transport and not because of asymmetric absorption of shortwave radiation as a result of hemispheric asymmetries in albedos (Yoshimori

Fig. 4.4 Estimates of the atmospheric heat transport across hemispheres (from the NCEPR1 atmospheric reanalysis) and hemispheric mean net radiation at the Top of the Atmosphere (TOA; ERBE and CERES satellite data), and ocean heat transport across hemispheres (computed as residual). The error bars in the estimates of all fluxes are ±0.1PW. (Reproduced from Marshall et al. 2014)

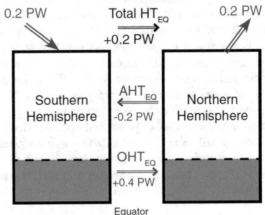

and Broccoli 2009; Frierson et al. 2013; Marshall et al. 2014). Other studies have shown that when AMOC is weakened in a coupled model simulation, the ITCZ shifts equatorward (Zhang and Delworth 2005), or when NH is cooled implicitly by imposing a southward ocean heat transport, the ITCZ moves southward with the enhancement of northward atmospheric heat transport (Broccoli et al. 2006).

4.2   Regional ITCZ

Although the perception of ITCZ being zonally symmetric is common, there is a significant zonal asymmetry in the ITCZ. For example, the seasonal variation of the ITCZ between the oceans and that over the continents is glaring. Over the oceans, the seasonal variations of the ITCZ in terms of their meridional migration from one season to the other are much weaker than that over the continental regions (Fig. 4.6).

4.2.1   Eastern Pacific ITCZ

The Tropical Eastern Pacific ITCZ displays a clear asymmetric feature about the equator with the ITCZ residing north of the equator along with the warm SST for the most part of the year (Fig. 4.7). The one season where it is symmetric about

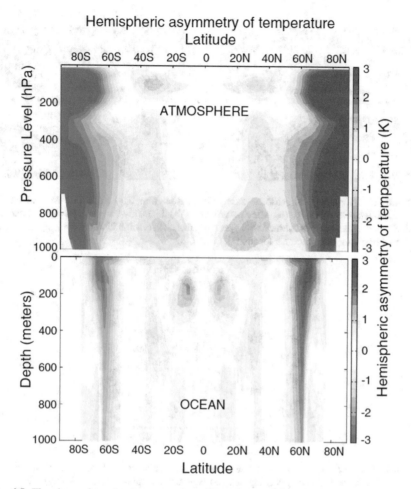

Fig. 4.5 The observed zonal mean asymmetry of temperature (°C) in the atmosphere and ocean from the NCEP-NCAR reanalysis and the World Ocean Atlas, respectively. The asymmetry of temperature (T) is computed as $\frac{[T(\phi)-T(-\phi)]}{2}$, where ϕ and $-\phi$ are corresponding latitudes in the opposite hemispheres. (Reproduced from Marshall et al. 2014)

the equator is in the boreal spring (MAM) season, when there is a double ITCZ straddling the equator with warm SSTs on either side of the equator (Fig. 4.7). The seasonal cycle in the equatorial surface winds over Eastern Pacific is most visible in the meridional component of the wind (Fig. 4.7). Although the surface winds in the Equatorial Eastern Pacific remain southerly throughout the year, it is strongest in the summer-fall seasons when the ITCZ is strongest and located farthest from the equator. This maximum in the southerly winds causes corresponding changes in upwelling, wind stirring, and evaporation that results in the appearance of the coldest temperatures in the Equatorial Eastern Pacific (a.k.a. the cold tongue) in the late summer-fall season (Fig. 4.7). Similarly, in March, the southerlies are weakest

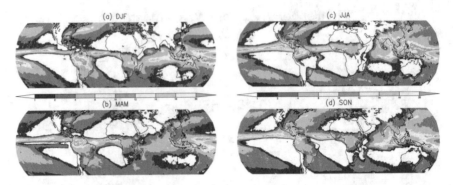

Fig. 4.6 The climatological seasonal mean precipitation (mm/day) from TRMM3B42 for (**a**) December-January-February, (**b**) March-April-May, (**c**) June-July-August, and (**d**) September-October-November seasons overlaid with a latitudinal peak in precipitation at each longitude (black line; as a proxy for the latitudinal location of the ITCZ)

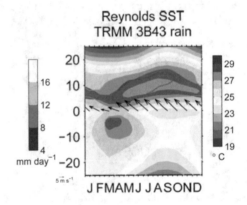

Fig. 4.7 The observed climatological seasonal cycle of zonally averaged (140°W–90°W over eastern Pacific) SST (colored shading and red contour) and rainfall (gray shading) with corresponding observed equatorial surface wind vectors. The ordinate shows latitude (from 25°S to 25°N), and the abscissa shows the 12 months of the year. (Adapted from de Szoeke and Xie 2008; © American Meteorological Society. Used with permission)

during the year when the Equatorial Eastern Pacific SST is warmest. Subsequently, in the spring (MAM) and in the summer (JJA) seasons, the warmest SSTs appear north of the equator co-located with maximum rainfall (Fig. 4.7).

In a novel set of coupled ocean-atmosphere modeling experiments, Philander et al. (1996) established that the asymmetry in the Tropical Eastern Pacific climate is a result of the geometry of the west coast of the North and South American continents. The northwest-southeast slant of the west coasts of the North and South American continents offers an asymmetry with the alongshore winds along the South American coast causing upwelling while the surface winds are nearly normal to the coast, north of the equator. The air-sea feedback then amplifies this asymmetry, making the SST warmer north of the equator than south of the equator.

In contrast, when the western coast of the Americas is altered to become aligned with the meridians while preserving the land area in each latitude, Philander et al. (1996) showed that the ITCZ remained essentially at the equator in the Tropical Eastern Pacific Ocean.

4.2.2 Eastern Atlantic ITCZ

Just as in the Eastern Tropical Pacific Ocean, the location of the ITCZ in the Eastern Tropical Atlantic Ocean is north of the equator throughout the year. Similarly, the equatorial wind in the Eastern Atlantic has a southerly component throughout the year, which causes an upwelling and elevates the thermocline at and south of the equator but causes downwelling and deepens the thermocline north of the equator. Philander et al. (1996), again using the coupled ocean-atmosphere model, showed that the hemispheric asymmetry in land area apparent in the West African bulge is the primary cause for the asymmetry in the location of the ITCZ in the Eastern Atlantic Ocean. This West African bulge, to the north of the Gulf of Guinea (~5°N), attains a very high land surface temperature than that of the surrounding ocean, resulting in winds that are like land-sea breeze. These cross-equatorial southwesterly winds toward the West African bulge cause upwelling along Southwestern Africa, resulting in an asymmetrical distribution of warm SST to the north and cold SST to the south of the equator. This asymmetry develops in the Eastern Atlantic Ocean despite relatively symmetrical solar radiation about the equator.

4.2.3 Indian Ocean ITCZ

Philander et al. (1996) showed the asymmetrical modes forced by the asymmetry of the continental mass and the coastlines are amplified in regions of the shallow ther-mocline in the equatorial oceans. However, as the depth of the thermocline increases as over the Tropical Indian and Western Pacific Oceans, these asymmetrical modes are ineffective. Philander et al. (1996) posit that these zonal variations in the thermocline depth are a result of the different wind systems in the Tropical Ocean basins. For example, in the Tropical Pacific and Tropical Atlantic Oceans, easterly trade winds prevail, which drive the warm surface waters westward, shoaling the thermocline in the east. In the Equatorial Indian Ocean, however, westerlies prevail on account of the large asymmetry in the continental distribution across the equator, resulting in relatively far less east-west variation in the depth of the thermocline. The ITCZ in the Indian Ocean shows large seasonal migrations compared to the other two tropical ocean basins. In fact, the annual mean ITCZ in the Indian Ocean resides well south of the equator (Figs. 4.1a, b). This is because even in the boreal summer season when the continental Asian monsoon dominates, it just diminishes the oceanic ITCZ in the Tropical Indian Ocean (Fig. 4.6c).

4.2.4   The South Pacific and South Atlantic Convergence Zones

In the austral summer season, three highly zonally asymmetric features develop in the SH that correspond to the SPCZ, the SACZ, and the less significant SICZ (Fig. 4.8). These convergence zones are oriented approximately in the northwest-southeast direction, and they have no counterparts in the NH. The ITCZ is more zonal in the NH. The SPCZ stretches diagonally from New Guinea in the Equatorial Western Pacific Warm Pool to 30°S and 120°W. The SACZ extends from the Amazon over southern Brazil towards the Southwestern Atlantic Ocean. The SICZ (between South Africa and Madagascar) in contrast is less prominent and does not extend as far into the middle latitudes as SACZ or SPCZ.

Van der Wiel et al. (2015) examined the spectra of OLR variability in the SPCZ and SACZ and found that ENSO variations at interannual scales dominate the variability. Furthermore, intraseasonal variability associated with the MJO is also found to dominate the SPCZ but not the SACZ. However, spectral peaks at 0.1 cycles day^{-1} (synoptic variations) are found in both the SACZ and the SPCZ. The diagonal rain pattern of the SACZ and the SPCZ is linked to tropical-extratropical interactions, and often these convergence zones serve as graveyards of fronts that transgress from higher to lower latitudes. This is best illustrated in the composite of the OLR, upper-level vorticity, and wind anomalies shown in Fig. 4.9. The Rossby waves (which owe their existence to gradients in potential vorticity) propagate from the subtropical jet stream region equatorward. These Rossby waves manifest as wave trains of cyclonic and anticyclonic vorticity, clearly visible at upper levels

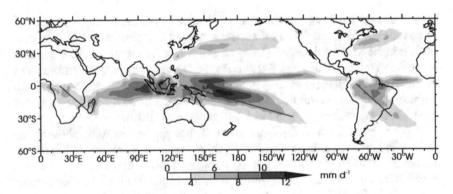

Fig. 4.8 The observed climatological November through April mean rainfall (mm/day) from CMAP. The red solid lines indicate the position of the South Indian, South Pacific, and South Atlantic Convergence Zones. (Reproduced from Van der Wiel et al. 2015)

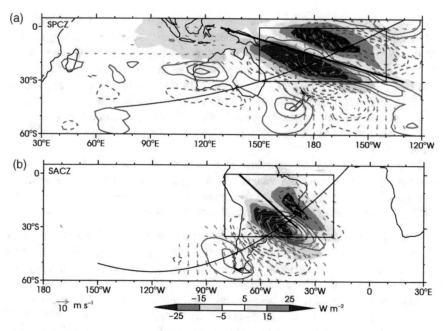

Fig. 4.9 The observed composite anomalies of OLR (shaded), 200 hPa vorticity anomalies (red contours with zero contour omitted and negative contours dashed), and 200 hPa anomaly wind vectors for (**a**) SPCZ and (**b**) SACZ. The composites were created over a time period covering 1979/1980 to 2012/2013 by conducting an EOF analysis of the OLR anomalies in the rectangular box and recognizing convective events when the time series of the principal component of the first EOF exceeded one standard deviation and reached maxima centered over a 10-day period of the convective event. The bold diagonal line represents the mean convection axis, and the curved line is the approximate wave propagation path. (Reproduced from Van der Weil et al. 2015)

(e.g., 200 hPa), but extend down to the surface as weaker anomalies (Van der Wiel et al. 2015). These vorticity anomalies appear initially as quasi-circular. But as these anomalies are refracted equatorward, toward the upper tropospheric westerlies, they get deformed owing to the shear on the northern edge of the jet and acquire a northwest-southeast diagonal orientation (Van der Wiel et al. 2015).

These Rossby waves can trigger convection in the SACZ and SPCZ regions. Following quasi geostrophic dynamics, the poleward flow ahead of the cyclonic anomaly in the wave train is associated with ascent and decreased static stability, which can trigger convection in the presence of warm SST in the tropics. Although it may be noted that convection is in quadrature with vorticity anomalies in the wave, with convection ahead of the cyclonic anomaly. This convection and associated latent heat release force additional ascent and upper-level divergence. But due to vortex stretching, anticyclonic tendency develops, which acts to weaken the initial cyclonic anomaly. Van der Wiel et al. (2015) suggests that the difference between the SPCZ and the SPCZ is that the negative feedback of the vortex stretching is stronger in the former than the latter. This leads to the dissipation of the ITCZ in the SPCZ and its further propagation downstream of the triggered convection in the SPCZ. In the SPCZ, the negative feedback of vortex stretching is strong enough to dissipate the wave and stop the propagation within a day (Matthews 2012). In the SACZ, the divergence forced by the convection is not strong enough, and therefore, the wave propagates beyond the mean convection axis (Van der Wiel et al. 2015). Several observational studies have confirmed this observed sequence of events of equatorward-propagating midlatitude disturbances setting up a diagonal rainfall band in the SPCZ and SACZ regions (Kiladis and Weickmann 1992a, b; Kiladis 1998).

In a series of AGCM experiments that examined the sensitivity of zonal asymmetry in SST, absolute values of SST, global orography, and continental configuration (the presence of the Australian and South American continents) on SPCZ, Van der Wiel et al. (2016) concluded the following:

- The zonal asymmetry of SST is very important to obtain the slope and strength of the SPCZ. Weakening the zonal asymmetry (akin to El Niño conditions) resulted in the SPCZ becoming more zonal and weaker. In contrast, strengthening the zonal asymmetry (akin to La Niña conditions) resulted in the SPCZ becoming stronger and diagonal. In these experiments the upper-level dynamical atmospheric forcing did not change. However, when the zonal asymmetry of the SST was reduced, the moisture transport into the SPCZ by the subtropical high, west of the Andes, weakened considerably.
- In the experiments where the absolute values of SST were changed but asymmetry was preserved, it was shown that colder SST make SPCZ lose its diagonal orientation and also weakened it considerably; warming the SST had the opposite effect on the SPCZ, making it stronger and more diagonal.
- Removing the orography or the continents of Australia and South America had very little impact on the SPCZ.

4.3   ITCZ as a Spawning Zone of Tropical Cyclones

A very large fraction (nearly 80%) of the global TCs form near or within the ITCZ (Gray 1979). The synoptic waves, which sometimes give ITCZ an undulating character, can be identified in satellite imagery by cloud patterns that have an inverted V pattern (Frank 1969). In some instances, such undulations result in the ITCZ spawning several TCs in clusters, which is often referred to as ITCZ breakdown (Hack et al. 1989; Guinn and Schubert 1993). Thereafter, the TCs translate to higher latitudes, while the ITCZ tries to reform as an elongated band of convection. These synoptic waves were first recognized by Riehl (1945) as easterly waves and associated their importance to tropical cyclogenesis. Ever since, easterly waves have been identified in the Atlantic Ocean and West Africa (Carlson 1969), in the Eastern and Western Pacific Oceans (Yanai 1961; Tai and Ogura 1987), in the Indian Ocean, and in the South China Sea (Saha et al. 1981). These easterly waves appear in the lower troposphere, typically with wavelengths of 1500–4000 km with speeds ranging from 5 to 8 ms^{-1} and with a period of 2–5 days.

In the context of West Africa, Burpee (1972) indicated that the combined barotropic and baroclinic instability is the source of these easterly waves. These easterly waves in West Africa ride on the AEJ. The West African region is somewhat special in that it produces an AEJ at around 600 hPa. The existence of the AEJ is a consequence of the thermal wind balance resulting from the strong positive temperature gradient to the south of the Sahara Desert. The vertical placement of the strong AEJ at ~600 hPa is also because at this level, the meridional temperature gradient is ~0 while it is positive below the jet and negative above the jet (Thorncroft and Blackburn 1999). The synergistic energy interaction between the AEJ and the African easterly waves leads to the growth of the latter at the expense of the former. The combined barotropic and baroclinic instability manifests as sign reversal of the meridional gradient of the time mean potential vorticity or a vanishing meridional gradient of the potential vorticity within a channel (or latitude band) in the lower troposphere (~700 hPa). The basic premise here is that the wave grows when the basic (or mean) state is unstable. Mathematically, the necessary condition for the existence of combined barotropic-baroclinic instability warrants

$$\frac{\partial \overline{\xi}_p}{\partial y} = \underbrace{\left(\beta - \frac{\partial^2 \overline{u}}{\partial y^2} \right)}_{\text{Term 1}} - \underbrace{\frac{\partial}{\partial p} \left(\frac{f_0^2}{\overline{\sigma}} \frac{\partial \overline{u}}{\partial p} \right)}_{\text{Term 2}} = 0 \qquad (4.6)$$

where $\overline{\xi}_p$ is the time mean potential vorticity, $\beta = \frac{\partial f}{\partial y}$, $\overline{u}$ is the time mean zonal wind, f_0 the is constant value of f, and $\overline{\sigma} \equiv$ static stability $= -\frac{R\overline{T}}{p\overline{\theta}} \frac{\partial \overline{\theta}}{\partial p}$. Terms 1 and 2 of Eq. 4.6 represent the contribution from barotropic dynamics and baroclinic dynamics, respectively.

The necessary condition for barotropic instability is

$$\beta - \frac{\partial^2 \overline{u}}{\partial y^2} = 0 \qquad (4.7)$$

or

$$\frac{\partial}{\partial y} \left(\frac{\partial \overline{u}}{\partial y} - f \right) = \frac{\partial \overline{\xi}_a}{\partial y} = 0 \qquad (4.8)$$

In other words, for barotropic instability to occur, Equation 4.8 suggests that the time mean absolute vorticity ($\overline{\xi}_a$) must reach a minimum or a maximum value somewhere within the channel. Alternatively, barotropic instability is dependent on the horizontal shear of the flow. It should be noted, however, that the second meridional derivative of the zonal wind should exceed β for barotropic instability to occur. β serves as a stabilizing influence on barotropic instability, especially if you have broad, weak zonal jets, which may display a maxima and minima of absolute vorticity in the meridional edges of the jet but may not be sufficient to cause barotropic instability. Therefore, Eq. 4.8 represents a necessary but not a sufficient condition for barotropic instability.

In the tropics, the presence of horizontal temperature gradients and the resulting thermal wind-related vertical wind shear is usually not sufficient to support free baroclinic instability (Goswami et al. 1980). Even over West Africa, where there are strong meridional thermal gradients near the southern border of the Sahara, free baroclinic modes do not exist. However, the barotropic term can itself contribute alone to the change of sign of $\frac{\partial \bar{\xi}_p}{\partial y}$. But in areas of wave growth, we see the combined barotropic and baroclinic terms contributing to the change of sign of $\frac{\partial \bar{\xi}_p}{\partial y}$, like in areas over West Africa. Several studies have shown that the barotropic term dominates in the mean flow instability over West Africa relative to the baroclinic term in Eq. 4.6 (Rennick 1976; Krishnamurti et al. 1979).

In mid-oceanic regions in the tropics, the lack of underlying meridional temperature gradients suggests that there must be other effects for the growth of these waves. Similarly, several tropical cyclogenesis events in the Eastern Pacific basin are traced to easterly waves originating from West Africa (Avila 1991). Although it has been observed that the strong African easterly waves can maintain their structure while propagating across the Atlantic into the Eastern Pacific (Shapiro 1986), more typically, they weaken as they pass over the cold Eastern Atlantic waters (Carlson 1969). Molinari et al. (1997) noted for an active TC season over the Eastern Pacific, there are strong zonal gradients in the time mean potential vorticity from West Africa to the Eastern Pacific that also modulates $\frac{\partial \bar{\xi}_p}{\partial y}$. For example, in the Eastern Atlantic, they found no sign reversal in $\frac{\partial \bar{\xi}_p}{\partial y}$. But over the Caribbean Sea and the Eastern Pacific (in the Gulf of Tehuantepec), they found the time mean flow was unstable with areas of reversal of the sign of $\frac{\partial \bar{\xi}_p}{\partial y}$. Molinari et al. (1997) suggested that wave growth in the Eastern Pacific can be triggered by enhanced convection in the ITCZ in the presence of an unstable mean state (where the meridional gradient of the seasonal mean potential vorticity displays a sign reversal in a channel). Furthermore, they also suggest that the presence of a preexisting but weak easterly wave in the presence of an unstable mean state can lead to the reinvigoration of the wave growth and downstream cyclogenesis. Alternatively, Ferreira and Schubert (1997) indicate that the mean state instability could be produced by the latent heat release in the ITCZ that can lead to the breakdown of the ITCZ into tropical disturbances. The deep convection in the lower troposphere produces cyclonic potential vorticity anomaly that has a sign reversal of $\frac{\partial \xi_p}{\partial y}$ on its poleward side and thereby satisfying the necessary condition for combined barotropic and baroclinic instability.

As Krishnamurti et al. (2013) note, there is mathematical difficulty in the linear stability analysis of the combined barotropic-baroclinic instability problem. The two terms (barotropic and baroclinic) in this combined instability problem are inseparable. Alternatively, one can approach this problem of understanding their individual roles by conducting an energy budget. This is done by adapting the zonal mean and eddy forms of the potential and kinetic energy budget equations of Lorenz (1955). In some studies, these easterly waves have been regarded as eddies relative to the zonal mean (e.g., Hseih and Cook 2007) and in others relative to the time mean (e.g., Alaka and Maloney 2014). The time rate of change of eddy kinetic energy (k_e) is given by

$$\frac{\partial k_e}{\partial t} = -\overline{\bar{u}\frac{\partial k_e}{\partial x}} - \overline{\bar{v}\frac{\partial k_e}{\partial y}} - \overline{\bar{\omega}\frac{\partial k_e}{\partial p}} - \overline{u'\frac{\partial k_e}{\partial x}} - \overline{v'\frac{\partial k_e}{\partial y}} - \overline{\omega'\frac{\partial k_e}{\partial p}} \underbrace{- \left(\overline{u'v'}\right)\frac{\partial \bar{u}}{\partial x}}_{\text{Term A}}$$

$$- \left(\overline{u'v'}\right)\frac{\partial \bar{u}}{\partial y} - \left(\overline{u'\omega'}\right)\frac{\partial \bar{u}}{\partial p} - \left(\overline{u'v'}\right)\frac{\partial \bar{v}}{\partial x} - \left(\overline{u'v'}\right)\frac{\partial \bar{v}}{\partial y} - \left(\overline{u'\omega'}\right)\frac{\partial \bar{v}}{\partial p}$$

$$- \frac{R}{p}\overline{(\omega'T')} - \frac{\partial \overline{(\phi'u')}}{\partial x} - \frac{\partial\left(\overline{\phi'v'}\right)}{\partial y} - \frac{\partial \overline{(\phi'\omega')}}{\partial p} - D \tag{4.9}$$

Equation 4.9 can be rewritten in vector format as

$$\frac{\partial k_e}{\partial t} = \underbrace{-\overrightarrow{\bar{V}}.\nabla k_e}_{\text{Term A}} \underbrace{-\overrightarrow{V'}.\nabla k_e}_{\text{Term B}} \underbrace{-\left(\overrightarrow{V_H'}.\left(\overrightarrow{V'}.\nabla\right)\overrightarrow{V}_H\right)}_{\text{Term C}} \underbrace{-\frac{R}{p}\overline{(\omega'T')}}_{\text{Term D}} \underbrace{-\nabla.\left(\overrightarrow{V'}\phi'\right)}_{\text{Term E}} - D \tag{4.10}$$

The time rate of change of time mean zonal eddy available potential energy $(\overline{p}_e)$ is given by

$$\frac{\partial p_e}{\partial t} = -\frac{\gamma}{T}\left(\overline{T'Q'}\right) - \frac{\gamma c_p}{T}\left(\overline{u'T'}\right)\frac{\partial \overline{T}}{\partial x} - \frac{\gamma c_p}{T}\left(\overline{v'T'}\right)\frac{\partial \overline{T}}{\partial y} - \frac{\gamma c_p}{T}\left(\overline{\omega'T'}\right)\frac{\partial \overline{T}}{\partial p} + \frac{R}{p}\overline{(\omega'T')} + R \tag{4.11}$$

Similarly, Eq. 4.11 can be rewritten in vector format as

$$\frac{\partial p_e}{\partial t} = \underbrace{-\frac{\gamma}{T}\left(\overline{T'Q'}\right)}_{\text{Term 1}} \underbrace{-\frac{\gamma c_p}{T}\overrightarrow{V'}T'.\nabla\overline{T}}_{\text{Term 2}} + \underbrace{\frac{R}{p}\overline{(\omega'T')} + R}_{\text{Term 3}} \tag{4.12}$$

where $k_e = \overline{\frac{u'^2+v'^2}{2}}$, $p_e = c_p\gamma\frac{\overline{T'^2}}{2T}$, $\overrightarrow{V}$ is the 3D wind vector, $\overrightarrow{V}_H$ is the horizontal wind vector, Q is the diabatic heating, ϕ is the geopotential height, R is the specific gas constant (= 287.058 $Jkg^{-1}K^{-1}$), c_p is the specific heat at constant pressure (= 1004 $JKg^{-1}K^{-1}$), $\gamma \equiv$ inverted static stability= $\frac{\Gamma_d}{\Gamma_d-\Gamma}$, $\Gamma_d \equiv$ dry static stability= $\frac{g}{c_p}$, $\Gamma \equiv$ Environmental lapse rate= $\frac{\partial \overline{T}}{\partial z}$, D is the dissipation of eddy kinetic energy due to friction and sub-grid scale processes, R is residual in the eddy available potential energy equation, overbar is time mean, and prime is deviation about this time mean.

Interpreting the terms on the RHS of Eqs. 4.10 and 4.12 is important to understand the importance of the various forms of instability.

Term A of Eq. 4.10: Advection of eddy kinetic energy by time mean wind
Term B of Eq. 4.10: Advection of eddy kinetic energy by anomalous winds
Term C of Eq. 4.10: It represents the barotropic energy conversion from mean kinetic energy to eddy kinetic energy in the presence of horizontal wind shear; barotropic disturbances derive their energy from the mean flow; in the case of the African easterly waves in West Africa, this term describes the energy transfer to eddy kinetic energy from the African easterly jet due to the presence of the horizontal shear; for barotropic disturbances to grow that derive energy from the zonal mean jet like the African easterly jet, they must tilt orthogonal to $\frac{\partial \bar{u}}{\partial y}$ This implies that the wave disturbances like the AEW, which propagate westward, would need to have a southwest-northeast tilt and be on the cyclonic shear side of the low-level jet to be able to draw the energy from the mean flow (Fig. 4.10). Alternatively, the easterly waves can tilt in the northwest-southeast direction when they are on the anticyclonic shear side of the low-level jet to draw the energy from the mean flow (Fig. 4.10). In the case of mid-latitude disturbances, they tilt in the same direction as (or parallel to) $\frac{\partial \bar{u}}{\partial y}$, and so the disturbance loses energy to the time mean flow by barotropic instability. Thus, barotropic instability is important for the maintenance of the mean flow in the middle latitudes.

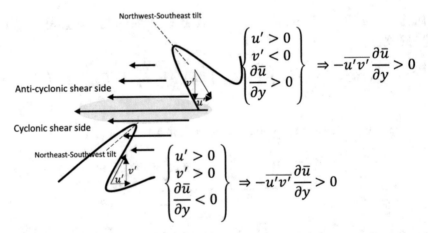

Fig. 4.10 A schematic showing the favorable northeast-southwest tilt and northwest-southeast tilt of the easterly wave on the cyclonic and anticyclonic shear side of the jet core (shaded; the African easterly jet) in the NH. The dotted arrows show the zonal (u') and the meridional (v') components of the easterly flow. The signage of Term A of Eq. 4.9, a dominant term under this scenario, is also illustrated

Term D of Eq. 4.10: It is the baroclinic energy conversion term. If warm air rises and/or relatively cool air descends, then eddy available potential energy is converted to eddy kinetic energy; in the case of the AEWs in West Africa, this term maintains the African easterly jet from the strong temperature gradient between the Gulf of Guinea and Sahara.

Term E of Eq. 4.10: It is the convergence of perturbation geopotential flux from local convergence.

Similarly, interpreting the RHS terms of Eq. 4.12, we can say the following:

Term 1 of Eq. 4.12: It is the generation term of eddy available potential energy; it is positive when diabatic heating and temperature covary.

Term 2 of Eq. 4.12: It represents the energy conversion from mean available potential energy to eddy kinetic energy; this term results from perturbation temperature flux being directed down the mean temperature gradient.

Term 3 of Eq. 4.12: It is like Term D of Eq. 4.10 but with the opposite sign. This term represents the energy that p_e loses to k_e.

Chapter 5
The Western Pacific Warm Pool

Abstract The feature of the warm sea surface temperature in the tropical Western Pacific is introduced with its importance to the atmospheric general circulation. The maintenance of this warm pool is then discussed that emphasizes the role of the atmosphere and the ocean in sustaining this feature.

Keywords Warm pool · Outgoing longwave radiation · Greenhouse effect · Sverdrup vorticity balance · Indonesian throughflow · Ocean thermostat theory · Atmospheric thermostat theory · Betts-Ridgway model

5.1 What Is the Significance of the Warm Pool?

The Western Pacific Warm Pool, which spans the western waters of the Equatorial Pacific to the Eastern Indian Ocean, is almost four times the size of the continental United States with an area $>30 \times 10^6$ km^2 (Fig. 5.1). It extends about 9000 miles in the zonal direction along the equator and 1500 miles in the meridional direction, and since it spans into the Eastern Indian Ocean, it is also often referred to as the Indo-Pacific Warm Pool. The temperature of the warm pool ranges anywhere from 81 °F (27 °C) at the edges of the warm pool to 86 °F (30 °C) in the center. The warm pool also shows a strong seasonal cycle apparent from the variations of the spatial extent of the warm pool and its overall location about the equator (Fig. 5.1). It may be noted that there is no defined isotherm as such to demarcate the warm pool, although 28 °C is commonly used to define the warm pool (Fig. 5.1).

This warm pool extends to almost 200 m below the ocean surface (Fig. 5.2) and remains warm (>28 °C) throughout the year. However, a large portion of the seas in the warm pool (e.g., parts of Flores Sea, Banda Sea, Celebes Sea, Sulu Sea, and Molucca Sea) are relatively shallow (≤200 m), which then allows the upper ocean temperatures to remain comparatively warm throughout the year from the impinging atmospheric fluxes under a background of very light surface winds (≤~3 m/s). As a consequence of this immense heat energy stored in the warm pool, it supports overlying strong atmospheric thunderstorm activity with clouds reaching heights of approximately 15 km and producing a tremendous amount of latent heat in the

© Springer Nature Switzerland AG 2023
V. Misra, *An Introduction to Large-Scale Tropical Meteorology*, Springer Atmospheric Sciences, https://doi.org/10.1007/978-3-031-12887-5_5

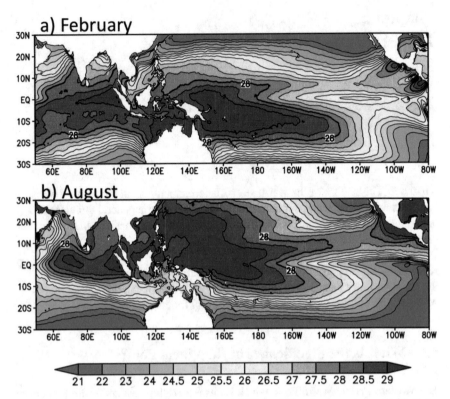

Fig. 5.1 The observed climatological annual mean SST that delineates the warm pool in the Indo-Pacific region with temperatures >28 °C during two contrasting months of the year, namely, (**a**) February and (**b**) August. The SST is from OISSTv2 and the climatology is computed over 1987–2010

atmosphere from the condensation of water vapor in the clouds. Webster and Lukas (1992) showed that the maxima of the total diabatic heating (including contributions from latent, radiative, and sensible heating) co-locate with the warm pool, setting up atmospheric global heating gradients. Therefore, this region of strong ascent over the warm pool, brought about by strong convergence of air and moisture, which is in fact regarded as the ascending cell of the Walker Circulation, leads to heavy precipitation (yielding an average of 3 m/year; Webster and Lukas 1992) and is often called the steam engine of the atmospheric general circulation.

Despite the large seasonal changes in the spatial extent of the warm pool (Fig. 5.1), it can also be argued that the SST in the core of the warm pool barely exhibits a seasonal cycle as it continues to remain very warm throughout the year. It is therefore safe to suggest that it is the warm water volume that undergoes significant changes (Figs. 5.1 and 5.2). For example, the centroid of the warm water volume migrates meridionally from ~10°S to 10°N from the SH summer to the NH summer, which is largely dictated by the corresponding seasonal variations of the

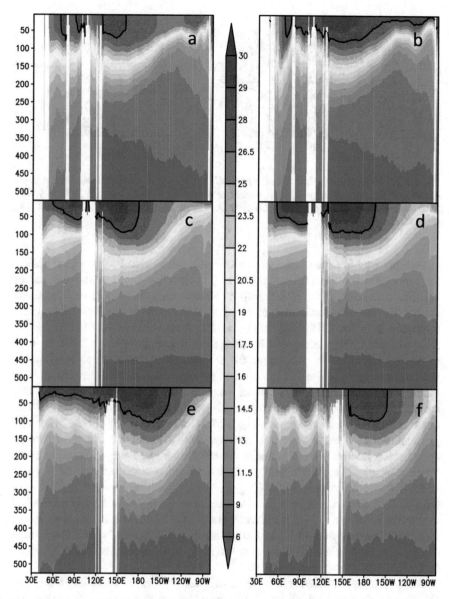

Fig. 5.2 Longitude-depth cross sections of ocean temperatures (°C) from the surface to a depth of 525 m for (**a, c, e**) February and (**b, d, f**) August at (**a, b**) 10°N, (**c, d**) equator, and (**e, f**) 10°S. The 28 °C isotherm is indicated by the bold black contour line

net atmospheric heat flux. But the zonal variations of the warm water volume are less affected by the seasonal cycle and more influenced at the inter-annual scales by variations like ENSO.

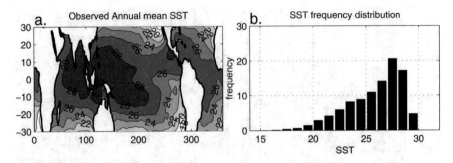

Fig. 5.3 (**a**) The climatological annual mean SST (°C) and the corresponding (**b**) frequency distribution of SST shown in (**a**) as a percent area (so that the sum over all SSTs is 100%). (Reproduced from Clement et al. (2006); © American Meteorological Society. Used with permission)

Another way to perceive the large warm pool in the Indo-Pacific region besides the ways shown in Figs. 5.1 and 5.2 is to plot the climatological area covered by adjacent isotherms in the global tropical oceans (Fig. 5.3b). Figure 5.3b shows the distribution in percent area occupied by the various isotherms of the annual mean climatological SST shown in Fig. 5.3a. It should be noted that in making Fig. 5.3b, one must compute the area between the adjacent isotherms of SST for each individual month and then compute the climatological annual mean to avoid the potential spreading of the SST distribution across seasonal variability. In Fig. 5.3b, the percent area occupied in the annual mean SST range of 26 °C to about 29 °C is the highest, after which the area drops significantly. It is important to recognize that the area between two isotherms in Fig. 5.3b is an inverse proxy for the mean SST gradient. For example, a large area between two isotherms means that they are far apart, which implies a weak gradient. Therefore, the large area occupied between the isotherms of 27 and 28 °C in Fig. 5.3b is a reflection of the vast expansive region of the warm pool with very weak SST gradients. On the other hand, the rapid drop of the area for SSTs warmer than 28 °C is a result of the asymptotic approach toward zero of the areal coverage of isotherms warmer than 30 °C. It is rare to have annual mean SSTs warmer than 30 °C. In a broad sense, the global SST spatial distribution can be described as a large body of warm SSTs in the tropics bounded by strong SST gradients in the higher latitudes.

Many studies have shown that global weather and climate variability are highly sensitive to changes in the warm pool's temperature, size, and volume (Bjerknes 1969; Palmer and Mansfield 1984; Webster and Lukas 1992; Trenberth et al. 1998; Meinen and McPhaden 2000; Hoerling and Kumar 2003). They not only affect atmospheric variations within the warm pool but also affect remote climates through modulation of the atmospheric waves. The variations of the warm pool are central to ENSO variability, affect the Asian Monsoon variations, and have a substantial influence on the intraseasonal variations and typhoons that develop over the Western Pacific region.

5.2 The Atmospheric Thermostat Theory for the Warm Pool

A commonly used way of representing the relationship between deep convection and SST is a scatter plot between OLR and SST between the latitude bands of 40°S to 40°N (Fig. 5.4). This scatter plot has a total of 40,000 data points (Lau et al. 1997). The bulk OLR-SST relationship is represented by the solid line in Fig. 5.4 that connects the mean OLR values (shown by open circles in Fig. 5.4) for every 0.5 °C SST bin. The scatter plot in Fig. 5.4 has the following three important features:

- An increasing OLR with increasing SST from 18 to 26 °C with a slope of ~ + 2.5 W/m²/°C (suggesting clear skies in this SST range, although deep convection could occur in this range).
- A steep decrease in OLR with SST ranging between 26.5 and 29.5 °C (suggesting deep convection in this SST range, although the large scatter of the low OLR suggests the temporal and spatial variabilities of deep convection).
- A steep increase in OLR at SSTs warmer than 29.5 °C (suggesting isolated regions of hot spots in regions of forced subsidence).

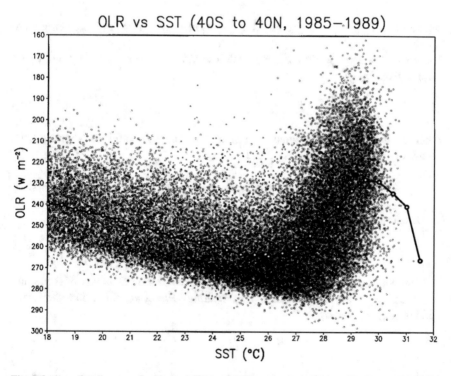

Fig. 5.4 A scatter diagram of collocated SST and OLR grid points for the tropical oceans (40°S–40°N). The overlaid solid line with open circles represents the OLR as a function of SST bin with a bin width of 0.5 °C. (Reproduced from Lau et al. (1997); © American Meteorological Society. Used with permission)

It is this threshold of around 26.5 °C for deep convection that intrigued many scientists and purported a whole host of theories including the atmospheric thermostat theory.

The atmospheric thermostat theory for maintenance of the warm pool gained significant attention at the time when Ramanathan and Collins (1991) proposed this idea. The cirrus clouds are produced over the warm pool because of the sustained deep convection. Using observed (satellite-based) estimates of radiation fluxes over an El Niño event (of 1987), Ramanathan and Collins (1991) indicated that the highly reflective cirrus clouds over the warm pool act like a thermostat to shield the ocean surface from solar radiation and thereby regulate the SST to never exceed ~305 K (or ~ 32 °C). The cirrus clouds they suggested form in response to a 'super greenhouse effect', which refers to the greenhouse effect increasing with surface temperature (T_s; in this case, the SST over the warm pool) at a rate that exceeds the rate at which radiation is emitted from the surface ($= \sigma T_s^4$; the Stefan-Boltzmann constant $= 5.67 \times 10^{-8}$ Wm^{-2} K^{-4}).

The total greenhouse effect of the atmosphere and clouds (G) is given by

$$G = E - F \qquad (5.1)$$

where $E = \sigma T_s^4$ is the energy emitted by the surface and F is the radiation emitted to space.

The atmospheric portion of the greenhouse effect or clear sky greenhouse effect (G_a) is given by

$$G_a = E - F_c \qquad (5.2)$$

where F_c is the radiation emitted to space under clear sky conditions. Therefore, the cloud longwave forcing (C_l) is given by

$$C_l = G - G_a = F_c - F \qquad (5.3)$$

Or using Eqs. 5.1 and 5.3, we have

$$F = E - G = E - G_a - C_l \qquad (5.4)$$

The clouds, by way of their higher albedo, increase the reflectivity of the planet and thereby reduce the planetary solar heating. Therefore, the cloud shortwave forcing (C_s) is given by

$$C_s = S(1 - A) - S_c \qquad (5.5)$$

where A is the column albedo, S is the solar radiation at the top of the atmosphere, $S(1 - A)$ is the solar radiation absorbed in the column, and S_c is the clear sky solar absorption. The net radiation at the TOA, H, is given by

$$H = S(1 - A) - F = C_s + S_c - E + G_a + C_l \qquad (5.6)$$

Equation 5.6 can be interpreted as the absorbed solar radiation under clear sky (S_c) heats the surface offset by the cloud shortwave forcing (C_s), wherein, $C_s < 0$. The surface emits energy as longwave radiation (E) that is partially trapped in the atmosphere given by $G_a + C_l$.

Ramanathan and Collins (1991) attempted to understand the variations of the terms in Eq. 5.6 with respect to the surface temperature using observations. So to evaluate the contributions of the response of the net radiation at the TOA to SST, we can write Eq. 5.6 as

$$\frac{dH}{dT_s} = \frac{dC_s}{dT_s} + \frac{dS_c}{dT_s} - 4\sigma T_s^3 + \frac{dG_a}{dT_s} + \frac{dC_l}{dT_s} \qquad (5.7)$$

Ramanathan and Collins (1991) found in the Tropical Pacific (10°N to 10°S) when T_s is ~300 K that $\frac{dG_a}{dT_s} \sim 6$ to 9 Wm^{-2} K^{-1} and $\frac{dE}{dT_s} = 4\sigma T_s^3 \sim$ 6.1 Wm^{-2} K^{-1}. This observational fact of $\frac{dG_a}{dT_s} \geq \frac{dE}{dT_s}$ is referred to as the super greenhouse effect, wherein the warm tropical ocean has lost the negative feedback between temperature and outgoing longwave radiation, which radiates out the excess surface heat to space. In essence, Ramanathan and Collins (1991) observed that at the warmest temperatures, the total greenhouse effect (G) increases significantly, and longwave emission at the surface (E) decreases with increasing surface temperature (T_s). Locally, this leads to instability, which is counterbalanced by the sharp increase in the cloud shortwave forcing (C_s). Although C_s and C_l are observed to be strongly anti-correlated, large negative values of C_s are correlated with warmer waters. Ramanathan and Collins (1991) diagnosed $\frac{dC_s}{dT_s} \sim -22$ Wm^{-2} K^{-1}. This led Ramanathan and Collins (1991) to posit: 'as the tropical oceans warm, the rapid rise of G_a with SST leads to an unstable feedback. The warming continues until the clouds become thick enough to shield the ocean from the solar radiation and arrest further warming. This thermostat effect of the cirrus anvils controls the maximum value of SSTs'.

This atmospheric thermostat theory had some serious implications. For example, one of the takeaways of this theory is that surface temperature will not warm appreciably with global warming unless the greenhouse gases increase significantly (by almost an order of magnitude!). This theory had such profound implications that the National Science Foundation and the Department of Energy in the United States funded an exclusive month-long field study to examine this issue. This exclusive field experiment was conducted in March 1993 in the Tropical Pacific, which was called the CEPEX whose mission was to directly measure the quantities to test the atmospheric thermostat theory.

As previously noted before the atmospheric thermostat theory was proposed, surface evaporation was considered to be the fundamental limiting mechanism on SST (Newell 1979; Graham and Barnett 1987). But the atmospheric thermostat theory challenged this idea because evaporation moistens the column and therefore further increases the super greenhouse effect and that water vapor is imported to the warmest oceanic regions and thus evaporation tends to be weak. Observations from CEPEX and TOGA-COARE indicated that cooling effects due to shortwave forcing of clouds play as significant a role as evaporation in cooling the Tropical Western Pacific Ocean with the former, however, being larger in magnitude (Waliser 1996). Furthermore, the atmospheric thermostat theory does not recognize that SST and convection are not only determined locally but are part of a system involving the whole atmospheric general circulation (Wallace 1992; Fu et al. 1992; Waliser 1996). The modifications to the atmospheric circulation by the SST anomaly, which is at a length scale larger than that of the SST anomaly, affect the local negative cloud feedback. For example, the temperature and moisture of the converging air (determined by processes remote to the local SST in the warm pool region) will determine the amount of cloudiness and local rate of evaporation. The observational study of Hartmann and Michelsen (1993) showed that the shortwave forcing of clouds is more sensitive to large-scale convergence than to SST as positive cloud albedo anomalies over convective regions are nearly compensated by albedo decreases elsewhere in the tropics.

In essence, the atmospheric thermostat theory undermines the contributions of the large-scale atmospheric general circulation and its role in modulating convection and thereby cloud radiative forcing. Nonetheless, the atmospheric thermostat theory serves as a good pedagogical tool to begin understanding the role of clouds in climate.

5.3   A Holistic Approach to Understanding the Regulation of the Warm Pool

Clement et al. (2006) have provided a very systematic analysis for the regulation or maintenance of the warm pool. This work is also illustrative of the scientific approach to understanding a phenomenon by using a hierarchy of models from a reduced dimensionality like their use of one-dimensional climate model to a full-blown coupled ocean-atmosphere general circulation model to assess the impact of individual processes of what, in reality, are inter-woven processes. The fundamental question that is raised is why is the distribution of the SST as shown in Fig. 5.3? More finely put, why is the maximum area occupied by the isotherm slightly cooler than the warmest SST in Fig. 5.3? Alternatively, these questions are framed to ask what maintains the Western Pacific Warm Pool.

Recognizing that the warm pool resides over the equator, it is reasonable to assess the sole contribution of the incoming solar radiation on the distribution of the SST. The solar insolation is maximum at the equator and decreases as the cosine of the latitude in the annual mean. Therefore, the solar insolation at the top of the atmosphere is relatively large for $\sim 10°$ on either side of the equator (Figs. 5.5a, b). If the SST were simply regarded as a linear function of this solar insolation at the

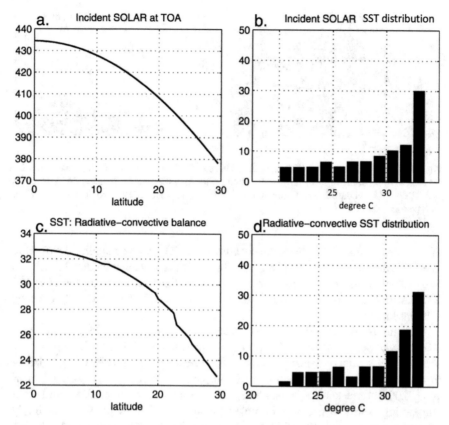

Fig. 5.5 (**a**) The annual mean incident solar radiation (Wm^{-2}) at the top of the atmosphere as a function of latitude used in the one-dimensional Betts-Ridgway model and (**b**) the corresponding frequency distribution. (**c**) The equilibrium SST (°C) as a function of latitude computed from the Betts-Ridgway model in response to the imposed solar radiation shown in (**a**) and (**d**) the corresponding frequency distribution of SST. (Reproduced from Clement et al. (2006); © American Meteorological Society. Used with permission)

top of the atmosphere (Fig. 5.5a), then it results in a very skewed distribution of the SST with the warmest SST occupying the largest area (Fig. 5.5b), which is unlike the observed SST distribution (Fig. 5.3b).

It is possible that local air-sea processes, convection, and radiation can alter this relationship between SST and the solar insolation that is ignored in the illustration shown in Fig. 5.5b. In order to include these processes, Clement et al. (2006) adopted a 1D climate model developed for the tropics following Betts and Ridgway (1989). This 1D climate model solves for SST, humidity, and temperature throughout the troposphere for given solar insolation, atmosphere, and ocean heat transports. Clement et al. (2006) ran this model for each degree of latitude from 30°S to the equator, specifying zero atmospheric and oceanic heat transports with solar insolation prescribed as a unique function of latitude (Fig. 5.5a) to isolate the impact

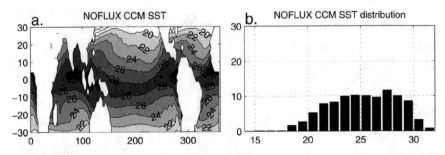

Fig. 5.6 (**a**) The annual mean SST from the AGCM (CCM3) in which ocean heat transports are set to zero everywhere and (**b**) the corresponding frequency distribution (percent area) occupied by SST (°C). (Reproduced from Clement et al. (2006); © American Meteorological Society. Used with permission)

of local processes. The Betts-Ridgway climate model yielded a similar distribution of SST as the solar insolation (Fig. 5.5c, d), which suggests that the local processes (e.g., radiative-convective processes) have no significant impact on the observed SST distribution and the homogenization of tropical SST than is already done by solar insolation.

Clement et al. (2006) proceeded with running a full AGCM coupled to a *mixed layer ocean model* with no oceanic heat transport but which allows for local air-sea thermodynamic processes. As a result of the processes included in this configuration, the distribution of the tropical SST (Fig. 5.6a, b from Clement et al. 2006) from such a model integration results in a rather flat distribution (suggesting nearly equal area occupied by many isotherms of SST that imply strong horizontal gradients of SST), which is quite different from the distribution of SST shown in Fig. 5.5. So primarily the contention is that the homogeneity of the tropical SST enforced by solar insolation and other local processes is nearly eliminated by the introduction of the atmospheric circulation. Simply put, the results in Fig. 5.6 suggest that the meridional gradient of the solar insolation at the top of the atmosphere drives a zonally symmetric atmospheric circulation (e.g., Hadley Circulation) that diverges heat from the equator to the poles and in the process alters the meridional distribution of relative humidity and surface winds and in the process changes the SST distribution to that shown in Fig. 5.6, which is far different from the local adjustment case shown in Fig. 5.5.

In order to exclusively assess the impact of the atmospheric heat transport on the SST distribution, Clement et al. (2006) prescribed a reasonable atmospheric heat transport as a function of column mean temperature in the Betts-Ridgway model run at each latitude, which gave rise to maximum heat transport at the equator and decreasing to zero at 30°S. To start with, they tested the Betts-Ridgway model by setting the atmospheric heat transport to zero and then examine the SST distribution (Fig. 5.7a, b). In this experiment, the column mean temperature (tcr) and surface wind speed were held constant. The distribution of SST from this experiment (Fig. 5.7b) restores the skewed distribution of SST obtained from solar insolation (Fig.

5.5b). However, the skewed distribution of SST in Fig. 5.7b is dissimilar from the observed SST distribution in Fig. 5.3b, which shows that the maximum area is occupied by the isotherm that is slightly cooler than the maximum SST.

In the next experiment, the atmospheric heat transport was restored in the Betts-Ridgway model. In this experiment, the heat transport is a maximum at the equator and tapers to zero at 30° latitude (Fig. 5.7c). The inclusion of atmospheric heat transport in the Betts-Ridgway model resulted in homogenizing the SST by cooling the warmest SST and producing an SST distribution with a maximum area occupied by SST values slightly below the maximum SST (Fig. 5.7d). However, the range of the SST distribution in Fig. 5.7d is rather narrow compared to the observed range (Fig. 5.3b). Nonetheless, from Fig. 5.7d, it can be concluded that atmospheric heat transport tends to homogenize the tropical SST.

As a next step, Clement et al. (2006) altered the relative humidity distribution in the Betts-Ridgway model. One of the consequences of the Hadley Circulation is the desiccation of the moisture in the atmospheric column of the subtropics owing to the implied subsidence over these latitudes. In the Betts-Ridgway model, the alteration of the relative humidity above the trade wind inversion is achieved by modifying a parameter called T_{crit}, which then alters the specific humidity above the trade wind version (q_T) given in the model as $q_T = q_s(T_{crit}, \theta_e)$, where q_s and θ_e are saturation specific humidity above the trade wind inversion and subcloud layer equivalent potential temperature, respectively. In the previous experiments with the Betts-Ridgway model, T_{crit} was set to 266 K. But now T_{crit} is made to vary with latitude. The resulting SST distribution from the inclusion of this meridional gradient of relative humidity is that it causes an expansion of the range of SST (Fig. 5.7e, f) compared to having just the local processes (Fig. 5.5a, b) or even including atmospheric heat transport (Fig. 5.7c, d). This change in the distribution of the SST is because of the reduced greenhouse effect in the subtropics from the relatively clearer skies that results in the cooling of the subtropical SSTs. However, the SST distribution in Fig. 5.7f continues to remain negatively skewed unlike the flat SST distribution observed with the full AGCM integration in Fig. 5.6b.

The final process to be included in the Betts-Ridgway model in order to understand the SST distribution from the AGCM integration was to include the meridional gradient of the surface wind speed. Therefore, the surface wind speed was made to vary from 5 ms^{-1} at the equator to about 9 ms^{-1} at 30°S, which is unlike the previous experiments where the surface wind speed was kept constant at all latitudes at 6.7 ms^{-1}. The inclusion of this varying surface wind speed with all of the other processes restored the SST distribution to be flat (Fig. 5.7g, h) and with a range comparable to that obtained from the AGCM integration (Fig. 5.6b). In the subtropics where the surface wind speed is higher, the atmospheric boundary layer tends to move closer to the ocean surface temperature in order to reduce the air-sea temperature gradient. A moist and warm atmosphere would therefore radiate more to space, and to restore equilibrium at the air-sea interface, SST will cool. On the other hand, the low wind speeds at the equator would warm the SSTs and thereby reduce the homogeneity of the SST in the tropics.

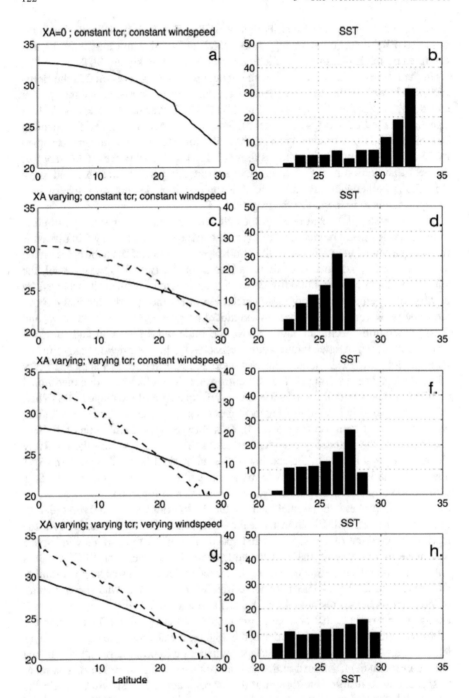

Fig. 5.7 The results from the one-dimensional Betts-Ridgway model showing (**a**) SST (°C) as a function of latitude for local balance with atmospheric heat transport (XA) set to zero and with column mean temperature (tcr) and surface wind speed held constant and (**b**) the corresponding

Obviously, the SST distribution shown in Fig. 5.6b from the AGCM coupled to a mixed layer ocean model is far from the observed SST distribution (Fig. 5.3b). This raises the potential for investigating the remaining component of the climate system, which is the ocean transport. We are deliberately ignoring the role of the clouds here for pedagogical reasons. However, it should be abundantly clear from the discussion in the previous subsection that clouds indeed play a very important role in regulating the SST in the warm pool. In order to understand the contribution of the ocean transport, we need to introduce two concepts: Ekman and Sverdrup transports.

In the boundary layer, the balance of forces is Coriolis (f), pressure gradient ($\overrightarrow{\nabla p}$), and friction ($\overrightarrow{F}$). A boundary layer with this balance of forces is often called the Ekman layer, after Vagn Walfrid Ekman. This boundary layer appears both at the bottom of the atmospheric and oceanic columns (at the ocean floor) where friction acts on flow over uneven/rough surfaces. Additionally, the boundary layer also appears in the upper part of the oceanic column, where the frictional drag acts on the atmospheric surface winds over the ocean surface. The horizontal momentum equations in the boundary layer can be written as

$$fu = \frac{-1}{\rho}\frac{\partial p}{\partial y} - F_y \qquad (5.8)$$

$$fv = \frac{1}{\rho}\frac{\partial p}{\partial x} + F_x \qquad (5.9)$$

where ρ, F_y, and F_x is the density of air, and friction terms in zonal and meridional momentum equations, respectively. Or in vector form, the above two equations could be written as

$$f\overrightarrow{V} = \hat{k} \times \frac{1}{\rho}\overrightarrow{\nabla p} + \hat{k} \times \overrightarrow{F} \qquad (5.10)$$

Fig. 5.7 (continued) frequency distribution of SST. (**c**) SST (°C, solid line, left axis) and atmospheric heat transport (Wm^{-2}, dashed line, right axis) as a function of latitude for the case of interactive atmospheric heat transport with T$_{crit}$ (see text) and surface wind speed held constant and (**d**) the corresponding frequency distribution of SST. (**e**) SST (°C, solid line, left axis) and atmospheric heat transport (Wm^{-2}, dashed line, right axis) as a function of latitude for the case of interactive atmospheric heat transport with T$_{crit}$ varying with latitude and surface wind speed held constant and (**f**) the corresponding frequency distribution of SST. (**g**) SST (°C, solid line, left axis) and atmospheric heat transport (Wm^{-2}, dashed line, right axis) as a function of latitude for the case of interactive atmospheric heat transport with T$_{crit}$ and surface wind speed varying with latitude and (**h**) the corresponding frequency distribution of SST. (Reproduced from Clement et al. (2006); © American Meteorological Society. Used with permission)

The boundary layer wind can be split into geostrophic ($\vec{V}_g$) and ageostrophic ($\vec{V}_{ag}$) components:

$$\vec{V} = \vec{V}_g + \vec{V}_{ag} \tag{5.11}$$

where

$$\vec{V}_g = \hat{k} \times \frac{1}{\rho f} \tag{5.12}$$

and

$$\vec{V}_{ag} = \frac{\hat{k} \times \vec{F}}{f} \tag{5.13}$$

The surface wind blowing over the ocean surface imparts motion to the surface water measured as shear force per unit contact area or more commonly referred to as surface wind stress. This wind stress is parameterized as

$$\vec{\tau} = \rho C_D \vec{V}_{10m}^2 \tag{5.14}$$

where C_D and $\vec{V}_{10m}$ are dimensionless wind drag coefficient and winds at 10 m above the surface, respectively. In the Ekman layer of the ocean extending from the surface to a certain depth (Ekman depth), the stress exerted by the surface wind gives the ocean current a velocity that is deflected to the right in the NH or to the left in the SH due to Coriolis force. However, because of the frictional drag, this deflection of the ocean current is at 45° to the force exerted over the layer. Therefore, as one moves downward through each successive layer of the Ekman boundary layer, the currents become slower consistent with the reducing wind stress and get deflected at an angle (45°) to the right of the current in the NH or left of the current in the SH in the layer immediately above (Fig. 5.8). The depth of the Ekman layer is determined when friction ceases or becomes negligible. Although the direction of the current is different within each layer, the theoretical average direction of the 'Ekman transport' is taken to be 90° to the right in the NH or to the left in the SH of the surface wind direction.

The friction force ($\vec{F}$) in Eq. 5.10 is given by

$$\vec{F} = \frac{1}{\rho} \frac{\partial \vec{\tau}}{\partial z} \tag{5.15}$$

Substituting Eq. 5.15 in Eq. 5.13 and putting it in scalar form, we get

$$-fv = \frac{1}{\rho} \frac{\partial \tau_x}{\partial z} \tag{5.16}$$

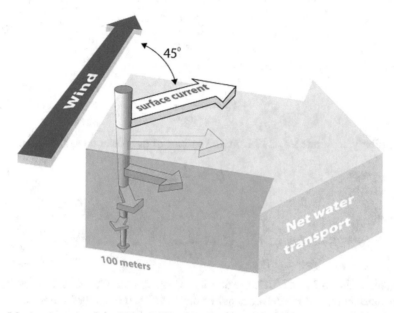

Fig. 5.8 A schematic of the Ekman drift with the top layer (or the ocean surface) driven by atmospheric wind and the subsequent layers below are moved by friction. In each succeeding layer from the surface, the ocean currents are moving at a slower speed and deflected at an angle (45°) to the right in the NH (or left in the SH) of the current in the layer immediately above until it reaches the bottom of the Ekman layer (in this example, at 100 m) where the effect of friction is no longer felt. The theoretical average direction of the Ekman flow in the Ekman layer is taken to be at right angles (90°) to the right (in NH) or the left (in SH) to the prevailing surface wind. (Reproduced from https://oceanservice.noaa.gov/education/tutorial_currents/media/supp_cur04e.html)

$$fu = \frac{1}{\rho}\frac{\partial \tau_y}{\partial z} \tag{5.17}$$

The mass continuity equation in the x, y, z coordinate system is given by

$$\frac{\partial u}{\partial x} + \frac{\partial v}{\partial y} + \frac{\partial w}{\partial z} = 0 \tag{5.18}$$

Now, integrate Eq. 5.18 from the surface of the ocean up to the depth of the Ekman layer (z_e) and substitute $U_e = \int_{z_e}^{0} u\, dz$, $V_e = \int_{z_e}^{0} v\, dz$, and $W_e = \int_{z_e}^{0} w\, dz$ to give

$$\frac{\partial U_e}{\partial x} + \frac{\partial V_e}{\partial y} = W_e \tag{5.19}$$

Likewise, integrate Eq. 5.16 from the surface of the ocean up to z_e to give

$$-fV_e = \frac{1}{\rho_o}\tau_x \tag{5.20}$$

Or

$$V_e = -\frac{1}{f\rho_o}\tau_x \tag{5.21}$$

Similarly, integrate Eq. 5.17 from the surface of the ocean up to z_e to give

$$fU_e = \frac{1}{\rho_o}\tau_y \tag{5.22}$$

or

$$U_e = \frac{1}{f\rho_o}\tau_y \tag{5.23}$$

Here, in Eqs. 5.20 and 5.22, the density of water in the Ekman layer is assumed to be constant (ρ_o) and therefore is pulled out of the integral. Substituting for V_e and U_e from Eqs. 5.21 and 5.23, respectively, into Eq. 5.19, we get

$$\frac{1}{\rho_o f}\left[\frac{\partial \tau_y}{\partial x} - \frac{\partial \tau_x}{\partial y}\right] = W_e \tag{5.24}$$

or in vector form:

$$W_e = \hat{k}.\left(\nabla \times \frac{\vec{\tau}}{\rho_o f}\right) \tag{5.25}$$

Equation 5.25 suggests that the upwelling or downwelling in the Ekman layer is equal to the curl of the wind stress. So a positive wind stress curl causes divergence in the Ekman layer and upward Ekman pumping (or upwelling), and a negative wind stress curl causes convergence and downward Ekman pumping (or downwelling). This can also be interpreted by using the right-hand thumb rule in the NH: Wrap the fingers of your right hand in the direction of the flow of the curl (clockwise or anticlockwise) and the direction at which your right thumb points will give the direction of the Ekman pumping (on whether it is upward or downward). Likewise, in the SH, one would use the left hand to ascertain the direction of the implied Ekman pumping from the curl of the wind stress.

Now to understand Sverdrup dynamics of the oceans, we make use of the vorticity equation:

$$\frac{\partial \eta}{\partial t} + u\frac{\partial \eta}{\partial x} + v\frac{\partial \eta}{\partial y} + w\frac{\partial \eta}{\partial z}$$
$$= \underbrace{-\eta\left(\frac{\partial u}{\partial x} + \frac{\partial v}{\partial y}\right)}_{A} \underbrace{- \left(\frac{\partial w}{\partial x}\frac{\partial v}{\partial z} - \frac{\partial w}{\partial y}\frac{\partial u}{\partial z}\right)}_{B} + \underbrace{\frac{1}{\rho^2}\left(\frac{\partial P}{\partial y}\frac{\partial \rho}{\partial x} - \frac{\partial P}{\partial x}\frac{\partial \rho}{\partial x}\right)}_{C} \tag{5.26}$$

where η is the absolute vorticity ($=\xi + f$). Terms A, B, and C in Eq. 5.26 capture the effect of horizontal divergence on vorticity, the transfer of vorticity between horizontal and vertical components (also called the twisting or the tilting term), and the effect of baroclinicity (also called the solenoidal term), respectively.

In considering the large-scale flow of the oceans, one can neglect the vertical advection of vorticity on the LHS of Eq. 5.26 and the twisting and solenoidal terms on the RHS side of Eq. 5.26. Furthermore, at large scales in the subtropical oceans, $\xi \ll f$. Therefore, $\eta \approx f$. So revisiting Eq. 5.26 after considering these factors, we get

$$u\frac{\partial f}{\partial x} + v\frac{\partial f}{\partial y} \approx -f\left(\frac{\partial u}{\partial x} + \frac{\partial v}{\partial y}\right) \tag{5.27}$$

Using mass continuity equation and knowing $\frac{\partial f}{\partial x} = 0$, Eq. 5.27 can be rewritten as

$$v\frac{\partial f}{\partial y} = f\frac{\partial w}{\partial z} \tag{5.28}$$

or

$$v\beta = f\frac{\partial w}{\partial z} \tag{5.29}$$

Equation 5.29 is also called the Sverdrup vorticity balance eq. A useful interpretation of this equation is that when v is northward or southward, then Eq. 5.29 suggests there is compensating convergence (i.e., $\frac{\partial w}{\partial z} > 0$) or divergence (i.e., $\frac{\partial w}{\partial z} < 0$), respectively. By the same argument, when the meridional flow is negligible as in the center of a subtropical gyre, the implied divergence from the Sverdrup vorticity balance equation would also be negligible.

Now integrate Eq. 5.29 from the bottom of the ocean (sea floor; z_o) to the base of the Ekman layer (z_e), which gives to:

$$\beta\int_{z_0}^{z_e} v dz = f\int_{z_0}^{z_e}\frac{\partial w}{\partial z}dz \tag{5.30}$$

But at the sea floor, $w = 0$ and $\int_{z_o}^{z_e} v dz = V_g \equiv$ vertically integrated meridional geostrophic transport. Therefore, Eq. 5.30 becomes

$$\beta V_g = f W_e \tag{5.31}$$

Now, substituting Eq. 5.25 into 5.31, we get

$$\beta V_g = f\left(\hat{k}.\left(\nabla \times \frac{\vec{\tau}}{\rho_o f}\right)\right) = \frac{f}{\rho_o}\left[\frac{1}{f}\frac{\partial \tau_y}{\partial x} - \frac{1}{f}\frac{\partial \tau_x}{\partial x} + \frac{\tau_x}{f^2}\frac{\partial f}{\partial y}\right] = \frac{1}{\rho_o}\left(\hat{k}.\left(\nabla \times \vec{\tau}\right)\right) + \frac{\tau_x \beta}{\rho_o f}$$

(5.32)

Substituting Eq. 5.21 in 5.32, we get

$$\beta V_g = \frac{1}{\rho_o}\left(\hat{k}.\left(\nabla \times \vec{\tau}\right)\right) - \beta V_e$$

(5.33)

or

$$\beta\left(V_g + V_e\right) = \frac{1}{\rho_o}\left(\hat{k}.\left(\nabla \times \vec{\tau}\right)\right)$$

(5.34)

$\beta(V_g + V_e)\equiv$ Sverdrup transport. So Eq. 5.34 indicates that the Sverdrup transport is the total meridional transport of β from the ocean surface to the bottom, which is proportional to the curl of the wind stress. On the other hand, the Ekman transport is described by Eqs. 5.21 and 5.23, which are proportional to the wind stress.

So now having some background on the Ekman and Sverdrup transports from the preceding discussion, we can go back to the discussion of the distribution of the tropical SST. In including the full 3D ocean circulation with a thermocline depth of 50 m, Clement et al. (2006) changed the SST distribution as shown in Fig. 5.9. It may be noted that the results in Fig. 5.9 are with the uncoupled ocean model imposed with observed wind stress and surface heat flux. The distribution in Fig. 5.9 is now skewed like the observations (albeit with a limited range of SST), exhibiting a far more homogenous SST in the tropics with the largest area occupied by isotherms that are slightly cooler than the warmest SST. The easterly wind stress imposed on the equatorial region causes upwelling in the Eastern Pacific, causing a cold tongue to form that is distinct from the warm pool in the Western Pacific (Fig. 5.9a). By varying the depth of the thermocline, one can, however, modulate the effect of the Sverdrup dynamics. As the thermocline is deepened uniformly, the warm pool expands eastward as the upwelled water in the Western Pacific is warm. As the warm pool expands and homogenizes the SST, the curl of the wind stress reduces in response to the weakening SST gradients. Therefore, by Eq. 5.34, the Sverdrup transport will weaken as the thermocline depth is uniformly increased. Likewise, as the thermocline depth is uniformly decreased, it would imply an increase in the Sverdrup transport from the consequent increase in the heterogeneity of the SST and curl of the wind stress. This is exactly observed when the thermocline depth was changed from 50 m to 100 m to 200 m (Fig. 5.10). As the thermocline depth was raised in the model experiments from 50 m to 200 m, the Sverdrup dynamics had a

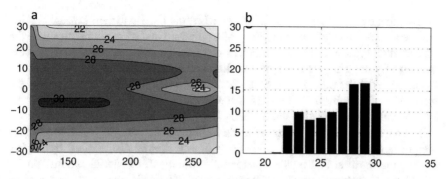

Fig. 5.9 The (**a**) spatial distribution and the corresponding (**b**) frequency distribution as a function of latitude of the annual mean SST from an ocean model (with full 3D ocean circulation included) with imposed atmospheric wind stress and heat flux. (Reproduced from Clement et al. 2006)

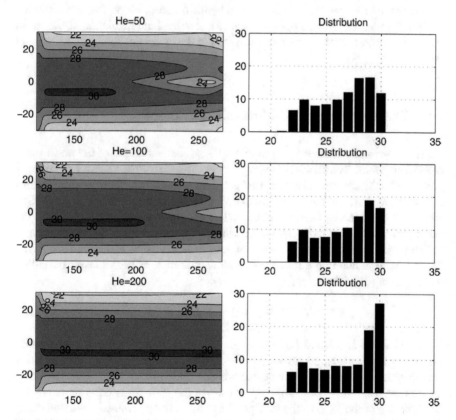

Fig. 5.10 The annual mean (left column) spatial distribution and (right column) corresponding frequency distribution of SST from the ocean model with imposed wind stress and heat flux where the depth of the thermocline at the eastern boundary is (top row) 50 m, (middle row) 100 m, and (bottom row) 200 m

diminishing effect on the SST distribution, while the Ekman dynamics were playing a more significant role in moving the water poleward and trying to homogenize the SST at all longitudes, with the consequence of the largest area being progressively occupied by the warmest isotherm. At shallower thermocline depths, the Sverdrup dynamics shift the peak in the SST distribution to slightly cooler SSTs than the warmest isotherm by bringing cold water from higher latitudes to the equatorial region, which is upwelled to the surface.

In conclusion, Ekman transport tends to homogenize the SST with the largest area occupied by the warmest isotherm (a highly skewed distribution) by way of cooling the warmest SST on the equator and warming off equatorial regions. On the other hand, Sverdrup dynamics shift the peak in the SST distribution toward slightly cooler values, making it less skewed by facilitating the upwelling of cooler waters in the equatorial region. Therefore, it may be said that the Sverdrup circulation helps in determining the geographical structure of the warm pool. In other words, the Western Pacific Warm Pool is also defined by the presence of the cool tongue in the Eastern Pacific, and therefore, one needs to understand the regulation of the warm pool in the entirety of the whole Tropical Pacific Basin and not just in the warm pool region.

5.4   The Surface Heat Budget of the Warm Pool

There are several studies that claim that the seasonal variations of the north-south movement of the warm water volume in the Tropical West Pacific are strongly governed by the corresponding air-sea heat flux variations while the zonal variations of the warm water volume are associated with interannual variations of ENSO, which are governed by ocean dynamics. The net atmospheric heat flux (N) on the ocean surface is the sum of sensible heat flux (S), latent heat flux (L), net shortwave radiation at the surface (SW), and net longwave radiation at the surface (LW). There is also in addition a relatively small contribution of heat transfer by falling precipitation (PF). Therefore,

$$N = S + L + SW + LW + PF \qquad (5.35)$$

The sensible and latent heat fluxes are enabled by the turbulent flow (whose horizontal and vertical scales are comparable and are bounded by the planetary boundary layer depth). These turbulent eddies enable mixing in the PBL, typically taking heat from the ocean surface to the top of the PBL while bringing cooler air from the top of the PBL to the surface. Since the enthalpy (sensible + latent heat) fluxes are generated by eddies that have a very wide range of spatial and temporal scales, these fluxes are estimated from statistical approximations:

$$S = -\rho_a c_{pd} \left(\overline{w'\theta'} \right) \qquad (5.36)$$

$$L = -\rho_a L_v \left(\overline{w'q'} \right) \tag{5.37}$$

where the overbar and prime are the time mean and deviations about this mean quantity and ρ_a, L_v, c_{pd}, w, θ, and q are the density of air, latent heat of evaporation (2.5×10^6 J/Kg), the specific heat capacity of dry air ($=1004$ J/kg/K), vertical velocity, potential temperature, and specific humidity in the PBL, respectively. The covariances, $\left(\overline{w'\theta'} \right)$ and $\left(\overline{w'q'} \right)$, though, can be measured through the eddy covariance method, which involves taking very high frequency measurements of w, θ, and q. But they are, however, commonly estimated by using the bulk aerodynamic method. The eddy correlation method is impractical because of the paucity of observations, especially in the open oceans, and inadequate sampling of the covariances for all ranges of weather conditions. This method is based on the premise that near surface turbulence arises from the prevalent vertical shear in the PBL and that S and L are proportional to the mean gradients of temperature and humidity in the PBL, respectively. Therefore, Eqs. 5.36 and 5.37 are written as

$$S = \rho_a c_{pd} C_H \left(|V_a| - |V_o| \right) \left(T_a - T_{surf} \right) \tag{5.38}$$

$$L = \rho_a L_v C_D \left(|V_a| - |V_o| \right) \left(q_a - q_{surf} \right) \tag{5.39}$$

where $|V_a|$, $|V_o|$, T_a, T_{surf}, q_a, q_{surf}, C_D, and C_H are the wind speed at 10 m above the surface, surface ocean current, air temperature at 2 m above the surface, surface temperature, specific humidity at 2 m above the surface, specific humidity at the surface (which is saturation specific humidity over ocean surface), surface drag coefficient, and heat exchange coefficient, respectively. It may be noted that fluxes directed from the ocean to the atmosphere are taken to be negative and fluxes directed from the atmosphere into the ocean surface as positive (a sign convention adopted here but should be checked at every instance when these quantities are presented because their interpretation can change dramatically!). Therefore, in the warm pool, where the surface temperature is warm, the sign of the fluxes in Eqs. 5.38 and 5.39 is going to be invariably negative (i.e., ocean surface is losing heat due to enthalpy fluxes). It should be noted that the speed of the surface ocean current is much smaller than that of the overlying surface wind.

The net shortwave flux at the surface is

$$SW = (1 - \alpha) SW_o \tag{5.40}$$

where α and SW_o are the shortwave surface albedo and downwelling shortwave flux at the surface. Here, in Eq. 5.40, the SW is going to be positive over the warm pool, although its magnitude can wildly fluctuate depending on the cloud cover (as discussed in Sect. 5.2). Similarly, the net longwave flux at the surface is

$$LW = LW_o - \epsilon\sigma T_{surf}^4 \tag{5.41}$$

where LW_o, ϵ, and σ are the downward infrared flux at the surface, surface emissivity, and Stefan-Boltzmann constant ($=5.67 \times 10^8$ Wm^{-2} K^{-4}), respectively. The last term on the RHS of Eq. 5.41 is the outgoing longwave radiation at the surface. The sign and magnitude of LW in Eq. 5.41 depend on the cloud cover (as discussed in Sect. 5.2). But Song and Yu (2013) found this term to be largely negative over the warmpool.

Heat transfer by precipitation occurs if the precipitation has a different tempera-ture than the surface temperature. This is given by

$$PF = \rho_w c_{pw} R \left(T_{wa} - T_{surf}\right) \tag{5.42}$$

where ρ_w, c_{pw}, R, and T_{wa} are the density of seawater ($=1022.4$ kgm^{-3}), the specific heat capacity of seawater ($=3994$ JKg^{-1} K^{-1}), rain rate, and *wet bulb temperature* at 2 m above the surface, respectively. The term 'PF' is often neglected as they are small when considering, say, monthly mean but could become an important term for shorter periods of averaging, especially over the warm pool where convection occurs very often from tall convective clouds. According to Fairall et al. (1996), this term, estimated from in situ measurements over the warm pool, is 2.5 to 4.5 Wm^{-2}.

The net heat flux, N, at the surface of the warm pool has been rather elusive despite several attempts at it (Hoffert et al. 1983; Newell 1986; Godfrey et al. 1991, 1998; Song and Yu 2013). It ranges from near zero (Hoeffert et al. 1983) to 100 Wm^{-2} (Godfrey et al. 1991). More recently, Song and Yu (2013) reported from intercomparing nine different flux products that the annual mean N over the warm pool is about 32 Wm^{-2} but with a standard deviation of 16 Wm^{-2}!

Song and Yu (2013) conducted a 'bubble analysis', which is a unique formulation to compute time mean heat budgets for an enclosed volume, such as the warm water volume bounded by an isotherm (28 °C). The heat budget equation of such a bubble is given by

$$\rho_w c_{pw} \underbrace{\left[(\{\theta\} - \theta_x)\frac{dV_l}{dt} + V_l\frac{d\{\theta\}}{dt} \right]}_{Term\ A} = \underbrace{\oiint_{A_s} N dS}_{Term\ B} - \underbrace{\oiint_{A_T} SW_{pen} dS}_{Term\ C} - \underbrace{\oiint_{A_T} Q_{dif} dS}_{Term\ D} \tag{5.43}$$

where V_l is the warm water pool volume defined by the 28 °C isotherm, $\{\theta\}\left(=\frac{1}{V_l}\iiint \theta dV_l\right)$ is the potential temperature averaged over the warm water volume, θ_x is the bounding temperature ($=28$ °C), A_s is the sea surface, A_T is the oceanic isothermal surface, and Q_{dif} is the total diffusive heat flux through the sides and base of the volume. Song and Yu (2013) regarded this warm water volume bounded by the 28 °C isotherm in the Tropical Western Pacific as an enclosed volume by neglecting the transport across the Malacca Strait (between the Malay Peninsula and the Indonesian island of Sumatra). The heat budget Eq. (5.43) of such a bubble has two main advantages: (1) it removes the advection terms across the isothermal boundary in the equation, which are prone to computing errors, and (2) for an enclosed volume surrounded by closed water, the diffusive term must be negative (or out of the volume or down the temperature gradient)—this reflects that on an annual mean basis, the warm pool gains heat from the atmosphere and loses heat to the surrounding colder waters by diffusive processes. Song and Yu (2013) computed all terms of Eq. 5.43 except Term D, which was computed as a residue. Since the sign of this Term D has to be negative, they used the sign of the residue term from the heat budget Eq. (5.43) as a powerful constraint to check the validity of the surface flux data. Although this validation checks the compatibility of an air-sea flux climatology with the temperature, it cannot determine the size of the error in the climatological fluxes.

Term A of Eq. 5.43 has contributions from changes in the pool volume and changes in the mean temperature of the pool. Term A is also referred to as 'pseudo heat content' because the volume or the bubble can change. Term B of Eq. 5.43 is the surface integral of the net heat flux on the surface of the volume. Term C of Eq. 5.43 is the penetrative shortwave radiative flux that penetrates across the isothermal surface to the colder water below, and Term D is the total (horizontal + vertical) diffusive heat flux across the isothermal surface, which is estimated as a residual. SW_{pen}, following Wang and McPhaden (1999), is computed empirically assuming that SW decays exponentially with a 25 m e-folding depth. This empirical relation is given by

$$SW_{pen} = 0.45 \times SW \times e^{-\gamma H} \tag{5.44}$$

where $\gamma = 0.004$ m^{-1}, H is the depth of the mixed layer, and SW follows from Eq. 5.40.

A summary of the comprehensive analysis of the datasets examined by Song and Yu (2013) are as follows:

- The average of all surface flux terms over the warm pool from all nine flux products examined is as follows:

  - $N = 28 \pm 7$ Wm^{-2}
  - $SW_{pen} = 17 \pm 2$ Wm^{-2}
  - $Q_{dif} = -11 \pm 1$ Wm^{-2}
  - $Term\ A = 0 \pm 6$ Wm^{-2}

- Comparison of the flux products with co-located in situ (buoy) measurements in the warm pool suggests that these flux products can be broadly classified as high- and low-bias net heat flux products; the annual mean of the four high-bias net heat fluxes is 49 Wm^{-2}, and the annual mean of the remaining five low-bias net heat fluxes is 18 Wm^{-2} over the warm pool.
- They found that the uncertainty in the net heat flux over the warm pool is largely caused by the varied values of SW and L from the nine different datasets they examined.

- The change in the pseudo heat content of the warm pool between November and May is dominated by the expansion and contraction of the pool volume, while the change between June and October is largely dictated by the cooling and warming of the warm pool water.
- The monthly variations of the pseudo heat content were in phase with the corresponding variations of N.
- The residual diffusive flux term computed from all nine flux products suggests that they are all physically consistent as they result in a down-gradient diffusive flux through the year, although their values range from $-36\,\mathrm{Wm^{-2}}$ to $-5\,\mathrm{Wm^{-2}}$. This diffusion is largely carried out by the vertical diffusive process for the volume bounded by the 28 °C isotherm (as it is determined by a simple scale analysis; they also found this scale analysis to be sensitive to the choice of the bounding isotherm as it dramatically reduces the size of the volume as you change the isotherm to, say, 28.5 °C or 29 °C).
- In the annual mean sense, the surface heat flux over the warm pool is balanced by the diffusive heat flux and the solar radiation penetration across the 28 °C isotherm.
- The seasonal variability of the heat diffusion through the warm pool volume is determined as ocean dynamics; for example, the vertical shear between the Pacific tropical jet and the equatorial undercurrent can enhance the vertical diffusivity coefficient and further facilitate the down-gradient vertical diffusion.

5.5   The Indonesian Throughflow

The ITF is a unique pathway that connects two major tropical ocean basins (the Pacific and Indian Oceans) with about 15 Sv (1 Sv $= 10^6$ $\mathrm{m^3s^{-1}}$; Fig. 5.11) of discharge in the throughflow. Fundamentally, in an annual mean sense, the ITF transports warm and fresh water from the Tropical Western Pacific to the Eastern Indian Ocean through the multitude of passageways (Lombok Strait [2.6 Sv], Ombai Strait [4.9 Sv], and the Timor Passage [7.5 Sv]). The ITF is primarily driven by the SLP difference between the Pacific and Indian Oceans as the easterly trade winds pile up the water in the Western Tropical Pacific. However, during El Niño conditions, the ITF flow through the Makassar Strait is weaker, and the thermocline is shallower. But the variations of the ITF flow through the Lombok Strait, Ombai Strait, and Timor Passage with ENSO variations are less clear because the transport through these passages is also related to the Indian Ocean variability (Sprintall et al. 2014).

Coupled ocean-atmosphere model experiments with and without ITF reveal that the ITF deepens the thermocline in the Indian Ocean to the west of Australia, resulting in an increase in SST in the Eastern Indian Ocean and a reduction of SST over the Western Pacific (Schneider 1998). This results in the relative shifting of the warm pool to the west and the corresponding westward shift of the ascending cell of the Walker Circulation by the inclusion of the ITF.

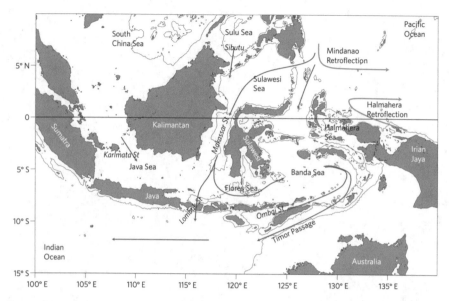

Fig. 5.11 The mean pathway of the Indonesian throughflow (red lines) and the contribution to the throughflow from the South Pacific (dashed red lines). (Reproduced from Sprintall et al. 2014)

The earlier understanding of the ITF was that it occurs within the warm near-surface layer with a strong seasonal variation driven by the seasonally reversing winds of the Northwest Indonesian Monsoon (in boreal winter) to the Southeast Indonesian Monsoon (in boreal summer). However, recent sustained observations indicate that the inter-ocean exchange primarily occurs at depths of about 100 m (within the thermocline) and exhibits variability from intraseasonal to decadal timescales (Sprintall et al. 2014).

Chapter 6
The Intraseasonal Variations

Abstract The intraseasonal variations that bridge weather and climate time scales in the range of 20–90 days are introduced in this chapter by way of illustrating their features from observations and reanalysis. The various theories proposed to explain their features are also presented. The operational detection of the propagating intraseasonal signal is described. The potential influence of the large-scale intraseasonal variations on extreme weather events is also discussed.

Keywords Madden Julian oscillation · Predictability of the second kind · Soil moisture · Supercloud clusters · Boreal summer intraseasonal oscillation · Rossby waves · Moist static energy · Moisture modes · Wind induced surface heat exchange · Wave-conditional instability of second kind · Tropical cyclones · Mixed Rossby and gravity waves · Kelvin waves

6.1 The Bridge Between Weather and Climate

A common adage used to draw the distinction between weather and climate is 'climate is what on an average we may expect, the weather is what actually we get' (Herbertson 1901). This essentially states weather is a state of the atmosphere at a particular instant, while climate is a statistical representation manifesting as the average of many atmospheric states. A more nuanced distinction is made when the weather prediction is claimed to be an initial value problem, while climate prediction is a boundary value problem. In other words, notwithstanding model development, improvements in weather prediction could be sought from improving the initial condition, while improvements in climate prediction happen to come when there are large anomalies of the boundary conditions (e.g., SSTA during strong ENSO years), which then condition the overlying atmosphere, although this separation of weather and climate is getting blurred with increasing emphasis on seamless prediction systems for weather and climate, which advocates the use of a unified modeling system (Link et al. 2017). So much so, that now decadal predictions are thought of as an initial value problem (that of initializing the deep ocean state; Meehl et al. 2014). Nonetheless, the yawning gap in the timescale between 10 days

© Springer Nature Switzerland AG 2023
V. Misra, *An Introduction to Large-Scale Tropical Meteorology*, Springer Atmospheric Sciences, https://doi.org/10.1007/978-3-031-12887-5_6

(weather) and 90 days (seasonal) is now targeted as the intraseasonal or subseasonal variations.

The growing interest in the intraseasonal oscillations came about from the discovery of the so-called MJO (Xie et al. 1963; Madden and Julian 1971) and the discovery of its multitude of impacts on extreme weather events across the world (Zhang 2013; Merryfield et al. 2020). The fact that useful weather prediction skill is limited to about 10 days before chaos in the atmosphere takes over somewhat limits numerical weather prediction (Lorenz 1969; Zhang et al. 2019). On the other hand, seasonal and longer-term predictions diverge from traditional weather prediction that target the 'predictability of the first kind' by targeting the so-called predictability of the second kind, leaving a gap between weather prediction and climate prediction. The predictability of the first kind refers to the information present in the initial conditions to which the model forecast is very sensitive. The predictability of the second kind refers to the information present in the boundary conditions, which can be assessed by the extent to which the model prediction is sensitive to it. Furthermore, the objective of climate prediction is not to predict the instantaneous weather conditions but a temporal average condition of the climate system. It is the ability of the intraseasonal oscillations to organize tropical convection on the subseasonal scales, which has lent the extended predictability beyond the weather prediction scale (Vitart 2017).

The intraseasonal predictions have shown tremendous potential from an end-user perspective. For example, many management decisions in agriculture, food and water security, disaster risk reduction, and public health safety reside in the subseasonal time range (between weather and seasonal timescales). In an enlightening application of the subseasonal forecasts, the Red Cross Red Crescent Climate Centre's International Research Institute for Climate and Society has proposed a 'Ready-Set-Go' concept for humanitarian aid and disaster preparedness. They use seasonal forecasts as a guidance to monitor subseasonal forecasts, update contingency plans, train volunteers, and test out the early warning systems to get the system 'Ready'. The subseasonal forecasts are used to alert volunteers and communities and even possibly get the aid material as close to the anticipated disaster-prone area to have the system 'Set'. The weather forecasts are then used to issue warnings, activate volunteers, distribute instruction to communities, and evacuate if need be ('Go').

The intraseasonal prediction is both an initial value and a boundary value problem that provides a likelihood for anomalous weather events on the subseasonal timescale. Over the years, the intraseasonal prediction has improved significantly (from improvement in initial conditions and numerical models and from the discovery of new sources of intraseasonal predictability) to provide useful information on the likelihood of anomalous weather events beyond the 2-week range, which has significantly helped in mitigating losses from such natural calamities.

Some of the recent success in intraseasonal prediction stems from sources in the stratosphere (Kidston et al. 2015). It is shown that events of sudden stratospheric warming (which are driven by Rossby waves from the troposphere) offer lagged impacts on sea level pressure, surface temperature, and precipitation anomalies over Eurasia and northeastern North America (Sigmond et al. 2013). Weather prediction models initialized during these sudden stratospheric warming events have produced useful skills out to 3–6 weeks in these regions (Domeisen et al. 2020). However, some limiting factors are that the predictability of these sudden stratospheric warming events is rather limited ranging from a few days to 2 weeks (Merryfield et al. 2020) and not all stratospheric warming events have a similar influence on the surface climate (Maycock and Hitchcock 2015).

The inertial memory of the root zone soil moisture, which can last for several weeks, can influence the overlying atmosphere by modulating the surface evaporation and the surface energy budget to cast an influence on air temperature and precipitation on subseasonal scales (Koster et al. 2010) both locally and in remote locations (Teng et al. 2019). Besides soil moisture, the thermal and radiative properties of the extent of snow cover (Orsolini et al. 2013) and vegetation states (Williams et al. 2016) are also found to affect subseasonal variations.

The upper ocean thermal structure, in particular the SST, is found to have a significant bearing on the intraseasonal variations in the tropics. For example, seasonal SSTA can modulate evaporation and atmospheric convection, which is then also associated with changes in atmospheric circulation. Tam and Lau (2005) find that the intraseasonal activity of the low-level wind is enhanced over the Central Pacific and reduced over the Western Pacific during warm ENSO events. They also find that the propagation of the MJO is slower during warm ENSO events.

6.2   The Structure of the Madden-Julian Oscillation

Madden and Julian (1971) originally discovered the MJO from a 10-year period of rawinsonde data from the Kanton Island (2.8°S and 171.7°W). They found a significant spectral coherence that appears between surface pressure, zonal winds, and temperature at various pressure levels over a broad time period range (12–100 days) that maximized between 41 and 53 days over Kanton Island in the Central Pacific Ocean. They found similar broadband temporal coherence in the rawinsonde data from other stations (e.g., Balboa in Panama at 9.0°N and 79.6°W, Singapore at 1.4°N and 103.9°E). Madden and Julian (1972) found a distinct spectral peak in the 40–50-day time range in surface pressure from stations on the east coast of Africa to the western half of South America but largely confined to ±10° of the equator. They also found evidence for eastward propagation at these timescales from the changing phase angles of the coherence of the zonal wind at 850 hPa at these various stations on intraseasonal timescales with the surface pressure at Kanton Island. The surface pressure in these equatorial stations with that over Kanton Island was coherent with 850 hPa zonal wind and nearly out of phase with temperature from 700 to 150 hPa. Similarly, the 850 hPa zonal wind was found to be coherent and out of phase with the zonal wind from 300 to 100 hPa. They did not find any significant spectral peak at these timescales in the meridional wind. These discoveries led Madden and Julian (1972) to propose a schematic showing the large-scale feature of the MJO (Fig. 6.1).

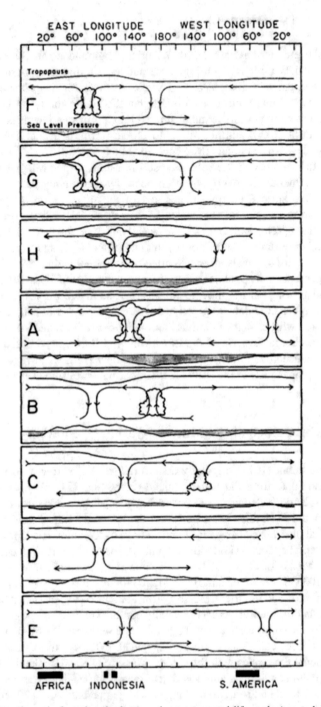

Fig. 6.1 The schematic featuring the large-scale structure and life cycle (top to bottom) of the Madden-Julian Oscillation (MJO) with longitude in the abscissa and pressure in the ordinate. The streamlines indicate the longitude-pressure anomalous zonal circulation associated with the deep convection (cloud symbol). The anomalous sea level and tropopause pressure level are shown as solid lines that bound the zonal circulation at the bottom and top, respectively. (Reproduced from Madden and Julian 1972. Published (1972) by the American Meteorological Society)

Figure 6.1 shows the eastward propagation of deep convection (active phase of the MJO) flanked by suppressed convection (inactive phase of the MJO) coincident with zonal overturning circulations that occur through the depth of the troposphere. In the lower troposphere (~850 hPa), Fig. 6.1 suggests anomalous westerly and easterly winds exist to the west and east of the convection, respectively. In the upper troposphere (~200 hPa), there is a phase reversal with easterlies and westerlies to the west and east of the deep convection, respectively. This eastward propagation is observed to have an envelope of deep convection and cloudiness around 10,000 km in scale (with an inactive phase of the MJO having larger zonal scales) with an average phase speed ranging from 4 to 8 ms^{-1}. This envelope of convection of the MJO is organized by equatorial waves and a broad spectrum of mesoscale convective systems. Figure 3.10 makes a subtle distinction of the large-scale MJO with majority of its spectral power in wave numbers 1 and 2 (roughly 12,000–20,000 km of the zonal extent measured by the positive and negative anomalies of cloud cover) from the relatively faster eastward-propagating waves. It may be noted that the zonal scale of convection is much less than that of the circulation, and as a result, the zonal wind has a spectral peak on the MJO timescale in zonal wave number 1, while precipitation has the spectral peak spread over zonal wave numbers 1–3 (Zhang 2005). Such zonal scales warrant only one fully developed MJO event in the tropics at a given time, although, on some occasions, two weak convective centers of the MJO with associated (weakly) anomalous circulations may exist such as a convection cluster initiated in the Indian Ocean while another cluster of convection is decaying in the Central Pacific Ocean.

The observations reveal that over the life cycle of an MJO event, the phase between the surface zonal wind anomaly and convection varies. For example, when the convection center of the MJO is in the Indian Ocean, it is more likely to have surface zonal westerlies to the west and zonal easterlies to the east of the convection. In contrast, when this convection center moves over to the Western Pacific Ocean, the surface westerlies are likely to prevail through the convection center. This led Rui and Wang (1990) to propose a schematic depiction of the large-scale circulation of the MJO (Fig. 6.2). Figure 6.2a depicts the scenario when the convection is over the Indian Ocean, wherein the 850 hPa westerlies to the west and easterlies to the east of the convection are associated with two subtropical cyclones and anticyclones, respectively. Similarly, at 200 hPa, the easterlies prevailing over the convection center are associated with two subtropical anticyclones. As the convection moves over to the Western Pacific Ocean (Fig. 6.2b), the easterlies at 200 hPa associated with two subtropical cyclones persist from Fig. 6.2a. But at 850 hPa, the easterlies ahead of the convection disappear with the subtropical anticyclone (Fig. 6.2b). In the meanwhile, weak easterlies associated with subtropical anticyclones over the Indian Ocean, west of the convection, begin to develop (Fig. 6.2b). The convection anomalies and the lower tropospheric circulation anomalies dissipate as the convection passes the Date Line and over the cooler waters of the Eastern Pacific Ocean. However, the circulation anomalies in the upper troposphere continue to traverse eastward as the convection dies out in the Central Pacific Ocean.

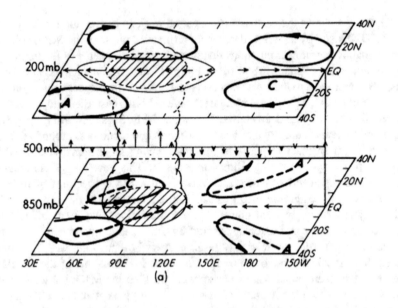

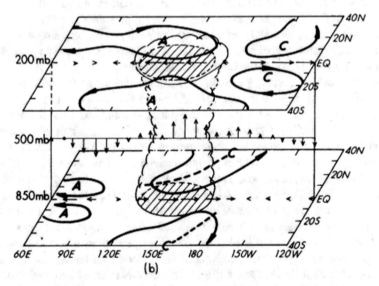

Fig. 6.2 A schematic of the large-scale circulation associated with the MJO when the convection center is over the equatorial (**a**) Indian Ocean and (**b**) Western Pacific Ocean. A and C refer to the anticyclonic and cyclonic centers of the circulation with dashed lines representing the trough and ridge lines. The zonal and vertical velocities in the zonal equatorial-pressure plane are also depicted. (From Rui and Wang 1990; © American Meteorological Society. Used with permission)

The MJO however is an amalgam of many different scales. Nakazawa (1986, 1988) identified a hierarchical structure within the convection envelope of the MJO.

This envelope comprises of convective systems moving in all directions, with the eastward propagation of the envelope characterized by new convective systems developing successively, each time to the east of the previous one. These convective systems (also called 'superclusters' or 'supercloud clusters') are identified as Kelvin waves with horizontal scales of 3300–6600 km, and timescales of <10 days (synoptic scale). The eastward phase speed of these Kelvin waves is observed to propagate more slowly over the Indian Ocean (12–15 ms^{-1}) compared to that over the West Pacific Ocean (15–20 ms^{-1}). These superclusters were further identified to comprise of even higher frequency westward-propagating disturbances (called 'cloud clusters') with periods of 2–5 days. These westward-propagating cloud clusters with a period of ~2 days were identified with $n = 1$ westward inertia-gravity waves (Takayabu 1994) and diurnal variation of convection (Chen and Houze 1997). The westward-propagating cloud clusters with a period of ~5 days were identified with equatorial Rossby and mixed Rossby-gravity waves (Wheeler and Kiladis 1999).

The broadband feature of the subseasonal variations has expanded the original timescale of 40–50 days to other ranges such as 30–50 days (e.g., Krishnamurti and Subrahmanyam 1982), 30–60 days (e.g., Weickmann et al. 1985), 10–20 days (e.g., Murakami 1976; Krishnamurti and Bhalme 1976), and 20–70 days (e.g., Sikka and Gadgil 1980). This diversity in the diagnosis of the timescales of the subseasonal variations is a result of the inherent broadband nature of the oscillation that is featured by the absence of a distinct spectral peak at a particular frequency and the differences from the various adopted spectral techniques, types of observations used, and the geographical location of the analysis in the different studies.

6.3   The Seasonality of the Intraseasonal Variations

The MJO signal is largely confined to the warm waters in the Equatorial Indian and the Western Pacific Oceans. Zhang (2005) also suggests that the effect of the warm SST on the location of the MJO can also be illustrated by the zonal dislocation of the MJO during ENSO events and the appearance of the MJO in the Eastern Pacific Ocean, just north of the cold tongue and adjacent to the Central American coast only in the boreal summer season, when the SST is sufficiently high. The convective envelope of the MJO is generally observed to be weak over the Maritime Continent (Kiladis et al. 2005) with possible explanations being that the strong diurnal variations of the convection over the land tend to compete with the MJO for moisture and energy, topography of the islands serves as a barrier to moisture flux convergence, and surface evaporation is comparatively reduced over the islands.

The primary peak season of the MJO is in the boreal winter season (coincident with the Australian summer monsoon), just south of the equator. A secondary seasonal peak of the MJO is in the boreal summer season (coincident with the Asian summer monsoon), when the strongest signals of the MJO are to the north of the equator. MJO has a single peak in the Eastern Pacific Ocean in the boreal summer.

Furthermore, at the equator over the Indian and the Western Pacific Oceans, MJO has a single peak in the boreal winter. The MJO signal migrates in latitude from 5°S–10°S in the primary peak season to 5°N–10°N in the secondary peak season, with the seasonal cross-equatorial migration being much stronger in the Western Pacific than in the Indian Ocean (Zhang 2005).

The MJO has a significant impact on the Australian monsoon rainfall variability. However, the intraseasonal variations of the Asian summer monsoon also share many of the features of the MJO but are also distinct from it owing to the seasonal modulation of the intraseasonal variations. There are a number of terminologies used for the intraseasonal variations of the Asian summer monsoon including ISO, BSISO, MISO, and even MJO. We will use ISO here. Just like the MJO, ISO also has a broad spectrum with a time period in the range of 30–50 days. ISO is associated with active and break periods of the Asian summer monsoon. But the propagation of the ISO is more complex. The ISO seems to propagate both in an eastward and a northward direction as illustrated in the schematic in Fig. 6.3. Just like the MJO, the convection in the ISO initiates in the Western Equatorial Indian Ocean, which propagates eastward toward the Central Indian Ocean (Fig. 6.3a). But roughly between the longitudes of 80°E and 130°E, the convection seems to propagate both eastward and northward. The off-equatorial disturbances shown in Fig. 6.3 are a result of the poleward moving Rossby waves emanating from the eastward propagation of the convection over the equator as a mixed Rossby-gravity wave. The summertime distribution of warm SSTs and abundance of low-level moisture and high wind shear favor convection north of the equator. The warm waters in the Eastern Equatorial Indian and Western Pacific Oceans provide a favorable environment to enhance convection and generate a Rossby wave response, which appears as lobes of cyclonic circulation on either side of the equator. As a result of the asymmetry of the seasonal monsoon and associated easterly vertical shear and moist static energy, the Rossby waves develop an asymmetric structure about the equator. Therefore, the Rossby waves generated by the ISO become asymmetric with the northern lobe of cyclonic flow getting stronger and showing a strong tendency for northward propagation (Fig. 6.3).

The ISO also exhibits differences over the course of the Asian summer monsoon season. During the onset months of the monsoon (May–June), most of the ISO variability occurs over the Indian sector, while in the final months of the monsoon, the ISO activity flares in the East Asian and West North Pacific regions. This evolution of the ISO with the monsoon seems to be consistent with the seasonal march of the warmest SSTs, which begin in the Northern Indian Ocean and eventually are found in Southeast Asia and the West North Pacific region. Based on these observations, Kemball-Cook and Wang (2001) arrived at the schematic shown in Fig. 6.3a, b. Stages 1–3 represent the initiation and eastward propagation of the ISO in Fig. 6.3a, b. The transition of the convection from the Indian Ocean to the Western Pacific Ocean is represented by Stage 4 (Fig. 6.3a, b). The Rossby waves are emitted from the Western Equatorial Pacific Ocean in Stage 5 and travel poleward in Stage 6 during the early monsoon period (Fig. 6.3a) and in Stages 7 and 8 in the late monsoon period (Fig. 6.3b). The dissipation of the cycle occurs in the

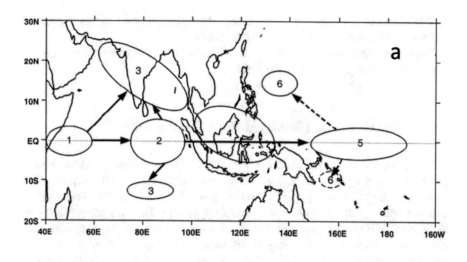

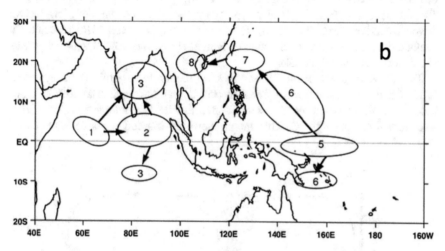

Fig. 6.3 Schematic of the BSISO for (**a**) May–June and (**b**) August–October time period. The ovals indicate convection with numbers indicating the evolution of the convection anomaly. The zonal arrows indicate eastward propagation, and the slanted/meridional arrows indicate poleward propagation of the convection due to Rossby waves generated from the equatorial convection. The dashed arrows indicate a low-amplitude signal. (Reproduced from Kemball-Cook and Wang 2001; © American Meteorological Society. Used with permission)

Stage 6 in the early monsoon period and in Stage 8 in the late monsoon period (Fig. 6.3b). The main differences in the ISO characteristics between the early monsoon (Fig. 6.3a) and late monsoon (Fig. 6.3b) periods are the following:

- Eastward displacement of the initiation phase of the ISO during the late monsoon period.

- The eastward propagation is somewhat discontinuous and jumps across the Maritime Continent in the late monsoon period.
- There is a latitudinal asymmetry in the late monsoon period because warm moist conditions is mostly north of the equator.

6.4 Theories of the Madden-Julian Oscillation

A persuasive theory for the MJO has been rather elusive. There are many theories proposed over the years, but they all have some limitations in explaining all of the features of the MJO. It is important to note from Fig. 3.7a that the MJO occupies a unique space in the wavenumber-frequency spectrum. The only wave that comes close to the MJO in the spectrum (Fig. 3.7a) is the eastward-propagating equatorial Kelvin wave that has a planetary scale zonal wind field resembling the MJO. But the convectively coupled Kelvin waves propagate much faster (10–15 ms^{-1}) than the MJO (5–10 ms^{-1}), and its spectral power does not lie along the Kelvin wave dispersion curve (Fig. 3.7a). Furthermore, the observed MJO resembles a combination of the convectively coupled Kelvin waves and Rossby wave responses to equatorial heating (Fig. 6.4).

The CISK theory prevailed over the MJO theory for a while (Lau and Peng 1987; Chang and Lim 1988). They theorized that the equatorial Kelvin wave becomes unstable when the associated convective heating interacts with the low-level moisture convergence. This theory however implied unstable wave-CISK

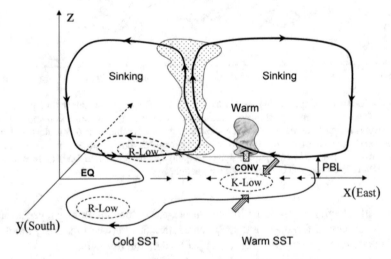

Fig. 6.4 A schematic of the vertical and horizontal structure of the MJO showing the streamlines and low-pressure anomalies associated with equatorial Kelvin (K-low) and Rossby (R-low) waves. The arrows indicate the wind vectors. (Reproduced from Wang 2003 under CC BY 4.0)

Kelvin modes that propagate at 16–19 ms^{-1}, which is comparable to the equatorial convectively coupled Kelvin waves and not the MJO (with an average phase speed of 5–10 ms^{-1}). Modifications were made to this wave-CISK theory by introducing the frictional effect on the moisture convergence in the atmospheric boundary layer to slow the propagation speed. By the inclusion of the boundary layer friction, Wang and Rui (1990) found that there was a coupling between the Moist Kelvin waves and the Rossby modes, which helps suppress the fast wave-CISK modes. The frictional wave-CISK theory posits that to the east of the convective heating, the low-level winds associated with the Kelvin wave turn equatorward due to friction, which produces low-level moisture convergence. A characteristic feature of the MJO is that the frictional convergence in the boundary layer is located to the east of the major precipitation region by a fraction of the wavelength, while the wave-induced convergence is in phase with major convection (Fig. 6.4). However, for the same reasons as discussed in Chap. 3, the wave-CISK theory has been criticized to be unphysical as well. This observation has been confirmed by examining several global data assimilation products (Hendon and Salby 1994).

There have been other modifications to the wave-CISK mechanism like varying the vertical heating profile of condensational heating. For example, Sui and Lau (1989) found that the bottom heavy heating profile tended to favor MJO-like modes, while Cho and Pendlebury (1997) claimed that the top heavy heating profile produced better MJO simulation. In a comprehensive study on this matter, Lin et al. (2004) made the following observations:

- TRMM data showed that stratiform precipitation contributes about 60% of the total anomalous intraseasonal precipitation at the time of maximum precipitation of the MJO. Furthermore, the vertical heating profile is top heavy at the time of maximum precipitation.
- Using data sets from TOGA-COARE, they showed that as the MJO passes over 155°E, the maximum condensational heating rises in altitude with time so that at the trailing edge of the rainy phase, the latent and radiative heating is from the stratiform anvils.
- In a review of wave-CISK literature, they find systematically that the phase speed of the MJO is found to decrease as the maximum heating rate reduces in altitude.
- They however suggest that the altitude of the peak latent heating profile is not the salient feature, but what vertical modes are excited by the heating profile is more pertinent. This is because waves of short vertical wavelength that are excited by either bottom heavy or top heavy latent heating profile propagate slowly. Therefore, they argue that relating MJO phase speed to vertical profile may be misleading.

Another popular theory in the context of the MJO is the WISHE theory (Yano and Emanuel 1991). So, unlike the wave-CISK theory where diabatic heating from convection is the destabilizing mechanism, the WISHE theory claims that condensational heating is compensated by adiabatic cooling. The WISHE theory posits that large-scale convection forces easterlies and westerlies to their east and west, respectively. Therefore, large-scale convection in the presence of mean low-level easterly winds will increase the easterlies and thereby increase evaporation to the east of the convection center. This will lead to an increase in the moist static energy and a zonal gradient in subcloud layer entropy (aka the quasi-equilibrium theory following Eq. 3.4), which subsequently leads to the eastward propagation of the wave. This theory has been criticized that over much of the Equatorial Pacific where the convection envelope of the MJO is observed, the mean low-level winds are extremely weak. Furthermore, observations indicate that evaporation anomalies are typically higher to the west rather than in the east of the convection anomaly (Jones and Weare 1996).

The MJO shares many of its features with the moisture modes (Adames and Maloney 2021) discussed in Sect. 3.7. For example, the precipitable water is in phase with precipitation anomalies of the MJO (Sobel et al. 2004). Similarly, Raymond (2001) showed from a toy model that when convection was made sensitive to moisture, then it displayed MJO-type variability more prominently. The relevance of the moisture mode theory to MJO is best understood from the moisture budget on an isobaric coordinate, given by

$$\frac{\partial \overline{q}}{\partial t} = -\overrightarrow{V}.\nabla \overline{q} - \overline{\omega}\frac{\partial \overline{q}}{\partial p} - \frac{q_2}{L} \tag{6.1}$$

The overbars in Eq. 6.1 refer to the area average, and q_2 follows from Eq. 3.2 in Chap. 3. In precipitating regions, the last two terms of Eq. 6.1 dominate over the moisture advection terms by nearly an order of magnitude (Adames and Wallace 2015). Upon applying the weak temperature gradient approximation (cf. Sect. 3.6) by defining vertical velocity that balances latent heat of condensation (ω_c) and vertical velocity that balances radiative heating (ω_R) and expanding on q_2, Adames and Maloney (2021) arrive at the following form of Eq. 6.1:

$$\frac{\partial L\overline{q}}{\partial t} = -\overrightarrow{V}.\nabla L\overline{q} - \overline{\omega}_c\frac{\partial \overline{h}}{\partial p} - \overline{\omega}_R\frac{\partial L\overline{q}}{\partial p} - \frac{\partial \overline{\omega'h'}}{\partial p} \tag{6.2}$$

where h is the moist static energy. The advantage of Eq. 6.2 over Eq. 6.1 is that the last two terms on the RHS of Eq. 6.1 that are large and tend to cancel each other in convective regions are replaced by terms on the RHS of Eq. 6.2 that are comparable to the moisture tendency (the LHS term of Eq. 6.2; Adames et al. 2021). The application of computing the moisture budget from Eq. 6.2 on the MJO reveals significant information on the propagation and maintenance of the oscillation (Fig. 6.5; Adames and Maloney 2021), like:

- Importance of cloud-radiative heating on the maintenance of the MJO.
- Horizontal advection of moisture is important for the propagation and maintenance of the MJO.

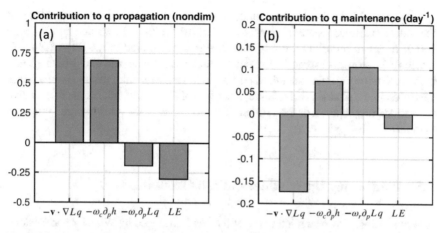

Fig. 6.5 The normalized contribution of the vertically integrated quantities on the RHS of Eq. 6.2 to (**a**) propagation and (**b**) maintenance of the MJO diagnosed from the ERA-Interim data. The overbars of the terms are not shown in the labels of the x-axis, and LE refers to the turbulent heat fluxes (last term of Eq. 6.2). The area average of these terms was computed over the domain of 20°N–20°S and 60°E–150°W. (Reproduced from Adames and Maloney 2021)

- Vertical advection moisture by convection is important in determining the propagation characteristics of the MJO.
- The surface fluxes are important for the propagation of the MJO.

Several modeling studies have shown that including coupled air-sea interaction improves the MJO simulation (DeMott et al. 2015). Although the physical pathway for this improved simulation is still under active investigation, DeMott et al. (2019) suggest that coupled models display steeper horizontal gradients of moisture that favor enhanced MJO propagation.

A theory for the ISO is explained in the previous section in the context of Fig. 6.3. Following Wang (2003), it should be noted that the eastward propagation of the tilted rainfall bands gives rise to a northward propagation component in the summer Asian monsoon domain. The moist Rossby Cells move west-northwest forming a northwest-southeast tilt of the rainfall band. In the absence of the mean flow, the Rossby waves would have propagated westward. But in the presence of the monsoon easterly vertical shear, the Rossby waves propagate northwestward. The Rossby wave induces a vertical velocity perturbation field that decreases meridionally on either side of the convection. The vertical shear creates positive and negative horizontal vorticity to the north and the south of the convection, respectively. This horizontal vorticity is twisted by the mean vertical velocity perturbation field to generate positive and negative vertical component north and south of the convection, respectively. This in turn results in convergence and divergence in the boundary layer north and south of the convection, respectively. The convergence to the north of the original convection spawns new convection. In the meanwhile, convection is suppressed south of the original convection.

Even for the ISO, the moisture mode theory has been invoked. It is suggested that the monsoonal mean state of moisture is critical for ISO (Lawrence and Webster 2002). In active monsoon periods, moisture is concentrated in the Bay of Bengal, resulting in a northeastward gradient of moisture. Karmakar and Misra (2020) indicate that the northwest-southeast tilt of the ISO is a result of the weaker moisture advection in the Arabian Sea relative to the Bay of Bengal, which results in a relative accelerated propagation of the ISO rainfall anomalies in the Arabian Sea compared to the Bay of Bengal.

6.5   Identifying the Madden-Julian Oscillation

A very succinct way of depicting the MJO was introduced by Wheeler and Hendon (2004). It is based on the RMM index, which is derived from a combined EOF analysis of daily anomalies in upper and lower level zonal wind and outgoing longwave radiation. The phases of the MJO labeled in Fig. 6.6 are determined by the principal components of the two leading EOFs, normalized by their standard deviation. Originally, as proposed in Wheeler and Hendon (2004), the EOF was conducted over the period of 1979–2001 using daily averaged values of OLR from NOAA polar-orbiting series of satellites (Liebmann and Smith 1996) and winds from NCEP-NCAR reanalysis (Kalnay et al. 1996). Wheeler and Hendon (2004) removed the seasonal cycle, the interannual variations associated with ENSO (by subtracting the component obtained by linear regression of the daily fields to the first EOF of Indo-Pacific SST), and a 120-day running mean on the three datasets before the combined EOF analysis was conducted. The two EOFs together explain about 25% of the variance of the original atmospheric fields and are well separated in their explained variance from the rest of the EOFs. The first EOF pattern shows enhanced convection (or negative OLR anomalies) over the Maritime Continent (~120°E). Furthermore, the first EOF showed low-level westerlies and upper-level easterlies to the west and low-level easterlies and upper-level westerlies to the east of the convection. The second EOF pattern shows enhanced convection over the Western Pacific Ocean (~150°E), with the low-level easterlies to the east of the convection being extremely weak and the winds in general being at quadrature to EOF1.

The composite pattern of the OLR for each of the phases of the MJO is illustrated in Fig. 6.7. The nominal time for transition between each of the numbered phases is about 6 days, which however can vary from event to event. From Fig. 6.7, one can easily trace the life cycle of a typical (canonical) MJO by following the large negative OLR anomalies that depict the convection center of the MJO as convection initiation phase (phases 1–3) over the Indian Ocean with low-level easterlies in the Indian Ocean and westerlies in the Pacific Ocean, followed by phases 4 and 5 when it passes over the Maritime Continent to phases 6 and 7 when it is over the Western Pacific Ocean and finally phases 8 and 1 when it is east of the dateline, circumnavigating the western hemisphere and starts a new cycle over. The amplitude of the MJO in Fig. 6.7 is determined by the distance of the dot from the center.

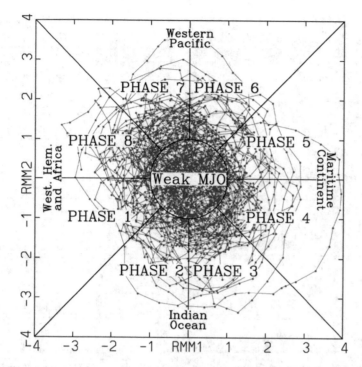

Fig. 6.6 Phase diagram of the real-time multivariate MJO (RMM) index for all available days in the December-January-February season from 1974 to 2003, with each dot representing a day. The eight phases of the MJO (shown in Fig. 6.7) are labeled. The distance of the dots from the center measures the amplitude of the MJO with points in the circle representing weak or inexistent MJO. (Reproduced from Wheeler and Hendon 2004; © American Meteorological Society. Used with permission)

Wheeler and Hendon (2004) determined that when the amplitude is <1, then the MJO is considered weak or inexistent and can be assigned a phase 0. It may be noted that the eastward propagation of the low-level wind anomalies is quicker than the OLR anomalies. Therefore, convection in phases 2 and 3 over the Indian Ocean is in near quadrature to the low-level winds, while convection in phase 7 over Western Pacific is wholly within the low-level westerlies.

The RMM index is also effective in the early part of the boreal summer season (May–June period). The composite canonical anomalies of OLR and wind anomalies for the eight phases of the ISO are shown in Fig. 6.8. These composites show eastward and northward movement of convection with the low-level easterly wind anomalies into India and the Bay of Bengal ahead of the convection in phases 1–3 and westerlies within and behind the convections in phases 3–5. For real-time application, the daily OLR fields and operational analysis of the winds (after removing their seasonal cycle, interannual component, and the running 120-day mean component) are projected over these two EOFs to yield the RMM index for the day.

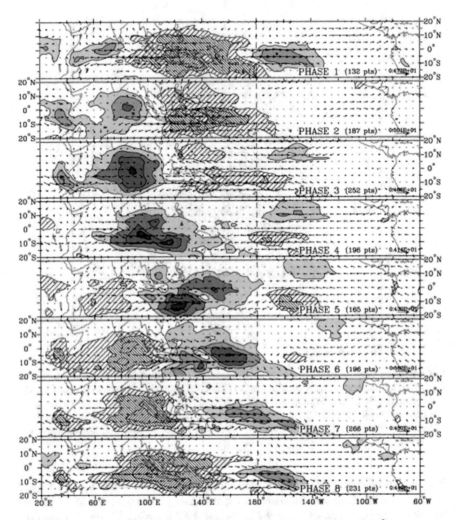

Fig. 6.7 The December-January-February composite of the OLR anomalies (Wm^{-2}) and 850 hPa wind anomalies based on the RMM index. (Reproduced from Wheeler and Hendon 2004). The negative OLR anomalies are shaded (done for anomalies of -7.5, -15, -22.5, -30 Wm^{-2}), and the positive OLR anomalies are hatched (done for anomalies of 7.5, 15, 22.5 Wm^{-2}). The magnitude of the largest vector is shown on the bottom right of each panel, and the number of days (points) in each phase of the MJO is also indicated in each panel. (Reproduced from Wheeler and Hendon 2004; © American Meteorological Society. Used with permission)

The RMM index for MJO has been criticized for amplifying the contributions of the zonal winds relative to OLR (Straub 2013; Wolding and Maloney 2015). In evaluating model performance of MJO, the RMM index rewarded models that had strong MJO signal in winds but weak signal in convection. Some of these weaknesses of the RMM index is attributed to the leakage of the power of some

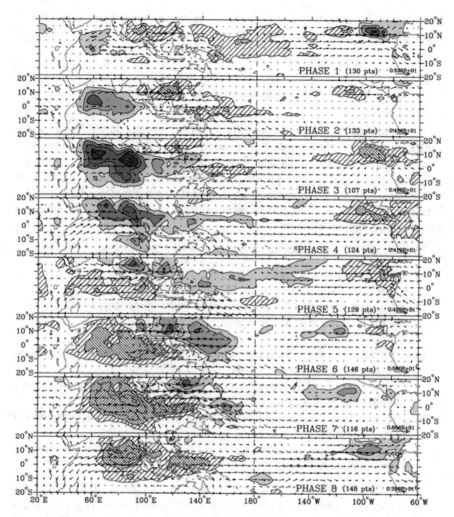

Fig. 6.8 The May–June composite of the OLR anomalies (Wm^{-2}) and 850 hPa wind anomalies based on the RMM index. (Reproduced from Wheeler and Hendon 2004). The negative OLR anomalies are shaded (done for anomalies of -7.5, -15, -22.5, -30 Wm^{-2}), and the positive OLR anomalies are hatched (done for anomalies of 7.5, 15, 22.5 Wm^{-2}). The magnitude of the largest vector is shown on the bottom right of each panel, and the number of days (points) in each phase of the MJO is also indicated in each panel. (Reproduced from Wheeler and Hendon 2004; © American Meteorological Society. Used with permission)

important modes of OLR at MJO timescales (Liu 2014). It is found that convection signal at MJO timescales in zonal wave numbers 2 to 5 is weak and could be sufficiently restored if more EOFs are retained. But incorporating more EOFs on a cartesian phase diagram like Fig. 6.6 is difficult. There have been several attempts to modify RMM index, which readers are encouraged to follow (e.g., Kikuchi et al. 2012; Ventrice et al. 2013; Liu et al. 2016).

6.6   The Impact of the Madden-Julian Oscillation

As stated earlier, one of the reasons for the rising interest in the intraseasonal variations is its near global impact on a variety of weather and climate events that often are associated with extreme weather or anomalies. Zhang (2013) argues that the universality of the impact of the intraseasonal variations should lead us to treat weather and climate as a continuum. In order to illustrate the global impact of the intraseasonal variations, it is best to illustrate them as a bulleted list, although by no means they are comprehensive:

1. The intraseasonal variations affect the monsoons:

    (a) A large fraction of the onset of the Indian Monsoon happens during the wet phase of the ISO (Karmakar and Misra 2019).
    (b) Extreme rain events in the ISM associated with tropical lows and depressions happen to occur during the wet phase of the ISO (Goswami et al. 2003; Karmakar et al. 2020).
    (c) MJO accounts for a majority of the onset of the Australian Monsoon (Wheeler and McBride 2012; Uehling et al. 2021).
    (d) MJO in phase 8 can cause rainfall in southern Brazil during austral summer to be heavier than normal by 50–75% (Marengo et al. 2012).
    (e) MJO in boreal summer has a large impact on the North American monsoon (Mo et al. 2012).
    (f) Over West Africa, intraseasonal variations account for 30% of the total monsoon rainfall (Janicot et al. 2011).

2. MJO affects the likelihood of extreme rain events: On a global scale, the occurrence of extreme rain is more likely by 40% when the MJO is present (in any one of the eight phases) relative to when it is absent (Jones et al. 2004).
3. MJO affects remote locations: by exciting Rossby wave trains that teleconnect to extra-tropics (Grimm and Silva Dias 1995) and forcing zonally propagating Rossby, Kelvin, and mixed RossbyGravity waves (Janicot et al. 2009).
4. MJO affects boreal spring tornadoes in contiguous United States: Thompson and Roundy (2013) indicate that in the boreal spring season, there are twice as many tornado outbreaks in the US when MJO is in phase 2.
5. MJO affects TC activity in all tropical ocean basins: MJOs tend to modulate vertical wind shear, low-level cyclonic vorticity, mid-level moisture, and synoptic disturbances that are considered key ingredients for TC development (Camargo et al. 2009).
6. MJO also affects climate anomalies:

(a) MJO also influences the Arctic oscillation (L'Heureux and Higgins, 2008) and the Southern Annular Mode (Carvalho et al. 2005).

(b) MJO modulates the amplitude of the North Atlantic Oscillation and the Pacific North American Pattern (Lin and Brunet 2009).

(c) There is evidence to suggest that MJO can affect many ENSO properties like its initiation, strength, period, and asymmetry (Zhang 2013).

(d) MJO is thought of as a possible mechanism to initiate Indian Ocean Dipole (Rao et al. 2008).

(e) The influence of the MJO is also seen in the ITF (Molcard et al. 1996) and the Wyrtki jet (a narrow surface eastward-flowing current in the Equatorial Indian Ocean; McPhaden 1982).

Chapter 7
El Niño and the Southern Oscillation

Abstract The most significant interannual variability phenomenon of our planet, namely the El Niño and the Southern Oscillation (ENSO), is discussed in this chapter starting with a definition to identify these events. The effort put to observe this phenomenon in the Equatorial Pacific along with the observed features of ENSO is discussed. The theories for the evolution of ENSO are also presented in this chapter.

Keywords El Niño and the Southern Oscillation (ENSO) · Southern Oscillation Index (SOI) · Buoys · Expendable bathythermograph · Upwelling · Bjerkness feedback · Teleconnection · Linear stochastic theory · Delayed oscillator theory · Recharge-discharge theory · Altimetry · Scatterometry · Rossby waves · Kelvin waves · Thermocline

7.1 ENSO Definition

Defining ENSO has remained far from definitive, although ENSO is a widely studied phenomenon and was discovered several decades ago (Walker and Bliss 1932). There are several instances when the varied concept of ENSO has come to the fore. For example, when the ENSO variations have to be distinguished from the anthropogenic climate variations (e.g., Compo and Sardeshmukh 2010) or when ENSO events tend to differ from the norm (Capotondi et al. 2015), then there is a particular focus on how ENSO is defined. A short summary of the various ENSO indices is indicated in Table 7.1. Although many of these indices are popular for their ease of computing and have their own particular merits, there is a strong sense of inadequacy in these indices that stem from the understanding that ENSO is far too complex to be represented by a single index. In fact, Walker and Bliss (1932) recognized that ENSO is a multivariate phenomenon when they used near surface air temperature and rainfall besides SLP to describe ENSO.

ENSO, as the name suggests is both an ocean and an atmosphere phenomenon and is regarded as a coupled air-sea phenomenon. Rasmusson and Carpenter (1982) in analyzing a combination of surface marine, satellite, and station observations

Table 7.1 The various ENSO-defining indices

	Popular name	Definition	Variable based on	References
1	Niño3.4 index	El Niño: Niño4-averaged SSTA exceeds one standard deviation and also exceeds the SSTA averaged over the Niño3 region La Niña: Niño3-averaged SSTA is below one standard deviation and also below the Niño4 region	SSTA	Kug et al. (2009) and Yeh et al. (2009)
2	El Niño Modoki index (EMI)	$EMI = \overline{SST}A_{CP} - \frac{1}{2}\overline{SST}A_{WP} - \frac{1}{2}\overline{SST}A_{EP}$ where $\overline{SST}A_{CP}, \overline{SST}A_{WP}, \overline{SST}A_{EP}$ are SSTA averaged over Central Pacific (10°S–10°N and 160°E–150°W), Western Pacific (125°E–145°E, 10°S–20°N), and Eastern Pacific (15°S–5°N and 110°W–70°W), respectively. This index was developed to emphasize the out-of-phase relationship of the SSTA of the Central Pacific from the Eastern and Western Pacific	SSTA	Ashok et al. (2007)
3	Trans-Niño index (TMI)	The difference between normalized SSTA between Niño4 and Niño1+2	SSTA	Trenberth and Stepaniak (2001)
4	Oceanic Niño index (ONI)	The Three-month running mean of SSTA exceeding ±0.5 °C over the Niño3.4 region for five overlapping seasons; the SSTA is based on a centered 30-year base period updated every 5 years	SSTA	Barnston et al. (1997)
5	Southern Oscillation Index (SOI)	Normalized SLP difference between Tahiti (in French Polynesia) and Darwin (in northern Australia) with a low pass filter (typically a running 11-month mean) to maximize the signal-to-noise ratio	SLP	Troup (1965), Trenberth (1984), and Allan et al. (1991)
6	Outgoing Longwave Radiation index (OLRI)	An index based on OLR anomalies over Eastern to Central Equatorial Pacific to separate El Niño events characterized by the presence or absence of OLR signal in this region as it was found to produce different remote impacts	OLR	Chiodi and Harrison (2013)
7	Multivariate ENSO index (MEI)	An EOF analysis conducted on a combination of variables over the Tropical Pacific	Surface winds, SST, surface air temperature, total cloudiness fraction	Wolter and Timlin (1993)

(continued)

Table 7.1 (continued)

	Popular name	Definition	Variable based on	References
8	Sea surface salinity index (SSSI)	SSSI is computed as the difference between the Western to Central Equatorial Pacific and Southeastern Pacific to the distinguish between the different El Niño types	SSS	Singh et al. (2011) and Qu and Yu (2014)
9	Warm water volume index (WWVI)	A proxy for the Equatorial Pacific content measured as the integrated warm water volume above the 20 °C isotherm between 5°N–5°S and 120°E–80°W	Sub-surface ocean temperature from TAO moorings, Argo floats, and XBTs	Meinen and McPhaden (2000); http://pmel.noaa.gov/elnino/upper-ocean-heat-content-and-enso

provided a detailed evolution of ENSO and identified the so-called Niño regions (1, 2, 3, and 4; Fig. 7.1). The Niño3.4 SST index (Table 7.1 and Fig. 7.1), by far the most popular ENSO index, was first introduced in Barnston et al. (1997), who identified this region that maximizes the ENSO signal among the SST-based indices. The SLP data for constructing the SOI (Table 7.1) is sometimes regarded as a benchmark to compare indices (Hanley et al. 2003). This is because SLP data was measured reliably over a very long period and is available from 1882 over Darwin and since 1935 from Tahiti. However, the SOI carries significant variability (or noise), which is not necessarily associated with ENSO and therefore must be craftily dealt with, using an appropriate low pass filter (Trenberth 1984).

Hanley et al. (2003) developed indices based on 5-month running mean of SSTA averaged over the various Niño regions and compared them with SOI to shed light on their ability to discriminate ENSO events. The running mean serves as a low pass filter. Hanley et al. (2003) classified events as an El Niño or a La Niña event if the index was above the 75th or below the 25th percentile from the SSTA time series, respectively. Similarly, they classified ENSO events based on the SOI time series using a 13-month running mean instead of a 5-month running mean. Any year that did not correspond to either El Niño or La Niña was defined as a neutral year. The comparison of these indices with SOI is shown in Table 7.2, where matching events with SOI are the diagonal values and the off-diagonal values represent false alarm or misses. This comparison indicates that the Niño1+2 index has the worst match with SOI for either El Niño or the La Niña events, while it does best in matching neutral events among the other Niño indices in Table 7.2. Similarly, Niño3.4 has the best matching for La Niña events with SOI, while for El Niño events, both the Niño3 and Niño4 indices are comparable.

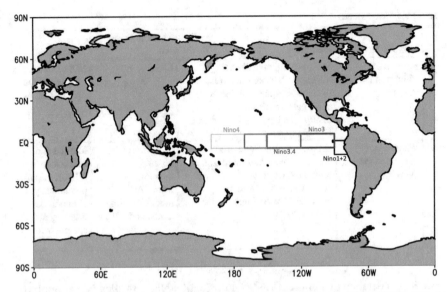

Fig. 7.1 The map outlining the boundaries of Niño1+2 (0°S–10°S and 90°W–80°W), Niño3 (5°N–5°S and 150°W–90°W), Niño3.4 (5°S–5°N and 170°W–120°W), and Niño4 (5°N–5°S and 160°E–150°W)

Table 7.2 Comparison of the SST-based indices over the various Niño regions with SOI for 83-year period (1911–1993)

	Index (based on 5-month running of SSTA)	Event	SOI (using 13-month running mean)			Matching percentage with SOI
			El Niño	Neutral	La Niña	
1	Niño3	El Niño	14	5	0	74%
		Neutral	5	34	5	77%
		La Niña	0	5	15	75%
2	Niño3.4	El Niño	14	6	0	74%
		Neutral	5	30	3	68%
		La Niña	0	8	17	85%
3	Niño4	El Niño	15	7	0	79%
		Neutral	4	31	4	70%
		La Niña	0	6	16	80%
4	Niño1+2	El Niño	12	3	0	63%
		Neutral	7	37	9	84%
		La Niña	0	4	11	55%

Adapted from Hanley et al. (2003); © American Meteorological Society. Used with permission

Despite the simplicity of some of these indices, one needs to realize that ENSO has a broad range of periodicity (~2–7 years), which reflects a complexity that cannot be easily represented by cursorily applying a filter (e.g., running a 5-month mean in case of some SST indices or a 13-month running mean in case of SOI) as it makes a tacit assumption that the SST variations related to ENSO reside within the chosen band of the filter and none of it outside of this band. The low-frequency variations of ENSO are widely recognized, and there is vigorous debate on the potential impact of climate change on ENSO. Furthermore, as observations have improved, reconstructions of past climate have been continuously refined with better techniques enabling discoveries on the subtleties of ENSO. These discoveries suggest that the core of the ENSO variations varies across the Niño regions while also questioning the thresholds used to define some of these indices (Kug et al. 2009; Song and Son 2018). Nonetheless, despite these issues, many of these indices indicated in Tables 7.1 and 7.2 remain popular and in current practice.

7.2 Observing ENSO

The global impact of ENSO is well known for a very long time, since at least Walker and Bliss (1932). Some of the major droughts in India, China, and Brazil toward the end of the nineteenth century that caused an estimated mortality of 30–50 million people had their origins associated with ENSO (Davis 2001; Cook et al. 2010; Mishra et al. 2019). Although colonial mindset that included apathy for the well-being of the native population, limited information, and knowledge on the evolving natural climate variations were also other overbearing factors that affected the response to and the prediction of famines (Roy 2016). Since ENSO affects climate worldwide and has a profound impact on human societies, considerable resources are devoted to observing, monitoring, and forecasting ENSO.

The first major initiative to observe the Equatorial Pacific was initiated through an international 10-year (1985–1994) program called the TOGA program (McPhaden et al. 1998). TOGA was initiated during a major El Niño event of 1982–1983 that appeared as a total surprise, which was neither predicted nor observed but had significant global impacts (Glantz et al. 1987). TOGA provided real-time measurements of some of the key oceanographic variables over the Equatorial Pacific, which included surface winds, SST, subsurface ocean temperatures up to a depth of 500 m, sea level, and a small number of moorings measured ocean current velocity. Furthermore, these moorings also measure atmospheric variables of surface air temperature and relative humidity. TOGA deployed the TAO array of moored buoys in the Equatorial Pacific Ocean (Fig. 7.2), a volunteer observing ship network of XBT measurements, a surface drifting buoy program, and an island and coastal tide gauge network . In addition, remotely sensed observations of surface winds, surface temperature, sea level, ocean color, and surface salinity from operational and research satellites complemented the in situ observations of the TOGA program.

It is important to understand that putting this TAO array of moored buoys in the Tropical Pacific Ocean was not an easy task given the strong surface currents prevalent in the region (Fig. 7.3). For example, as Fig. 7.3a shows, the eastward-flowing NECC between 5°N and 10°N, the westward-flowing NEC (~8°N–13°N)

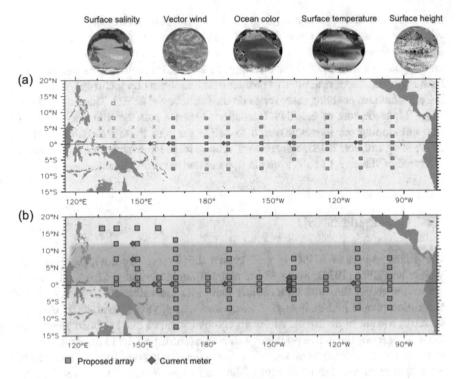

Fig. 7.2 The schematic of (**a**) the current TAO-TRITON array of moored buoys and (**b**) the proposed upgrade of the array by TPOS 2020. The current meter locations are shown in red diamonds, vacant TRITON sites are marked by x, and the proposed arrays of TPOS2020 is shown in large green boxes. In both TAO-TRITON and TPOS2020, an array of satellite and other in-situ systems (top row) further supplement the observations. (Reproduced from Chen et al. 2018)

and the SEC (~0°–5°N) on either side of NECC, and the eastward-flowing EUC have typical speeds of 50 cm/s, 40 cm/s, 20 cm/s, and 120 cm/s, at say around 170°E, respectively. The zonally averaged speeds of the zonal current are much lower (Fig. 7.3b). Dr. David Halpern of NOAA's PMEL in the 1970s developed the first moorings that were successfully deployed in the strongly sheared flows of the Tropical Pacific Ocean (Halpern 1987), which led to the deployment of the low-cost ATLAS mooring in TAO (Hayes et al. 1991). The data measured by the ATLAS system as daily averages of these variables (with some spots taking up to 10-min samples) is telemetered to shore via NOAA polar-orbiting satellites. As TOGA wound down in 1994, other national and international programs (e.g., US GOALS program, the World Climate Research Program's CLIVAR) supported this through 2010. Over the years, the TOGA-TAO array has provided invaluable data for forecasting and research, which has resulted in a large array of publications and understanding of ENSO and other aspects of variations of Tropical Pacific and has supported many other field programs in the Tropical Pacific.

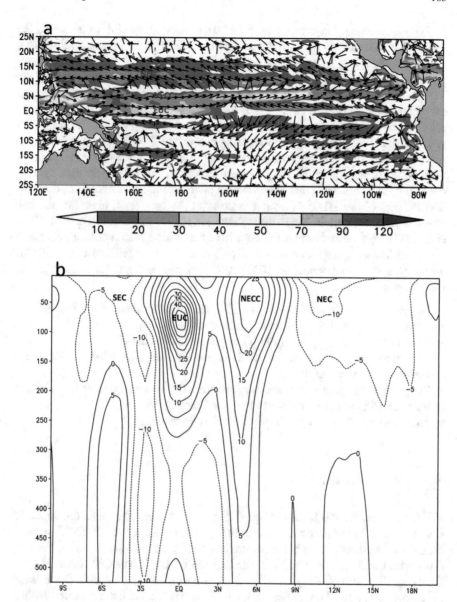

Fig. 7.3 (a) Annual mean climatology of surface ocean currents (cm/s) over the Tropical Pacific. (b) Cross section of the zonal mean (140°E–80°W) zonal currents (cm/s) across Tropical Pacific from SODAv3.4.1 reanalysis with eastward and westward currents contoured in solid and dotted contours, respectively. The names of the various currents is printed. The climatology was computed over a time period of 1980–2015

TOGA-TAO has also evolved over the years. In 2000, the JMA introduced the TRITON to extend TAO westward to become the TAO/TRITON array. Furthermore, Argo ocean profiling floats, ships of opportunity lines of *expendable bathythermograph*, and newer products from satellite remote sensing have supplemented the TOGA-TAO array to be now called the TPOS (Fig. 7.2b). Although these moorings are comparatively cheaper, their ~1 year lifetime makes their servicing a continuous effort, which is expensive. For example, McPhaden et al. (1998) indicate that 5.7 years of ship time was utilized in the 10 years of the TOGA program to service more than 400 TAO moorings deployed on 83 research cruises involving 17 ships from six different countries. Despite these expenses, the invaluable data coming out of this array has made this observational platform indispensable. But the recent deterioration of the TRITON array has been, at the very least, alarming and has renewed efforts to sustain TPOS through the TPOS2020 project (TPOS2014; Chen et al. 2018). With advances in technology in ocean observations such as autonomous platforms including Argo and gliders (Gould et al. 2004), satellite-derived salinity, *altimetry*, and *scatterometry*, the TPOS2020 project is being planned to meet the following objectives:

- To provide data to monitor and quantify the evolving state of the surface and subsurface ocean of the Tropical Pacific.
- To provide data in support of ENSO and other forecasting efforts
- To support the integration of satellite and in situ measurements including calibration and validation
- To provide observing system infrastructure for process studies to advance understanding of the climate system in the Tropical Pacific
- To maintain and extend the Tropical Pacific climate record.

7.3 ENSO Features

ENSO is often referred to as a coupled ocean-atmosphere phenomenon, which is best illustrated by the schematic shown in Fig. 7.4. Under normal or ENSO-neutral conditions, the thermocline is steeply inclined between the Western and the Eastern Equatorial Pacific Ocean, and the ascending cell of the Walker Circulation is confined over the warm pool region of warm SSTs over the Western Pacific Ocean with the associated low-level easterlies of the trade winds and the surface easterly wind stress over the Equatorial Ocean (Fig. 7.4b). During El Niño conditions, the zonal gradient in the depth of the thermocline is weaker than under normal conditions, with the thermocline being deeper in the Eastern Equatorial Pacific Ocean and shallower in the Western Equatorial Pacific Ocean (Fig. 7.4a). Furthermore, the warmer SSTs spread across Central Pacific during El Niño conditions, and the ascending cell of the Walker Circulation also stretches eastward with the warm SSTs over the Central Pacific, with associated weakening of the easterly trade winds and easterly wind stress that now appear as westerly anomalies (Fig. 7.4a).

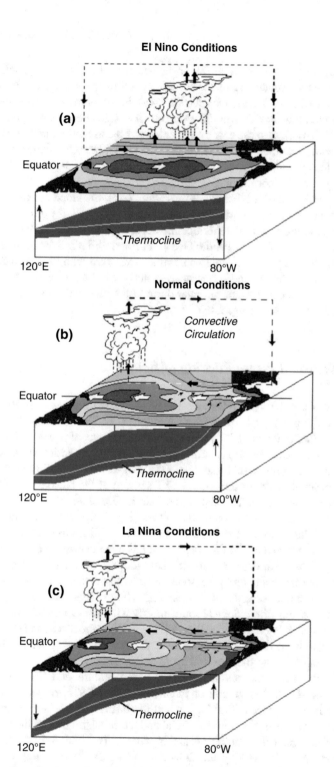

Fig. 7.4 A schematic of the overlying atmospheric (Walker) circulation, underlying thermocline, and SST of Tropical Pacific during (**a**) El Niño, (**b**) Normal, and (**c**) La Niña conditions. (Reproduced from https://www.pmel.noaa.gov/elnino/schematic-diagrams)

Accompanying these changes in the upper ocean of the Equatorial Pacific during an El Niño is the broadening of the ascending cell of the Walker Circulation from the Western to Central Pacific Ocean (Fig. 7.4a). Further the descending cells of the Walker Circulation also shift in the west over the Asian Monsoon region and over the Brazilian Amazon to the east on either side of the ascending cell in a typical El Niño. It should be noted that these changes in the Walker Circulation are noted in the anomalous circulation (obtained as the difference between the circulation in an El Niño year and the corresponding climatological circulation).

In contrast, during La Niña conditions (Fig. 7.4c), the slope of the thermocline becomes even steeper than under normal conditions (Fig. 7.4b) or during El Niño (Fig. 7.4a) conditions with the thermocline becoming deeper in the Western Pacific and shallower in the Eastern Pacific Ocean. Furthermore, the ascending cell of the Walker Circulation shifts and contracts further westward with the warm SSTs. The low-level easterly trades and the surface easterly wind stress anomalies over the Western Pacific Ocean become stronger during La Niña conditions (Fig. 7.4c) than under normal conditions (Fig. 7.4b).

7.3.1 Seasonal Phase Locking of ENSO

From the above discussion, it is clear that ENSO variability circumscribes features of the overlying atmosphere, sub-surface ocean, and the surface ocean of the Tropical Pacific. A feature of ENSO is its apparent seasonal phase-locking feature, which refers to peak variations of ENSO appearing in a particular season of the year. Phase locking refers to the season of the peak ENSO variability being in phase with the seasonal cycle (or the coldest SST season) of the background state of the Equatorial Pacific region. The Equatorial Pacific Ocean shows a distinct seasonal cycle in SST (Fig. 7.5a), in zonal wind stress (Fig. 7.5b), and in meridional wind stress (Fig. 7.5c). During boreal spring (MAM) and fall (SON) seasons, the SSTA tends to be warmer and colder than the annual mean in the Eastern Equatorial Pacific Ocean (Fig. 7.5a). Similarly, the meridional wind stress in the Eastern Equatorial Pacific Ocean displays a strong seasonal cycle with northerly anomalies in MAM and southerly anomalies in the SON season (Fig. 7.5c). These wind stress anomalies under the prevailing climatological southeasterly wind stress anomalies (cf. Fig. 4.7) suggest the weakening of the upwelling (cf. Eq. 5.25) in the MAM and strengthening of the upwelling in the SON seasons over the Eastern Equatorial Pacific Ocean, which is consistent with the seasonal cycle of SST (Fig. 7.5a).

An illustration of the seasonal phase-locking feature of ENSO is shown in Fig. 7.6. The standard deviation of monthly Niño3.4 SSTA peaks in November-December-January season (Fig. 7.6), when the background SST over the Eastern Equatorial Pacific Ocean is cold (Fig. 7.5a) and the zonal gradient of the thermocline depth in the Equatorial Pacific Ocean is strong. The Niño3.4 SSTA for some select warm and cold ENSO years show that indeed the largest amplitude over the evolution of the ENSO event is attained during the boreal winter months, consistent

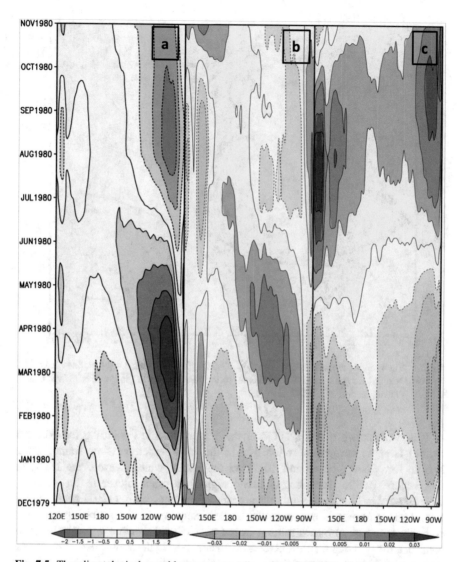

Fig. 7.5 The climatological monthly mean anomalies of (**a**) SST (°C), (**b**) zonal wind stress (N/m^2), and (**c**) meridional wind stress (N/m^2) after subtracting the corresponding annual mean over the Equatorial Pacific Ocean (120°E to 80°W). The climatology is computed over the period of 1979–2008 from NCEP-NCAR reanalysis. (Kalnay et al. 1996)

with the seasonal phase-locking feature of ENSO (Fig. 7.6). Hirst (1986) noted that the annual mean state of the Equatorial Pacific Ocean was too stable to support the onset of ENSO as a coupled ocean-atmosphere instability. Other studies have noted that the seasonality of the background state of the Equatorial Pacific Ocean creates a favorable state for ENSO to initiate through a coupled ocean-atmosphere instability (e.g., Philander 1983; Zebiak and Cane 1987; Tziperman et al. 1997).

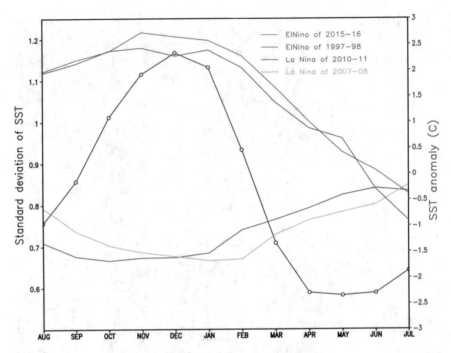

Fig. 7.6 The monthly mean standard deviation of SSTA (°C) over Niño3.4 (black line) and SSTA (°C) over Niño3.4 for some select ENSO events (colored lines). The SST is from ERSSTv5 (Huang et al. 2017) using period of 1975–2020

It is well known that the ITCZ in the Eastern Pacific (including the Niño3.4 region) is at its northernmost location in the July–September time frame, when it is in the vicinity of around 10°N. In the boreal winter season, the ITCZ in the Eastern Pacific (including Niño3.4 region) is much closer to the equator, with associated low-level wind convergence. Tziperman et al. (1997) contend that this wind convergence enhances the air-sea coupling in the following manner: convergence causes upward motion in the atmosphere, which eventually leads to condensation of the moisture and release of latent heat in the atmosphere. This latent heat release further engenders low-level convergence that accelerates the low-level wind and the wind stress in such a way as to increase evaporation off the ocean surface and upwelling of the cold waters to the surface of the ocean, respectively, and thereby cooling the SST further. On the other hand, when the ITCZ is further away from the equator during the boreal summer season, then the mean state of the SST and the atmospheric latent heating over the Eastern Equatorial Pacific Ocean are decoupled from each other (because there is net low-level divergence) at the equator, thereby reducing the air-sea coupling strength. Tziperman et al. (1997) turned off the seasonal cycle by prescribing the annual mean in SST, ocean upwelling, ocean currents, and low-level wind velocity except for wind divergence in their model and found that the ENSO variations in the model were stronger than the control integration and with respect to all other individual model integrations where the seasonal cycle of the other variables was exclusively included. Therefore, they termed this effect of the low-level wind divergence on ENSO as the first-order seasonal effect. It should be noted that the atmospheric condensational heating was parameterized as a function of large-scale convergence in their model. Tziperman et al. (1997) recognized the inconsistency in their approach of prescribing seasonal cycle of some background fields, while in others, it is not included given that these climatological fields are all interlinked and mutually dependent. But they insisted on using this approach to address the relative importance of the seasonal cycle of these variables on ENSO variations.

Tziperman et al. (1997) also found that although the influence of the seasonal cycle of the low-level wind divergence because of the seasonal migration of the ITCZ is dominant on the seasonality of ENSO, it cannot completely explain the seasonal phase-locking feature of ENSO. The model experiment that exclusively included only the seasonal cycle of wind divergence revealed that the ENSO simulated in the model was subtly different from the control integration, with some ENSO onsets in the latter occurring in months when the ITCZ was not necessarily closest to the equator. This led them to conclude that the background stability is set by additional factors besides the background low-level wind divergence. Tziperman et al. (1997) found that the seasonal cycle of SST in addition to the seasonal cycle of low-level wind divergence restored the ENSO variability in the model to very nearly to that of the control integration, suggesting the secondary effect of the seasonal cycle of SST on the seasonality of ENSO variations. Additionally, when the seasonal cycle of the upwelling was also introduced in addition to the seasonal cycle of low-level wind divergence and SST, then the ENSO variation was completely restored to that in the control integration, suggesting that seasonality of the upwelling in the Eastern Pacific Ocean has a tertiary impact on ENSO.

7.3.2   The Bjerknes Feedback Mechanism of ENSO

Time and again, ENSO is referred to as a coupled ocean (El Niño)-atmosphere (Southern Oscillation) phenomenon. The Peruvian fishermen in the sixteenth century observed that periodic warming of the SST, off their coast during December, caused considerable reductions in anchovy fisheries and seabird populations and so named it El Niño (little boy of Christ), and the opposite condition were named La Niña (little girl of Christ). But Bjerknes (1966, 1969) was the first to recognize the coupled ocean-atmosphere interactions of this Tropical Pacific variability. This coupled instability as proposed in Bjerknes (1969) functions in the following manner: suppose a cold SSTA appears over the Eastern Equatorial Pacific Ocean, then it raises the zonal SST gradient with the relatively warm Western Equatorial Pacific Ocean. This increased zonal gradient of SSTA will further strengthen the prevailing easterly trade winds, augmenting the cooling in the Eastern Equatorial Pacific through more wind-driven evaporation and by the inducement of stronger

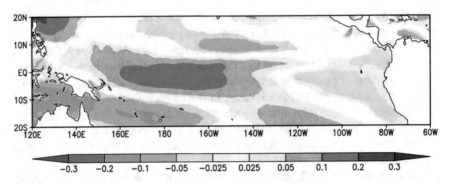

Fig. 7.7 The regression of the zonal wind stress anomaly (dynes/cm^2/°C) on the Niño3.4 SSTA computed from CFSR (Saha et al. 2010) over a period of 1979–2009. Only significant values at a 95% confidence interval according to the t-test are shaded

upwelling of cold waters (through its impact on the wind stress). The trade winds strengthen not only because of the direct result of the stronger zonal SST gradients and the associated pressure gradients within the PBL (Lindzen and Nigam 1987) but also because of an associated zonal pressure gradient imposed as a result of convection/vertical motion being confined further westward over the warm pool of the Western Pacific Ocean that will accelerate the easterlies further. A similar set of arguments could be made for the growth of warm SSTA in the Eastern Equatorial Pacific Ocean. This positive feedback between the atmosphere and the ocean is also referred to as the Bjerknes feedback mechanism of ENSO. It is clear from this discussion that the Bjerknes feedback mechanism fails to explain the initial appearance of SSTA in the Eastern Equatorial Pacific and is also inadequate to explain the phase change of ENSO from an El Niño to a La Niña or vice versa. The strength of the Bjerknes feedback mechanism is usually quantified by the linear regression of the zonal wind stress anomaly on to Niño3.4 SSTA (Fig. 7.7). The positive regression coefficients over the Equatorial Pacific in Fig. 7.7 suggest that a warm or a cold anomaly of the Niño3.4 SSTA (of say $|1|$ °C) implies a proportional (given by the slope of the linear regression) overlying westerly or easterly wind stress anomaly, respectively.

7.3.3   ENSO Spectrum

A characteristic feature of ENSO is its broad spectrum of periodicity ranging ~2–7 years (Fig. 7.8). This is typically quantified by the spectrum (as shown in Fig. 7.8) over the Niño3.4 region. This spectrum suggests that ENSO is aperiodic, i.e., an oscillation with an amorphous frequency in the range of 2–7 years. This aperiodicity of ENSO poses a huge challenge for its prediction and continues to elude some of the best models (Chen et al. 2020).

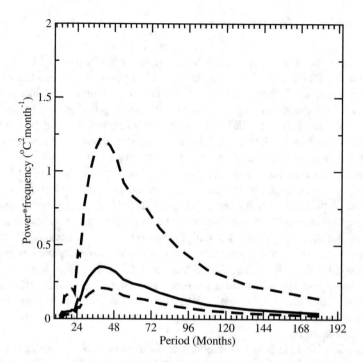

Fig. 7.8 The spectrum of Niño3.4 SST based on ERSSTv5 (thick solid line) over the period 1920–2010. The spectrum was generated after smoothing the time series with a 6-month running mean. The 95% confidence interval according to the chi-squared test is shown in dashed lines

Many studies indicate that the seasonal forcing could be a potential reason for the irregularity in the ENSO period. These studies suggest that ENSO oscillation can enter a nonlinear resonance with the seasonal forcing to produce ENSOs, irregularly. For example, at resonance, perfectly periodic behavior of ENSO is characterized in exact phase with the seasonal cycle. But several such resonances may coexist, for sufficiently nonlinear dynamics in which case the ENSO oscillator will jump irregularly between the different resonances as it is not able to prefer a single such resonance thereby, producing the observed ENSO irregularity. Tziperman et al. (1997) showed from their model integrations that prescribing the seasonal cycle of the critical variables like the low-level wind divergence is critical to get the broad range in the ENSO spectrum. In contrast, prescribing the annual mean of low-level wind divergence while prescribing the seasonal cycle of SST and upwelling simulated a rather acutely periodic ENSO.

7.3.4   ENSO Teleconnections

One of the reasons for ENSO's notoriety is the global impact it has, as shown schematically in Fig. 7.9. The near reversal of the surface temperature and precipitation anomalies between the two phases of ENSO causes a significant impact on the local climate from major wildfires with prolonged droughts (e.g., Janicot et al. 1996; Marengo et al. 2008) to severe flooding and cold snaps (Janowiak 1988; Ropelewski and Halpert 1989) that sometimes define climate extremes in the region. Many studies have reported that the monthly mean 1000 to 200 hPa tropospheric temperature anomalies over the global tropical strip of 20°N–20°S uniformly warm during El Niño from both eastward-propagating Kelvin waves and westward-propagating Rossby waves at different lead and lag times (Fig. 7.10). The reason for this uniform distribution of the tropospheric temperature anomalies in the tropics on timescales of a month or two (Fig. 7.10) is because the tropical atmosphere cannot maintain horizontal pressure gradients. As depicted in Fig. 7.9a, most tropical regions including the oceans and land area, remote from the Niño3 or Niño3.4 region, are warm and dry, except for Eastern Equatorial Africa and the Equatorial Pacific Ocean (Janowiak 1988; Misra et al. 2007). This is a consequence of the warming of the troposphere during El Niño events that lead to increased atmospheric stability.

To explain the ENSO teleconnections in the tropics, Chiang and Sobel (2002) divided the atmospheric response into three phases based on the evolution of ENSO. A growth phase, which is approximately −5 to 0 months, where negative months mean tropospheric temperature anomalies are leading Niño3 SSTA (Fig. 7.10), a mature phase (+1 to +5 months in Fig. 7.10), and a decay phase (~+6 months and onward in Fig. 7.10), where the Niño3 SSTA lead the tropospheric temperature anomalies. During the growth phase, moist convection is suppressed, while the SST slowly warms over the remote tropical oceans from the Niño3 region as a result of the increased stability imposed by the warming of the tropospheric wave propagation of the tropospheric temperature anomalies. Furthermore, the tropospheric temperature anomaly propagates to the surface in regions of deep and or shallow convection in the following manner:

(i) The PBL layer equivalent potential temperature undergoes an instantaneous adjustment as a result of the imposed tropospheric temperature anomalies because the free troposphere maintains a moist adiabatic lapse rate in active convective regions that is precisely locked to the equivalent potential temperature of the PBL (Arakawa and Schubert 1974). So, the boundary layer equivalent potential temperature adjusts to the imposed overlying tropospheric temperature anomalies by entraining the warm air from the free atmosphere into the PBL.

(ii) The surface fluxes act to bring the PBL and ocean surface layer in equilibrium since the PBL equivalent potential temperature is closely tied to SST through surface flux. But because of the high thermal inertia of the ocean mixed layer, the remote tropical oceans do not warm up instantly but with a lag of a

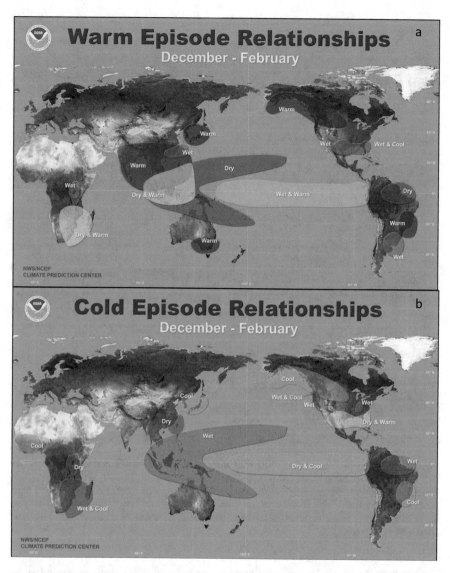

Fig. 7.9 The schematic of the global impacts of (**a**) El Niño and (**b**) La Niña on seasonal surface temperature and rainfall in the boreal winter season. (Reproduced from https://www.cpc.ncep.noaa. gov/products/precip/CWlink/MJO/enso.shtml#educational%20material)

few months. Over land, where the thermal inertia is comparatively less, the land surface temperatures come to equilibrium relatively quickly by the same mechanism.

Regarding teleconnection between Niño3 SSTA and precipitation, it is far more complex as it is more variable across a wide range of spatial and temporal scales.

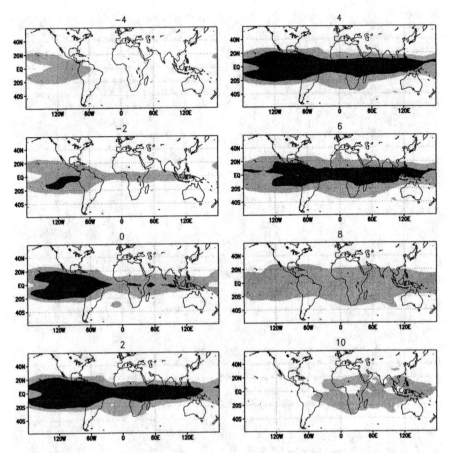

Fig. 7.10 The lag-correlation computed over the period of 1977–1999 between Niño3 SSTA and Microwave Sounding Unit (MSU) channel-2 measured 1000 to 200 hPa monthly mean tropospheric temperature anomalies with negative or positive value mentioned at the top of each panel, signifying that Niño3 SSTA lags and leads the tropospheric temperature anomalies by as many months. (Reproduced from Chiang and Sobel 2002; © American Meteorological Society. Used with permission)

Furthermore, the variations of precipitation are large only if the mean precipitation is relatively large. Therefore, the teleconnections are more seasonally dependent (e.g., on the latitudinal location of the ITCZ because of its seasonal migration). Nonetheless, Chiang and Sobel (2002) find that the precipitation response to tropospheric temperature anomalies is quite robust. They find that as they increase the depth of the ocean mixed layer, the precipitation response (as in the reduction of precipitation in the case of warm tropospheric temperature) becomes more significant. It may be noted that the thermal inertia of the upper ocean becomes larger as the depth of the mixed layer increases. Therefore, the disequilibrium between the atmospheric boundary layer and the ocean mixed layer becomes larger as the depth of the mixed layer is increased in the remote tropical oceans. So, effectively, the SST response becomes weaker in such remote tropical oceans with deep mixed layers as convection is drastically reduced from the imposed tropospheric temperature warming, which causes further decoupling of the surface from the free atmosphere. The warming of the SST that occurs in the remote tropical oceans with deep mixed layers is largely a result of a reduction in the latent heat flux as a response to the high boundary layer equivalent potential temperature.

Parhi et al. (2016) make a distinction of the precipitation anomalies over Western Sahel as a response to the growth and mature stages of ENSO. They argue that during the growth phase of ENSO, as the Tropical Atlantic SST is yet to warm, the moist static energy in the remote Tropical Atlantic Ocean builds up but is insufficient to overcome the stability from the warm tropospheric temperature anomalies. So, during this growth phase, which coincides with the first rainy (July-August-September) season of the year over the Western Sahel, it experiences a dry anomaly as a result of the remote ENSO forcing. But during the mature phase, when the ocean SSTs warm up in the Tropical Atlantic and are in equilibrium with the atmospheric boundary layer temperature, then there is a sufficient buildup of the moist static energy to overcome the stability of the overlying warm atmospheric column. Furthermore, the regional ocean, in this case the Tropical Eastern Atlantic, now serves as a moisture source for convection, with significant moisture advected into the boundary layer over Western Sahel during its second rainy season of October-November-December being sufficient to overcome the overlying stability from the warm tropospheric temperature anomalies to generate a wet seasonal anomaly during the mature phase of ENSO.

ENSO teleconnections are however not limited to just the tropical latitudes but also extend into higher latitudes (Fig. 7.9). This is also illustrated by the regression of the 200 hPa geopotential heights on the Niño3.4 SSTA, which shows a wave train of highs and lows extending across North America during the boreal winter season (Fig. 7.11a). These are a result of the stationary Rossby waves excited by the anomalous diabatic heat release from the convection in the Niño3.4 region. Likewise, a similar wave train is excited in the boreal summer season in the SH (Fig. 7.11b). These teleconnection patterns in the midlatitudes generated by ENSO variations are seasonally dependent because the subtropical jets in both hemispheres serve as a waveguide for these extratropical Rossby waves (Lee et al. 2009).

7.3.5   ENSO Diversity

Although it was well known that no two ENSOs are alike, a renewed interest in the diversity of ENSO has come about from some interesting observations made during some relatively recent ENSO events that differed from the traditional ENSO events (Fig. 7.12). In fact, Ashok et al. (2007) called these new types of El Niños as El Niño 'Modoki' (a Japanese word that means 'similar but different'), and others refer to them as CP El Niños while the traditional El Niños are called EP type. Figure 7.12

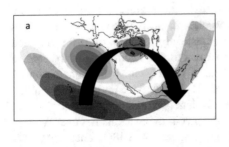
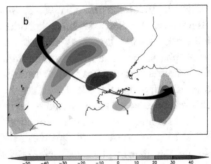

Fig. 7.11 The regression of the (**a**) DJF and (**b**) JJA 200 hPa geopotential height anomalies on the corresponding Niño3.4 SSTA. The solid arrows show the teleconnection pattern emanating from the Equatorial Pacific into (**a**) North American and (**b**) South American continents. Only significant values at a 5% significance level according to the t-test are shaded. This regression was conducted using CFSR (Saha et al. 2010) over the period of 1979–2008

highlights the essential difference between the CP and the EP type of El Niños, with the former type forming a peak SSTA in the Central Equatorial Pacific (~ in the Niño4 region) while the latter having its peak in the Eastern Equatorial Pacific (~ Niño1+2 and Niño3 regions). A more detailed description of the differences between the two types of ENSOs is listed in Table 7.3. However, it should be mentioned that La Niña events are not as distinct as the El Niños of the two types (Fig. 7.12), and therefore, most of the emphasis on ENSO diversity is focused on El Niños. Furthermore, Fig. 7.12 shows that peak SSTA associated with El Niños occurs over a wide range of longitudes in the Equatorial Pacific, and some of these events can be thought of as some linear combination of these two types of El Niños.

7.4 ENSO Theories

7.4.1 Linear Stochastic Theory

ENSO, by its name, was recognized to be basin-wide warming of the tropical Pacific (El Niño) associated with the overlying relaxation of the trade winds and a displacement of the Walker Circulation with the atmospheric convection shifting to Central Pacific from the Maritime Continent region. Although the importance of coupled air-sea feedback in ENSO is accepted, there is still debate of whether ENSO is a nonlinear and self-sustained oscillatory system (Zebiak and Cane 1987; Munnich et al. 1991; Tziperman et al. 1997) or a linear damped stochastically driven system (Penland and Sardeshmukh 1995; Kleeman and Moore 1997; Moore and Kleeman 1999a, b; Newman et al. 2011b). In the latter (hereafter linear theory of ENSO), it is assumed that without an external noise forcing (like weather

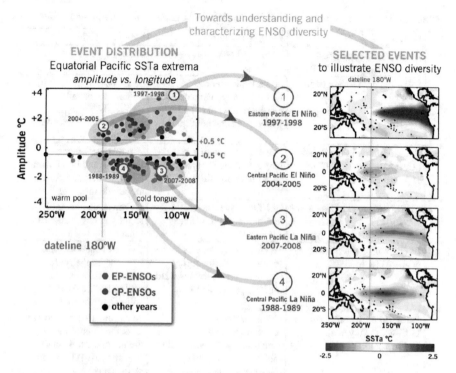

Fig. 7.12 (Left) The distribution of the amplitude of the extrema of seasonal November-December-January (NDJ) SSTA in the Equatorial Pacific (2°S–2°N and 110°E–90°W) as a function of longitude over the period of 1900–2013. The SSTA is computed as departures from the 1945 to 2013 climatology. The ENSO events before 1945 are identified in gray. EP-type events are identified when SSTA over the Niño3 region exceeds one standard deviation. CP-type events are identified when the standard deviation of the leading principal component of SSTA residual after removing the SSTA regression on the Niño index. (Right) The spatial patterns of SSTA for specific warm and cold events of either type of ENSO are illustrated with a shading interval of 0.25°C. Reproduced from Capatondi et al. (2015); © American Meteorological Society. Used with permission

variations), the ocean-atmosphere system will not produce ENSO-like events. In the former theories (hereafter nonlinear theory of ENSO), ENSO is described as an oscillation with irregular periods that arise naturally as a result of instabilities in the system (of the like described in Sect. 7.3.1).

Moore and Kleeman (1999a, b) offered further insight into this stochastic forcing to the coupled system. They suggested that the stochastic forcing enhances the seasonal to interannual variability by inducing the onset of new ENSO episodes and by disrupting developing or existing ENSO events. Furthermore, the ability of the stochastic forcing to enable seasonal to interannual variations depends on the following:

Table 7.3 Differences between Eastern Pacific (EP)- and Central Pacific (CP)-type El Niños

	Feature	EP-type El Niño	CP-type El Niño
1	SSTA peak region (Ashok et al. 2007; Kug et al. 2009)	Niño3 and Niño1+2 region	Niño4 region
2	SSTA pattern (Ashok et al. 2007)	SSTA peak over the Niño1+2 region and extends westward toward Niño3 with decreasing amplitude	SSTA peak over the Niño4 region and extends southeastward toward Niño1+2 with very weak amplitude
3	The amplitude of the SSTA peak (Ashok et al. 2007)	Are comparatively much larger than the CP type but attain peak amplitude in winter	Are relatively much weaker in amplitude than EP type but attain peak amplitude in winter
4	ENSO asymmetry (Kug and Ham 2011)	El Niños have a larger amplitude than La Niñas	La Niñas tend to be slightly stronger than El Niños
5	El Niño initiation (Newman et al. 2011a, b)	SSTA initially appears in the far eastern Pacific during spring and extends westward during summer and fall	SST anomalies extend from the northeastern subtropical Pacific to central Pacific during spring and summer
6	Paleoclimate reconstructions of ENSOs (Moy et al. 2002; Convoy et al. 2008)	Coral records suggest reduced interannual variability in the east Pacific mid-Holocene period (6000 years ago)	According to coral records from the Central Pacific, the interannual variance over the past 1000 years is statistically indistinguishable from that during the mid-Holocene period. This implies that ENSO variability may have been displaced westward in the mid-Holocene period
7	ENSO theory	The recharge-discharge process of the Equatorial Pacific oceanic heat content as in the recharge-discharge theory (Jin 1997a, b; Meinen and McPhaden 2000) is more influential in explaining the SST variations in the eastern Pacific related to ENSO where the thermocline is shallow	The advection of the large zonal temperature gradients by anomalous zonal currents as part an advective-reflective oscillator (Picaut et al. 1997) tends to be more effective to explain variations of CP-type El Niños due to the large zonal SST gradients near the edge of the warm pool
8	Teleconnections	Less North Atlantic TC activity, catastrophic floods in parts of Ecuador and northern Peru	Precipitations associated with Indian and Australian summer monsoons are more affected than by EP type; stronger SH winter storm activity and North Atlantic TC activity are observed

- The phase of the seasonal cycle of the background state: They found that by virtue of the seasonal cycle of the SST in the Eastern Equatorial Pacific Ocean, perturbations of the coupled system engendered by the stochastic forcing that develops in the late or early part of the year have a greater likelihood of developing into an ENSO event.
- The presence of nonlinearities in the system: They found that the growth of the perturbations to the system grew in the absence of nonlinearities. For example, during ENSO episodes when the system resides in a nonlinear regime, then the growth of the perturbations is inhibited. Moore and Kleeman (1999a, b) note two important nonlinearities, which include a threshold for moist static energy, below which deep convection anomalies cannot occur, and the depth of ocean thermocline, which could be either too deep or too shallow to affect SSTA. Both these nonlinearities are active in the Central and Eastern Equatorial Pacific Oceans during the mature phase of either El Niño or La Niña.
- The time series of the stochastic forcing: They found that if the phase of the noise forcing was changing continually, then the effect of stochastically forced perturbations will have interference with perturbations of the opposite sign that occur soon after, leading to weak or inexistent ENSO events. On the other hand, seasonal to interannual variations are significantly affected if the phase of the stochastic forcing remains consistent for a while even if the amplitude varies. For example, several El Niño events (e.g., 1976–1977, 1982–1983, 1986–1987, 1997–1998) were induced by a series of westerly wind bursts associated with the MJO.

The linear theory of ENSO has yielded very useful information on the evolution of ENSO. For example, Fig. 7.13 shows how SSTA patterns for EP- and CP-type ENSO events evolve from an initial anomaly pattern from 6 months before. In this example, shown in Fig. 7.13, the precursor (or initial) pattern was obtained from a linear inverse modeling approach of a fitted linear stochastically forced dynamical model to observations of SSTs, 20 °C isotherm (or thermocline) depth, and surface zonal wind stress for 1959–2000 (Newman et al. 2011b; Capotondi et al. 2015). These initial patterns clearly show the contrast in the region of the initial stochastic forcing with Eastern Pacific and Niño4 areas being important for the EP and the CP type of El Niños, respectively. Since these initial patterns were derived from a linear model, it may be noted that the patterns of the opposite sign will lead to cold events, although it should be noted that the differences in mature La Niñas are not as distinct as El Niños between the EP and CP types of ENSO (Fig. 7.12). This linear inverse modeling approach also reveals in Fig. 7.13c, d that the evolved mature states of ENSOs in either types appears as some linear combination of the evolution from the optimal initial patterns of EP and CP type of El Niños shown in Fig. 7.13a, b, respectively. In other words, Fig. 7.13c, d suggest that it is very hard to bracket a given ENSO event as purely EP- or a CP-type event.

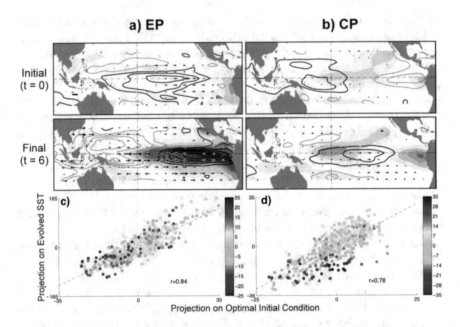

Fig. 7.13 Evolution of (**a**) EP-type and (**b**) CP-type El Niño events from initial (top panels of (**a**) and (**b**)) SSTA (shaded), thermocline depth (contoured), and zonal wind stress anomalies (vectors) derived from linear inverse modeling approach by fitting a linear stochastically forced dynamical model fitted to observed SSTs, thermocline depth, and zonal wind stress over a period 1959–2000. The bottom panels of (**a**) and (**b**) show the evolution of the SSTA, thermocline depth anomalies, and zonal wind stress anomalies after 6 months. The projection of the optimal initial condition vs the projection of evolved SST 6 months after, for (**c**) EP-type and (**d**) CP-type El Niño events. The color shading of the dots in (**c**) indicates the value of the projection on CP-type optimal initial condition, and likewise the shading of dots in (**d**) indicates the value of the projection on EP-type optimal initial condition. The spectra of colors in (**c**) and (**d**) suggest that EP and CP of El Niños evolve as some linear combination from the optimal initial conditions of the two types. Note that the ordinate of (**c**) is much larger in range than (**d**), suggesting the amplitude of EP-type El Niños are larger than those of CP type. (Reproduced from Capatondi et al. 2015; © American Meteorological Society. Used with permission)

7.4.2   Delayed Oscillator Theory

The Bjerknes feedback mechanism is one of the foremost theories to partially explain ENSO variations. But as described in Sect. 7.3.2, this feedback mechanism does not explain the initiation of ENSO or its phase change from, say, a warm phase (El Niño) to a neutral or cold phase (La Niña). But it nonetheless recognizes the importance of the air-sea interaction in the evolution of ENSO after the initial anomaly develops.

The delayed oscillator theory (Suarez and Schopf 1988; Battisti and Hirst 1989) proposes the important role of the equatorial waves in the ocean that supplement the growth of ENSO due to Bjerknes feedback mechanism while also providing a way for the reversal of the ENSO phases. Unlike the linear stochastic theory, the coupled ocean-atmosphere system produced ENSO-like oscillations under steady external (atmospheric) forcing. The instability in this mechanism comes in the way of Kelvin Modes that tilt the thermocline, enhancing the SSTA and further engendering the Bjerknes feedback.

By this delayed oscillator theory, a westerly wind stress anomaly over Central Equatorial Pacific (Fig. 7.14a), for example, excites a downwelling Kelvin wave in the upper ocean that propagates rapidly eastward. Simultaneously, a slower westward-moving Rossby wave from the place of the application of the wind stress anomaly is also generated (Fig. 7.14b). The downwelling Kelvin wave displaces the thermocline to make it deeper, which manifests as warm SSTA in the Eastern Equatorial Pacific where, climatologically, the thermocline is shallow (Fig. 7.14c–e). This warm SSTA grows further because of the Bjerknes (air-sea) feedback mechanism (Fig. 7.14f, g). In the meanwhile, the Rossby wave propagates westward toward the western boundary (Maritime Continent) as an upwelling wave, where the thermocline variations have little impact on the SSTA and local air-sea coupling. The Rossby wave that appears as an upwelling wave (with thermocline anomalies of the opposite sign compared to those of the eastward-downwelling Kelvin wave) is now reflected as an upwelling Kelvin wave from the western boundary (Fig. 7.14g) that reaches the Eastern Equatorial Pacific Ocean (Fig. 7.14h–j), interfering with downwelling Kelvin wave and eventually terminating the warm event (Fig. 7.14i, j). The name 'delayed' appears in the name of this theory to account for the time delay required for the Rossby waves to propagate westward, reflect at the boundary, and return as a Kelvin wave to the forcing region. It may be noted in Fig. 7.14f that the reflection of the eastward-propagating Kelvin wave as Rossby waves, at the eastern boundary, is not prominent. It would take nearly 3 years for these reflected Rossby waves from the eastern boundary to traverse the Pacific Basin to the western boundary. Furthermore, these reflected Rossby waves from the eastern boundary propagate much slower than the Rossby waves forced by the wind stress in the Central Pacific because they have much smaller equivalent depths. In addition, Fig 7.14 shows that once the Kelvin waves reach the eastern boundary, then they also propagate as coastal Kelvin waves, poleward in either hemisphere (Fig. 7.14e–j).

In the above scenario, if the westward-propagating Rossby wave is the first meridional mode, then the transit time for the first Rossby wave and the pair of (eastward downwelling and eastward upwelling) Kelvin waves to traverse the Pacific Basin and to reverse the ENSO phase as illustrated in Fig. 7.15 is way too short than the observed duration of El Niño (~9–12 months). Cane et al. (1990) then made the argument that the SSTA in the Eastern Equatorial Pacific grows as a result of the local air-sea feedback. Therefore, Cane et al. (1990) suggested that additional westward-propagating Rossby waves to the western boundary and the associated reflected Kelvin waves are required to reverse the oscillation. Therefore, in essence, the ENSO period (or duration of an event) in this delayed oscillator model is determined by the competition between the restoring ocean wave dynamics and local air-sea feedback instability in the Eastern Equatorial Pacific.

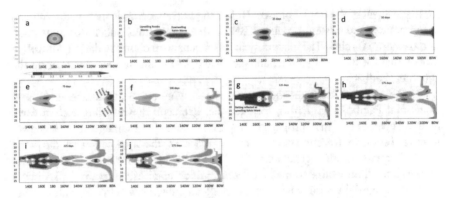

Fig. 7.14 (a) The initial westerly zonal wind stress (dynes/cm^2) anomaly imposed over the Central Equatorial Pacific Ocean in the Cane and Zebiak coupled model (Cane et al. 1986). (b–j) The evolution of the thermocline depth anomalies after the imposed zonal wind stress anomaly as a result of the equatorial-propagating Kelvin and Rossby waves. The positive and negative thermocline anomalies are suggestive of the downwelling and upwelling equatorial waves, respectively (Courtesy of David Dewitt, personal communication)

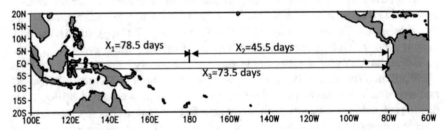

Fig. 7.15 A schematic illustration of the time it takes for the upwelling equatorial Rossby wave with a phase speed of 1 ms^{-1} to reach the western coast ($X_1 \approx \frac{61° \times 111000}{1 \times 86400} = 78.5$ days), downwelling equatorial Kelvin wave to reach the eastern coast ($X_2 \approx \frac{99° \times 111000}{2.8 \times 86400} = 45.5$ days), and upwelling Kelvin wave to reach the eastern coast from the west coast ($X_3 \approx \frac{160° \times 111000}{2.8 \times 86400} = 73.5$ days), when a perturbation is applied at the date line

Schneider et al. (1995), however, argued that time delays due to the equatorially trapped waves alone cannot determine the ENSO period. Schneider et al. (1995) suggested that excitation of higher meridional mode Rossby waves (or off-equatorial Rossby waves) can lead to long delays that are consistent with the ENSO period. However, the delayed oscillator theory as originally proposed involved only the gravest (first meridional mode) Rossby wave. Schneider et al. (1995) from their modeling experiments found that the thermocline anomalies in the western boundary were small, and time delays due to equatorially trapped modes alone could not produce the observed period of ENSO.

Kirtman (1997) nicely illustrates the role of the off-equatorial Rossby waves (with higher meridional mode) in determining the period of ENSO variations in a coupled model. Kirtman (1997) used the Zebiak and Cane (1987) ocean model coupled with a very simple statistical model of the atmosphere, where the structure of the wind stress anomaly is determined by linearly regressing on the Niño3 SSTA as illustrated below:

$$\tau_x = \alpha\,(i,\,j) \times \text{SSTA} \tag{7.1}$$

$$\tau_y = \beta\,(i,\,j) \times \text{SSTA} \tag{7.2}$$

The Zebiak and Cane (1987) ocean model is a linear, *reduced gravity two-layer model*, which simulates thermocline depth anomalies, depth-averaged baroclinic currents, and surface wind-driven currents. It also includes a prognostic equation for SST that has horizontal and vertical advection by both mean and anomalous currents and anomalous surface heat flux to damp the SSTA, with a prescribed seasonal cycle of the ocean currents, temperature, and thermocline depth. Kirtman (1997) conducted two sensitivity experiments (Narrow and Broad) in addition to the Control experiment with this coupled model. These experiments were multiyear integrations of the coupled model over the Tropical Pacific, wherein the control integration yielded realistic ENSO simulations. For the sensitivity experiments, the structures of α and β in Eqs. 7.1 and 7.2 were modified in such a way that the meridional extent of the zonal wind stress anomaly was modulated (Fig. 7.16), respectively. Therefore, the Broad and the Narrow experiments have extended and reduced meridional extent of the zonal wind stress anomalies about the equator (Fig. 7.16a), and the implied wind stress curl from these anomalies (Fig. 7.16b) also shows a similar modulation of its meridional extent. As a result, in the Broad experiment, the off-equatorial (or higher meridional mode) Rossby waves are excited relative to the Narrow experiment. The time-longitude cross section of the thermocline anomalies over the Equatorial Pacific from the three model experiments (Fig. 7.17) clearly shows that the ENSO period is modulated by these changes to the wind stress. For example, in the Broad, Control, and Narrow experiments, the ENSO periods are 9 years, 5 years, and 3 years, respectively. Kirtman (1997) argues that it is the excitation of the relatively slower westward-moving off-equatorial Rossby waves that is responsible for increasing the ENSO period in the broad experiment. In a novel set of additional experiments, Kirtman (1997) filtered out the off-equatorial Rossby waves in all three (Control, Broad, and Narrow) experiments. As a result, the ENSO period in these novel experiments became nearly identical at around 2 years for the originally Broad, Control, and Narrow experiments (Fig. 7.18), which ascertained the importance of the off-equatorial Rossby waves in determining the timescale of ENSO in the coupled model.

The off-equatorial Rossby waves are higher meridional mode Rossby waves that propagate much slower than the equatorial Rossby waves. For example, if the gravest and the off-equatorial Rossby wave at, say, 7°S were excited simultaneously in the Central Pacific, the gravest Rossby wave would reach the western boundary in about 18 days before the off-equatorial Rossby wave. The slower phase speed of the off-equatorial Rossby wave results in a phase lag between the thermocline anomaly

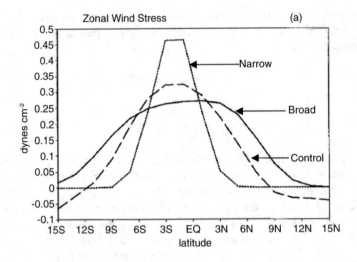

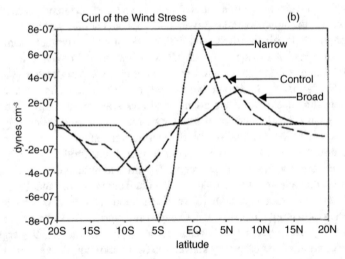

Fig. 7.16 The meridional structure of (**a**) zonal wind stress anomaly and (**b**) the corresponding wind stress curl at the date line. Control, Broad, and Narrow refer to the name of the three experimental integrations conducted with the coupled model by Kirtman (1997). The wind stress anomaly is based on a linear regression on Niño3 SSTA, which is set at 1.35 °C. (Reproduced from Kirtman 1997; © American Meteorological Society. Used with permission)

at the equator and the off-equatorial latitudes. As a consequence, the amplitude of the reflected Kelvin wave from the western boundary is considerably reduced when the impact of the off-equatorial Rossby waves is included. It may be noted that the amplitude of the Kelvin wave at the western boundary (*A*) is given by

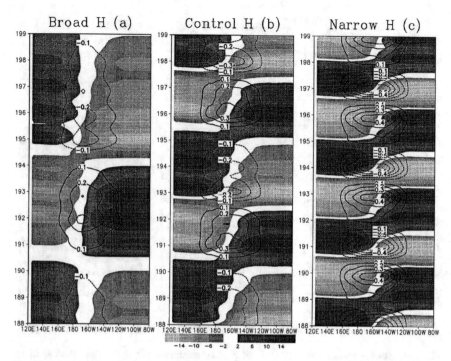

Fig. 7.17 Time-longitude cross sections of the thermocline depth anomalies (shaded; in m) superposed with the corresponding zonal wind stress anomaly over Equatorial Pacific for (**a**) Broad, (**b**) Control, and (**c**) Narrow coupled model experiments. The ordinate denotes time in years. (Reproduced from Kirtman 1997; © American Meteorological Society. Used with permission)

$$A\,(Y_S, Y_N) = -2 \int_{Y_S}^{Y_N} u_R\,(x, y)\,\mathrm{d}y \qquad (7.3)$$

where $u_R(x, y)$ is the zonal velocity due to all of the Rossby waves at the western boundary (x) between the latitudes of Y_S (= 7°S) and Y_N(= 7°N). This reduced amplitude of the Kelvin wave then suggests that it is less efficient in reversing the sign of the thermocline anomalies and SSTA in the Eastern Pacific, and thus, a longer oscillation period ensues. The experiments that produced the results in Fig. 7.18a reveal that the ENSO period was modified in the Broad experiment because the higher meridional Rossby waves were carrying the dominant westward signal. By the filtering of the off-equatorial Rossby waves in the original Broad experiment, the signal (of the thermocline anomaly) striking the western boundary was damped, and thereby, the amplitude of the reflected Kelvin waves was increased, which consequently resulted in the decrease of the ENSO period from around 6 years in Fig. 7.17a to about 2 years in Fig. 7.18a.

In summary, the experiments conducted by Schneider et al. (1995) and Kirtman (1997) suggest that the thermocline displacements at the Western Equatorial Pacific

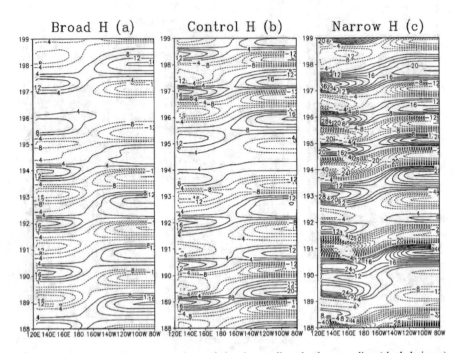

Fig. 7.18 Time-longitude cross sections of the thermocline depth anomalies (shaded; in m) superposed with the corresponding zonal wind stress anomaly over Equatorial Pacific after the off-equatorial Rossby waves are filtered out in (**a**) Broad, (**b**) Control, and (**c**) Narrow coupled model experiments. The ordinate denotes time in years. (Reproduced from Kirtman 1997; © American Meteorological Society. Used with permission)

are critical to the evolution of ENSO. The wind stress anomalies excite equatorial and off-equatorial Rossby waves over Central Pacific that cause thermocline displacements that propagate westward. It may be noted that the off-equatorial Rossby waves cause the largest thermocline displacements off the equator but are very strongly correlated to thermocline displacements at the equator and therefore set the value of the thermocline displacement at the equator of the western boundary. Since the east-west thermocline slope in the Equatorial Pacific is in equilibrium with the wind stress, the thermocline anomalies at all longitudes adjust instantaneously (at Kelvin wave timescale) to maintain the same slope with the new thermocline displacements at the western boundary as a result of the Rossby waves impinging on the western boundary.

An application of this theory is demonstrated in Fig. 7.19. In Fig. 7.19a, c, we show the regression of the zonal wind stress anomalies on the Niño3.4 SSTA between two arbitrary coupled ocean-atmosphere Climate Models A and B. In the case of Model A, the zonal wind stress anomaly over the Central Pacific is relatively narrow (Fig. 7.19a). This results in the Niño3.4 SSTA spectrum of Model A showing variability at comparatively high frequency at ~36 months (Fig. 7.19b). On the other hand, the zonal wind stress anomaly over Central Pacific for Model B

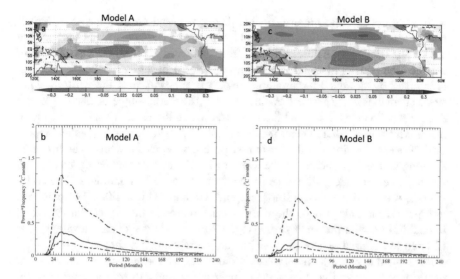

Fig. 7.19 The regression of the zonal wind stress (dynes/cm²) on the corresponding Niño3.4 SSTA from (**a**) Model A and (**c**) Model B. The spectrum of the Niño3.4 SSTA from (**b**) Model A and (**d**) Model B with the blue vertical line indicating the frequency of the peak power in the models. The dashed lines in (**b**) and (**d**) denote the 95% confidence interval of the spectrum by the chi-squared test

is comparatively broader (Fig. 7.19c), which results in the Niño3.4 SSTA spectrum displaying peak periodicity at relatively lower frequency of ~52 months (Fig. 7.19d).

7.4.3 Recharge-Discharge Theory

Recharge-discharge theory of ENSO (Jin 1997a, b) suggests that the discharge process due to Sverdrup transport is the negative feedback to counter the positive feedback mechanism of Bjerknes. In its simplest form, this theory suggests that warm water volume (defined above the 20 °C isotherm between 5°N and 5°S, 120°E to 80°W following Meinen and McPhaden 2000) builds up in the Equatorial Pacific prior to El Niño (recharge phase), and then the *ocean heat content* is discharged to higher latitudes during El Niño (discharge phase). As the thermocline displacement occurs in the Equatorial Pacific, the Bjerknes feedback mechanism kicks in as the SSTA develops and changes the gradients that affect the overlying winds and the overall Walker Circulation until the oscillation begins to transition. This theory claims that the meridional transport in the middle of the Equatorial Pacific Ocean (away from the lateral boundaries) are close to being in Sverdrup balance with the wind stress curl. But these transports are not compensated by the transports of the western boundary currents and therefore lead to gradual (low-frequency) changes in the warm water volume near the equator associated with ENSO. The basic premise of this theory is based on the quasi-equilibrium maintained between the equatorial zonal wind stress and zonal mean thermocline depth. The oscillatory nature of ENSO from this theory is from the disequilibrium between the equatorial zonal wind stress and zonal mean thermocline depth that is initiated by sustained perturbations (e.g., the appearance of the westerly wind stress anomalies in the Western Pacific Ocean). The theory suggests that the Central Pacific zonal wind stress, zonal thermocline tilt, and the Eastern Pacific SST are in phase with one another. Therefore, when westerly wind stress anomalies appear over Western Pacific (and are sustained by Bjerknes mechanism), which causes shoaling of the thermocline in the Western Pacific and deepening in the Eastern Pacific, a compensatory discharge of the warm water to high latitudes affected by the poleward meridional Sverdrup transport is initiated (Fig. 7.20a). The poleward transport is minimized once the warm water volume at the Equatorial Pacific is in equilibrium with the wind stress (Fig. 7.20b), which marks the transition from El Niño to La Niña, where the zonal mean thermocline has shoaled closer to the surface. At this point, as the easterly wind stress anomalies appear and grow (at maturity of La Niña; Fig. 7.20c), the compensatory equatorward transport of warm water is initiated by the Sverdrup transport that eventually leads to the deepening of the zonal mean thermocline (Fig. 7.20d) when the wind stress anomalies have minimized once again, marking the transition to El Niño.

Although the modulation of the warm water volume was well known historically (Wyrtki 1985; Cane et al. 1986), the lack of subsurface ocean measurements in the Equatorial Pacific limited the efforts to quantify the recharge of and discharge from Equatorial Pacific. The TOGA-TAO array ameliorated this limitation and helped in quantifying the recharge and discharge phases of the ENSO cycle (Meinen and McPhaden 2000). Typically, this is quantified by either conducting an EOF analysis of the Tropical Pacific thermocline anomalies or doing a lead/lag regression of the thermocline depth anomalies on Niño3 SSTA (Fig. 7.21). Both yield very similar answers. In doing the EOF analysis, the first EOF explains 28% of the total variance and represents the see-saw pattern of thermocline anomalies between the East Pacific and the West Pacific (Meinen and McPhaden 2000), which is very similar in pattern to the correlation pattern shown in Fig. 7.21a. The second EOF mode explains 21% of the total variance (Meinen and McPhaden 2000) and is again very similar to the correlation pattern shown in Fig. 7.21b. The second EOF pattern (or Fig. 7.21b) represents the recharge-discharge of the warm water volume of the Equatorial Pacific and leads the first EOF by about nine months (Meinen and McPhaden 2000). It is this leading pattern of EOF that makes the recharge-discharge theory more endearing to prognostic applications. Conceivably, one could monitor the warm water volume in the Equatorial Pacific to anticipate either the development or the demise of an El Niño, almost nine months in advance. It has been however noted that the changes to the recharge-discharge process in CP-type El Niño events have reduced the effectiveness of warm water volume as a predictor of such events (Hendon et al. 2009; McPhaden 2012). Lengaigne et al. (2004) find that if westerly wind bursts occur when the Tropical Pacific upper ocean heat content is larger than normal (i.e., in the recharged state), then the likelihood of EP-type El

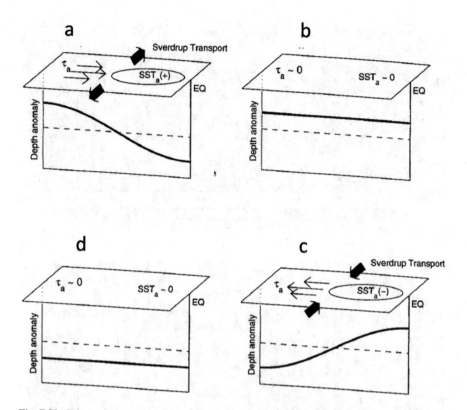

Fig. 7.20 Schematic representation of the recharge-discharge theory of ENSO. (**a**) Warm, (**b**) transition from warm to cold, (**c**) cold, and (**d**) transition from cold to warm phases of ENSO. The dashed line represents the zero line of the thermocline depth anomalies, and the black lines are the thermocline depth anomalies. (Reproduced from Meinen and McPhaden 2000; © American Meteorological Society. Used with permission)

Niño increases. However, Fedorov et al. (2014) find that when the upper ocean heat content in the Tropical Pacific is near normal, then westerly wind burst events may result in CP-type El Niños.

7.4.4 Advective-Reflective Oscillator Theory

This is another theory to describe ENSO (Picaut et al. 1997), which empha-sizes the role of anomalous zonal currents in the Equatorial Pacific modulating the zonal extent of the warm pool. According to this theory, it recognizes the upwelling, westward-propagating Rossby and downwelling, eastward-propagating Kelvin waves that originate from the Central Pacific where the wind stress anomalies are imposed. These downwelling Kelvin waves serve as a positive feedback mech-

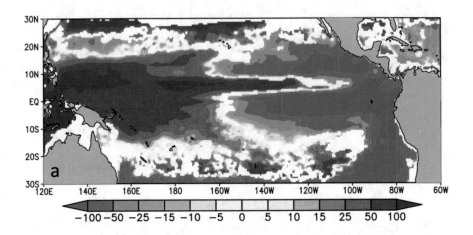

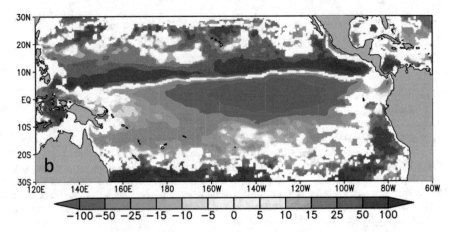

Fig. 7.21 (**a**) The contemporaneous regression of thermocline depth anomalies (m/°C) on the Niño3.4 SSTA. (**b**) The regression of the thermocline depth anomalies (m/°C) on Niño3.4 SSTA anomalies with Niño3.4 SSTA lagging by 9 months. The thermocline depth and Niño3.4 SST anomalies are from SODA3.4.1, and the regression is computed over the period of 1980–2015. The thermocline depth is diagnosed as the depth of the 20 °C isotherm. Only significant values at a 95% confidence interval according to the t-test are shaded

anism with the eastward anomalous zonal currents advecting warm Western Pacific waters eastward. However, as these waves reach their respective boundaries, they reflect as downwelling Rossby wave in the eastern boundary and upwelling Kelvin wave in the western boundary that tends to push the warm pool back to its original position in the western boundary by way of advection by anomalous westward zonal currents. In addition, Picaut et al. (1997) also recognize the role of the mean zonal currents in the Equatorial Pacific in the advection of the eastern boundary of the warm pool. In contrast to the delayed oscillator theory, this theory lays less emphasis on the vertical advection and entrainment for setting up the coupled ENSO system, and more emphasis is laid on the zonal advection. Furthermore, this theory adds more significance to the equatorial wave reflection on the eastern boundary in contrast to the delayed oscillator theory which attaches more importance to wave reflection in the western boundary. Picaut et al. (1997) argue that in their model, they can simulate ENSO-like oscillation with very little western boundary reflection of equatorial waves, while they found it impossible to obtain these oscillations with even moderate changes to the boundary reflection in the eastern boundary compared to their control integration.

7.4.5 The Western Pacific Oscillator Theory

Weisberg and Wang (1997) proposed this theory, which emphasized the role of the western Pacific Ocean in the evolution of ENSO. According to this theory, an atmospheric Rossby wave response to the anomalous condensational heating during ENSO over West-Central Pacific creates a pair of cyclones on either side of the equator, which generates westerly wind anomalies on the equator. These westerly wind anomalies act to deepen the thermocline, warm the SST, and provide a positive feedback to the SST growth in the East Pacific. In the meanwhile, the off-equatorial cyclones, which are west of the condensational heating, raise the thermocline through Ekman pumping and cool the SST and raise the SLP in the Western Pacific. As El Niño matures, the off-equatorial anomalous anticyclones now induce easterly wind anomalies in the Western Pacific that cause upwelling, raise the thermocline, and cool the SST eastward, allowing the system to oscillate.

7.4.6 The Unified Oscillator Theory

With all of these several theories of ENSO, which are capable of producing ENSO-like oscillation, Wang (2001) recognized that more than one theory could operate in nature. Wang (2001) proposed the unified oscillator theory that combines coupled air-sea interactions of anomalous SST in the Eastern Pacific, zonal wind stress anomalies in the Central and Western Pacific, and the off-equatorial thermocline depth anomalies in the Western Pacific to explain ENSO. Alternatively, it may be noted that the other, earlier, ENSO theories discussed in this section are posited as special cases of the unified oscillator theory.

7.5 ENSO Asymmetry

Two aspects of ENSO asymmetry need attention: (1) the differences in the amplitude of the SSTA between El Niño and La Niña events and (2) the differences in the

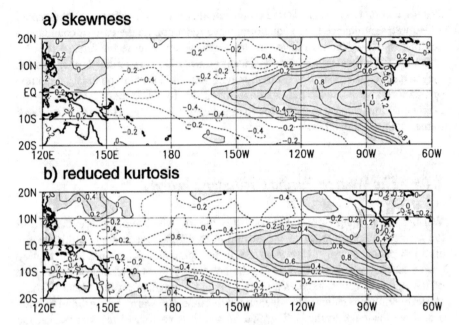

Fig. 7.22 The spatial distribution of (**a**) skewness (S) and (**b**) kurtosis (K) of the observed monthly mean SST over the period of 1950–1997, where $S = \dfrac{\frac{1}{N}\sum_{i=1}^{N}(x_i-\bar{x})^3}{\left(\sqrt{\frac{1}{N}\sum_{i=1}^{N}(x_i-\bar{x})^2}\right)^3}$ and $K = \dfrac{\frac{1}{N}\sum_{i=1}^{N}(x_i-\bar{x})^4}{\left(\sum_{i=1}^{N}(x_i-\bar{x})^2\right)^2} - 3 - 1.5S^2$, x is SST, and N is the total sample size. (Reproduced from Burgers and Stephenson 1999)

duration of El Niño and La Niña events. Additionally, Kessler (2002) points out that the transitions from La Niña to El Niño are slower than those from El Niño to La Niña. In terms of the seasonal cycle, both El Niño and La Niña typically develop in the late spring-summer period of the year and reach their peak toward the boreal winter season. This seasonal phase-locking feature of ENSO variations peaking during the equatorial cold season when the SST gradients are the strongest suggests the importance of the Bjerknes feedback mechanism for both El Niño and La Niña.

Historically, El Niño has drawn more attention than La Niña because of the more severe climate impacts of the former compared to the latter. However, the climate impacts of the strong 1988–1989 La Niña event brought researchers to recognize that El Niño and La Niña are complementary phases of ENSO. Clear evidence of the ENSO asymmetry is shown in Fig. 7.22, which shows the skewness and kurtosis of Tropical Pacific SSTA. It is seen that both skewness and kurtosis decrease in magnitude from East to West Pacific. In the East Pacific, the skewness is positive (suggesting that the positive deviation about the mean is relatively large and frequent), while in the West Pacific, it is negative (suggesting that the

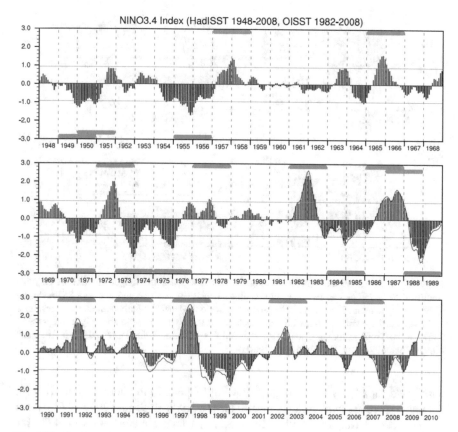

Fig. 7.23 The time series of the Niño3.4 SSTA from HadISST (color bars; 1948–2008) and OISST (black line; 1982–2008). The horizontal color lines indicate one standard deviation of the Niño3.4 index. (Reproduced from Okumura and Deser 2010; © American Meteorological Society. Used with permission)

negative deviation about the mean is comparatively large and frequent). Similarly, the kurtosis is negative in the Western Pacific (suggesting that the SST distribution has lighter tails and departures from the mean have smaller peaks), and kurtosis is positive in the Eastern Pacific (suggesting that the SST distribution has heavier tails and departures from the mean have higher peaks). In the Central Pacific (150°W–160°W), both skewness and kurtosis are near normal distribution values.

Alternatively, the ENSO asymmetry can also be explained by Fig. 7.23, which displays the time series of the Niño3.4 SSTA, where the two asymmetrical features of ENSO are highlighted: the peak amplitudes of SSTA during El Niño are generally higher than the complimentary La Niña phase, and the duration of La Niña is usually longer (~2 years) than El Niño (~<1 year). During this comparatively prolonged cold La Niña phase, the SSTA re-intensify each winter as evidenced by the sawtooth shape of the time series in Fig. 7.23. In contrast, a majority of the El Niños

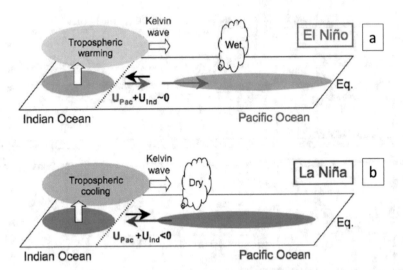

Fig. 7.24 A schematic diagram illustrating the mechanism for the longer duration of La Niña compared to El Niño. The pink and blue ellipses depict the positive and negative SSTA during El Niño and La Niña. The red and blue arrows indicate the surface wind anomalies as a response to the Equatorial Pacific SSTA, while the black arrow is the surface wind anomaly in response to SSTA in the Tropical Indian Ocean. (Reproduced from Okumura and Deser 2010; © American Meteorological Society. Used with permission)

terminated rapidly by the summer following its peak in the boreal winter season. Although there are two prolonged warm phases (1990–1995 and 2002–2005), none of the warm phases displayed Niño3.4 SSTA, which remained above 0.4 °C for more than 2 years.

The asymmetric feature of the varied duration of El Niño and La Niña is explained in the schematic shown in Fig. 7.24, following Okumura and Deser (2010). They attribute this asymmetric feature of ENSO to the influence of the Tropical Indian Ocean. As discussed earlier, in connection with the tropospheric temperature anomalies in Sect. 7.3.4, the basin-wide warming of the Indian Ocean results with the warm phase of the El Niño. This warming of the Tropical Indian Ocean causes a further increase in tropospheric temperatures (by moist adiabatic adjustment as the free atmosphere lapse rate is tied to the boundary layer equivalent potential temperature) and forces a baroclinic Kelvin wave into the Western Pacific during an El Niño (Fig. 7.24a). The lower troposphere easterly wind anomalies forced by this Kelvin wave over the Western Pacific counteract the prevailing anomalous westerly wind anomalies that mark the beginning of the termination of El Niño (following the Western Pacific oscillator mechanism). A similar mechanism prevails during La Niña, as the Kelvin wave forced by the basin-wide cooling of the Tropical Indian Ocean induces westerly wind anomalies over the Western Pacific (Fig. 7.24b). However, this westerly wind anomaly forced from the Indian Ocean is far smaller compared to the prevailing easterly wind anomalies in the Western

Pacific as a result of the large precipitation anomalies over the Western Pacific. Therefore, this enhanced duration of the zonal wind anomalies in the Western Pacific during La Niña results in their longer duration than El Niño. The reason for the disproportionate changes in precipitation anomalies between La Niña over the Western Pacific and El Niño precipitation anomalies over Central Pacific is because of the nonlinear relationship between precipitation and local SST (cf. Fig. 5.4). It may be noted that there is a large zonal shift in the rainfall anomalies from Central to Western Pacific as a result of the nonlinear dependence of atmospheric convection to SSTs. Small changes in absolute values of SST in the already very warm pool region of the Pacific during La Niña compared to relatively cooler SSTs in the Central Pacific lead to this disproportionate response in the precipitation and hence in the wind anomalies of the lower troposphere. Hoerling et al. (2001) suggest that SSTA and atmospheric deep convection response lack spatial asymmetry during weak (warm or cold) ENSO events. Similarly, warm ENSO events that do not develop until late summer or fall tend to be weaker and persist longer into the second year (Horii and Hanawa 2004). Sooraj et al. (2009) contend that the late onset of warm ENSO may delay and weaken the Indian Ocean SST response and thereby delay the remote negative feedback to the Western Pacific.

There are, of course, other theories to explain this asymmetry in ENSO. For example, Meinen and McPhaden (2000) report that positive anomalies in the warm water volume are associated with larger SSTA than negative anomalies of comparable magnitude in the warm water volume. Dommenget et al. (2012) suggest that the nonlinear response of the zonal wind stress anomalies to SSTA, with La Niña wind stress anomalies shifted further westward than El Niño wind stress anomalies, leads to asymmetry of ENSO. Su et al. (2010) on the other hand suggest that the nonlinear zonal and meridional temperature advection by ocean currents can lead to asymmetry of ENSO. Im et al. (2015) provide evidence to suggest that the dynamical ocean response (manifesting in thermocline displacements) per unit zonal wind stress is larger for warm compared to cold ENSO phases. Some of these issues are further discussed in the context of the symmetry between Atlantic Niño I and Atlantic Niña I in Sect. 8.3.3.

Chapter 8
Tropical Atlantic Variations

Abstract Tropical Atlantic Ocean is the smallest of the three tropical ocean basins. However, it has many distinct types of variations that manifest beyond the tropical latitudes of the Atlantic. The chapter discusses some of these features including the phenomenon of the Atlantic Niño, Benguela Niño, and the Atlantic Meridional Mode.

Keywords Benguela Niño · Atlantic meridional mode · Atlantic Niño · Nordeste · Angola-Benguela frontal zone · static stability · North-Atlantic Oscillation

8.1 A Broad Overview

The Tropical Atlantic variations are an amalgam of many distinct types of variations at interannual (periodicity of ~2–5 years), intra-decadal (periodicity of ~5–10 years), and inter-decadal (>10 years) temporal scales. Some of these have tropical origins, and others have fingerprints of higher latitude oscillation in the Tropical Atlantic. Nonetheless, it must be recognized that the Tropical Atlantic Ocean is unique in that it is a relatively small ocean basin compared to the Tropical Pacific Ocean and the Tropical Indian Ocean. The Tropical Atlantic Ocean is flanked by two large continents that have large hydroclimatic variations (e.g., the Sahel [or the sub-Saharan] region in West Africa and Amazon in South America), which has an overwhelming influence on the variations in the Tropical Atlantic Ocean (Chang et al. 2006).

The seasonal cycle dominates the Tropical Atlantic variations. The Equatorial Atlantic Ocean also has El Niño-like variability on the interannual scales often referred to by different names such as Atlantic Niño I, the Atlantic Equatorial Mode, the Zonal Mode, and the Equatorial Cold Tongue Mode (Merle 1980; Zebiak 1993). Off the coast of Angola in Benguela (~10°S), there is another El Niño-like oscillation, which is commonly called Benguela Niño or Atlantic Niño II (Shannon et al. 1986). The AMM is the cross-equatorial meridional gradient of SSTA in the Tropical Atlantic that displays variations at interannual and intra-decadal timescales

© Springer Nature Switzerland AG 2023
V. Misra, *An Introduction to Large-Scale Tropical Meteorology*, Springer Atmospheric Sciences, https://doi.org/10.1007/978-3-031-12887-5_8

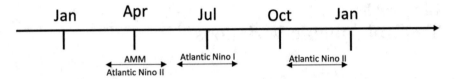

Fig. 8.1 A schematic of the time of the seasonal peaks of the various oscillations in the Tropical Atlantic Ocean as evidenced in the SSTA. The AMM and Atlantic Niño I have their peaks in the MAM and JJA seasons, respectively. The Atlantic Niño II is known to occur in either MAM or in October–November-December period

(Moura and Shukla 1981). The AMM is also referred to sometimes as the Atlantic Dipole Mode or the Gradient Mode or the Interhemispheric Mode.

Interestingly, all these oscillations have a distinct seasonality (Fig. 8.1). For example, as illustrated in Fig. 8.1, AMM and Atlantic Niño I have their peak variability in the MAM and JJA seasons, respectively. Atlantic Niño II variations are known to occur in MAM and sometimes later in the year in the October–December period.

8.2 Seasonal Cycle

The seasonal cycle in the Equatorial Atlantic is significant (Figs. 8.2, 8.3, and 8.4). The SST in the Eastern Equatorial Atlantic reaches their maximum and minimum in MAM and JAS seasons, respectively (Fig. 8.2a). In July and August, a distinct cold tongue forms, centered slightly south of the equator, which is called the eastern equatorial tongue. The seasonal cycle of the thermocline in the Equatorial Atlantic undergoes similar variations, with the Eastern Equatorial Atlantic deepening in the boreal spring season and shoaling in the JAS season. We find that similar seasonality with comparable magnitudes of the SSTA exists over the Eastern Equatorial Pacific Ocean (Fig. 8.2b). In the Equatorial Atlantic, the seasonal SSTA is more zonally oriented (Fig. 8.2a), unlike the weak westward propagation of the warm SSTA evident in the Eastern Equatorial Pacific Ocean (Fig. 8.2b). The seasonal cycle of the Equatorial Atlantic seasonal cycle of the zonal (Fig. 8.3a) and meridional (Fig. 8.4a) wind stress suggests the weakening and the strengthening of the equatorial southeasterly trade winds in the MAM and the JAS seasons, respectively. In the Equatorial Pacific Ocean, the seasonal cycle of the southeasterly trade winds is stronger in the eastern part of the basin (Figs. 8.3b and 8.4b), while the stronger seasonal variations of the zonal and the meridional wind stress are in the Western Equatorial Atlantic Ocean (Figs. 8.3a and 8.4a).

The ITCZ over the Atlantic is strongly tied to the warmest SST band, which implies that the ITCZ is closest to the equator during the boreal spring season. The Atlantic ITCZ reaches its northernmost position in September. Intriguingly, the convection in the ITCZ is stronger in July–August relative to the March–April time

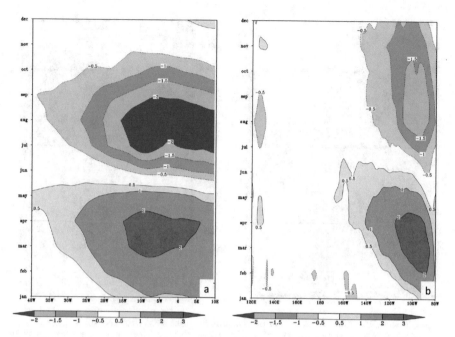

Fig. 8.2 Longitude-time cross-section of climatological observed SST (°C) over Equatorial (**a**) Atlantic and (**b**) Pacific Oceans (averaged between 2°S and 2°N)

frame even though the in situ SSTs are warmer in the latter period by ~1 °C. Some of this can be explained by the seasonality of the greater frequency of African easterly waves in the boreal summer season that can trigger convection in the relatively warm ocean.

The seasonal cycle exhibited by the Equatorial Atlantic Ocean is highly asymmetric. For example, at 10°W over Equatorial Atlantic, SST reaches 28 °C in March–April, evolving over a period of 7 months but drops to 23 °C in July–August in a time period of 3 months (Xie and Carton 2004). Xie (1994) argues that the meridional migration of the ITCZ, north of the equator, is the principal cause for the seasonal cycle in the Eastern Equatorial Atlantic and the Eastern Equatorial Pacific Oceans. The ITCZ modulates the southerly cross-equatorial winds, which intensify in the boreal summer/fall season and relax in the boreal spring season. Okumura and Xie (2004) indicate that the zonal wind in the Equatorial Atlantic is affected by the West African Monsoon in the east and by the seasonal cycle of the SST in the west. The rapid cooling of the equatorial cold tongue in the Eastern Atlantic is a result of the abrupt onset of the West African Monsoon, which causes rapid intensification of the southerly winds in the Gulf of Guinea, causing significant upwelling of cold waters slightly south of the equator and downwelling slightly north of the equator. The thermocline is known to shoal by about 60 m from April to August in the Equatorial Gulf of Guinea (Houghton 1983).

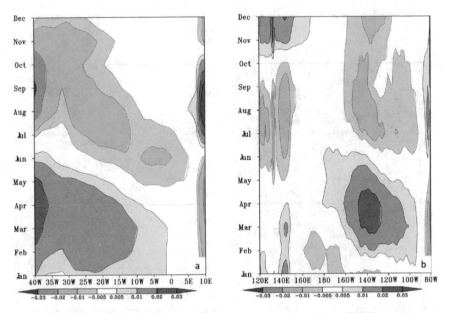

Fig. 8.3 Longitude-time cross-section of climatological observed (from NCEP R2) zonal wind stress (N/m^2) over Equatorial (**a**) Atlantic and (**b**) Pacific Oceans (averaged between 2°S and 2°N)

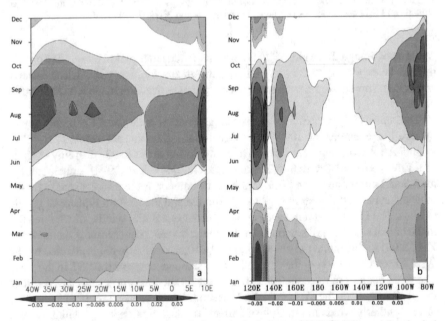

Fig. 8.4 Longitude-time cross-section of climatological observed (NCEP R2) meridional wind stress (N/m^2) over Equatorial (**a**) Atlantic and (**b**) Pacific Oceans (averaged between 2°S and 2°N)

8.3 Atlantic Niño I

8.3.1 The Structure of Atlantic Niño I

The Atlantic Niño I mode, just like the ENSO in the Equatorial Pacific Ocean, appears as a coupled ocean-atmosphere phenomenon (Zebiak 1993). Ruiz-Barradas et al. (2000) isolated this mode elegantly by conducting a five-variable (viz., ocean heat content, sea surface temperature, zonal and meridional wind stress, and atmospheric diabatic heating anomalies at 500 hPa) combined rotated EOF analysis, which resulted in producing the Atlantic Niño I mode that explained 4.9% of the total variance of the Tropical Atlantic Ocean (Fig. 8.5). The spatial pattern of this Atlantic Niño I mode shown in Fig. 8.5a is consistent with anomaly patterns one would obtain by compositing warm events of the Atlantic Niño I over several decades (e.g., 1963, 1966, 1968, 1973–1974, 1981, 1984, 1987, 1988; Ruiz-Barradas et al. 2000).

The description of a canonical Atlantic Niño I event is as follows: (a) The warm Atlantic Niño I event manifests with the warming of the SST in the eastern part of the basin (Fig. 8.5c), which is (b) coincident with the relaxation of the trade winds and development of westerly wind stress anomalies in the western part of the basin (Fig. 8.5b), and (c) a deepening of the thermocline in the eastern part of the basin and shoaling in the western Atlantic, which results in higher and lower heat content anomalies in the eastern and western equatorial Atlantic Ocean (Fig. 8.5a), respectively. The diabatic heating pattern in Fig. 8.5d associated with this mode indicates that the ITCZ is shifted southward of its climatological position and is closest to the equator. This latitudinal shift of the ITCZ is consistent with the warming of the surface waters that result in the disappearance of the equatorial cold tongue in the summer during an Atlantic Niño I event. The spectral analysis of the time series of the principal component shown in Fig. 8.5e yields a spectral peak at interannual scales of around 2 years and an interdecadal oscillation that reflects the nonstationary behavior of the Atlantic Niño I mode (Losada and Rodriguez-Fonseca 2016).

The Atlantic Niño I mode is seasonally phase locked, with its peak variance in the boreal summer season when the cold Equatorial Atlantic tongue with its 23 °C appears seasonally. Ruiz-Barradas et al. (2000) indicate that the Atlantic Niño I mode explains 7.5% of the total Tropical Atlantic variance, if the EOF analysis is conducted over the summer period only.

8.3.2 Theories for Atlantic Niño I

The various mechanisms to explain Atlantic Niño I are summarized in Fig. 8.6 following Lubbecke et al. (2018). Zebiak (1993) using a model setup for predicting ENSO was able to adapt it for Atlantic Niño I events, which clearly suggests

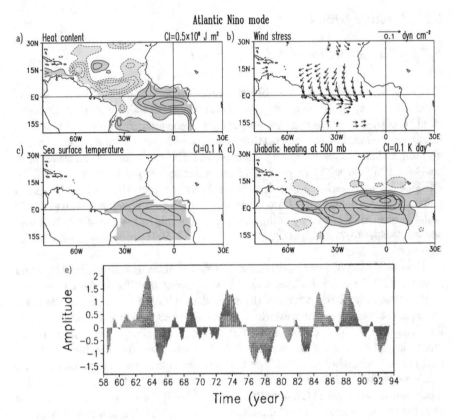

Fig. 8.5 One of the two dominant modes of observed Tropical Atlantic variability isolated from combined five-variables rotated principal component analysis is associated with Atlantic Niño I. The spatial EOF pattern associated with this Atlantic Niño I is shown for (**a**) ocean heat content ($\times 10^8$ Jm^{-2}), (**b**) wind stress (dynes/cm^2), (**c**) SST (°C), and (**d**) diabatic heating at 500 hPa (°C/day) along with the (**e**) time series of the corresponding principal component. (Reproduced from Ruiz-Barradas et al. 2000; © American Meteorological Society. Used with permission)

the applicability of the delayed oscillator theory for Atlantic Niño I (Fig. 8.6a). Keenlyside and Latif (2007) suggest that the three components of the Bjerknes feedback, namely, (1) the forcing of the surface wind anomalies in the western part of the basin (Fig. 8.6a), (2) the forcing of the thermocline anomalies in the Eastern Atlantic by the eastward-propagating Kelvin waves forced by the westerly wind stress anomalies in the Western Atlantic (Fig. 8.6a), and (3) the forcing of the SSTA in Eastern Equatorial Atlantic by the thermocline anomalies, are all active in the Equatorial Atlantic (Fig. 8.6b).

Richter et al. (2013) proposed an alternative theory to account for the noncanonical Atlantic Niño I where warm Niño I events occur in the presence of strong easterly wind anomalies in the western part of the basin. In such noncanonical events, Richter et al. (2013) claim that the advection of warm ocean temperature

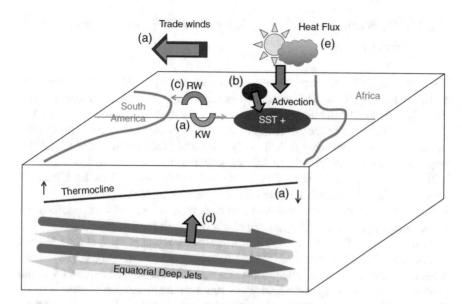

Fig. 8.6 A schematic representing the various proposed mechanisms for the genesis and evolution of the Atlantic Niño I event, which includes (**a**) the Bjerknes feedback mechanism including modulation of the trade winds, adjustment of the equatorial thermocline zonal slope through the propagation of Kelvin waves (KW), (**b**) meridional advection of upper ocean (top 50 m) temperature anomalies, (**c**) reflection of Rossby waves, (**d**) equatorial deep jets, and (**e**) net surface heat flux anomalies. (Reproduced from Lubbecke et al. 2018)

anomalies from Northern Tropical Atlantic is crucial for the development of Atlantic Niño I. A pre-existing, cross-equatorial SST gradient induces a wind stress curl anomaly that modulates the thermocline just north of the Equatorial Atlantic and consequently affects the upper ocean temperature anomalies that get directly advected to the equator (Richter et al. 2013; Fig. 8.6b). Alternatively, Foltz and McPhaden (2010) suggest that these thermocline anomalies induced by the wind curl anomalies north of the Equatorial Atlantic propagate westward as a Rossby wave, which then gets reflected as an equatorial Kelvin wave to trigger an Atlantic Niño I event (Fig. 8.6c).

Brandt et al. (2011a) suggested an Atlantic Niño I mechanism that was intrinsic to the ocean and independent of any atmospheric forcing. They indicated that Equatorial Atlantic deep jets with short vertical wavelengths are known to propagate their energy upward and affect the equatorial zonal current anomalies and SSTA (Fig. 8.6d). In contrast, Nnamchi et al. (2015) propose that atmospheric heat flux anomalies excited by stochastic atmospheric variations can explain a large fraction of the SST variability in the Eastern Equatorial Atlantic Ocean. However, Dippe et al. (2017) claim that ocean dynamics related to Bjerknes feedback is the dominant mechanism for Atlantic Niño I, and the warm bias over the Tropical Atlantic displayed by most climate models leads to an overestimation of the thermodynamical influence on Atlantic Niño I events.

8.3.3   The Symmetry Between Atlantic Niño I and Atlantic Niña I

We have learned from Chap. 6 that El Niños are stronger and shorter in duration and have their maximum SST located farther east than La Niñas. On the other hand, Atlantic Niño I and Atlantic Niña I are relatively far more symmetric (Fig. 8.7). For example, the composite anomalies in Fig. 8.7 show that the evolution and the demise of Atlantic Niño I and Atlantic Niña I and their amplitude are nearly identical and opposite along Equatorial Atlantic Ocean, which is also reflected by the negligible differences between the composites. In Fig. 8.7, we can observe for both Atlantic Niño I and Atlantic Niña I that initial SSTA appear along the Angola coast in April which reaches the cold tongue in May and peak in June–July with an amplitude of ~1 °C before they decay in August and nearly disappear by September.

Lübbecke and McPhaden (2017) contrasted the three components of the Bjerknes feedback mechanism between ENSO and Atlantic Niño I events (Fig. 8.8) by fitting a linear regression line separately for warm and cold events between SSTA in the eastern part of the basin and zonal wind stress anomalies in the western part of the basin (Fig. 8.8a, b), wind stress anomalies in the west leading to thermocline anomalies in the east (Fig. 8.8c, d), and eastern thermocline anomalies inducing in situ SSTA (Fig. 8.8e, f). Their comparison clearly showed that the slopes of the linear regression fit representing the three components of the Bjerknes feedback for Atlantic Niño I are comparable and statistically indistinguishable between the warm and cold Atlantic Niño I events. However, in the case of ENSO, while the first component of the Bjerknes feedback (between zonal wind stress in the west and SSTA in the east) shows symmetry (Fig. 8.8b), the other two components show strong asymmetry (Fig. 8.8d, f). In an interesting calculation, Lübbecke and McPhaden (2017) find that regression slopes of these Bjerknes feedback components become more symmetric for warm and cold ENSO events when the amplitude range of Niño3 SSTA is limited to those found in Atlantic Niño I. In other words, they suggest that the symmetry between Atlantic Niño I and Atlantic Niña I is related to their smaller amplitude of the SSTA.

8.3.4   The Impact of Atlantic Niño I

The Atlantic Niño I has local and remote impacts within and outside of the tropics. Locally, the Gulf of Guinea and the Guinea coast have drier summers because of the warm Atlantic Niño I event because the warm SSTs reduce low-level flow inland, while during Atlantic Niña I events, the land-ocean thermal contrast becomes larger and intensifies the southerly flow over the Guinean coast and the southwesterly flow into West Africa, which leads to wetter anomalies inland (Brandt et al. 2011a, b). Other studies have shown that Atlantic Niño I is associated with a dipole precipitation anomaly, with the region of Sahel having precipitation anomalies of

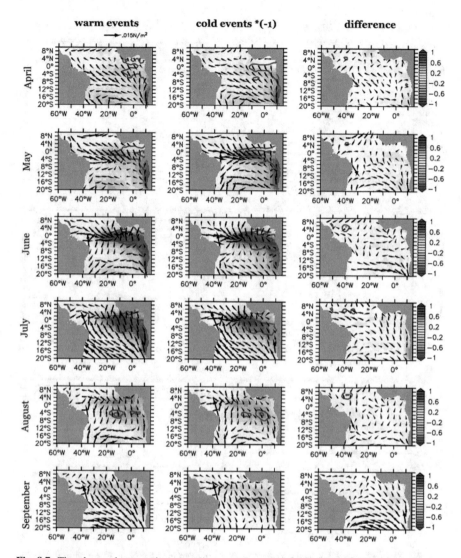

Fig. 8.7 The observed composite anomalies over the period 1958–2009 of SST (shaded), wind stress (vectors), and thermocline depth as measured by the depth of the 20 °C isotherm (blue contours for 3 m and 5 m) for (left column) warm Atlantic Niño I events, (middle column) cold Atlantic Niño I events or Atlantic Niña I events, and (right column) their corresponding difference between warm and cold Atlantic Niño I events. The anomalies in the panels of the middle column are multiplied by −1.0 to make the anomalies comparable to the corresponding panel in the left column. (Reproduced from Lubbecke and McPhaden 2017)

opposite sign than those over Guinea (Losada et al. 2010). There are several studies that suggest Atlantic Niño I has weak influence north of 10°N (Giannini et al. 2005), while other studies point to counteracting influence of SSTA of the Tropical Pacific

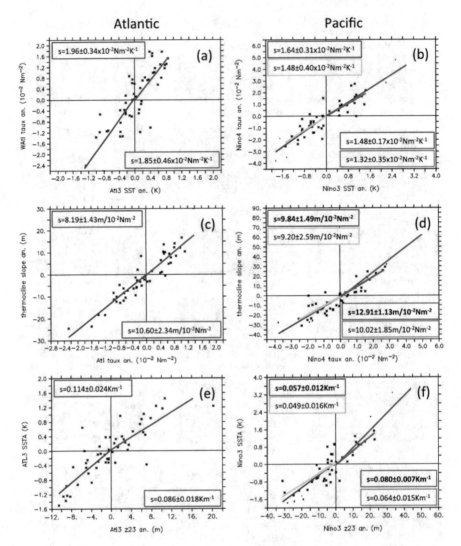

Fig. 8.8 The three Bjerknes feedback components are examined for (**a, c, e**) Equatorial Atlantic and (**b, d, f**) Equatorial Pacific Oceans. Linear regression between (**a**) contemporaneous May ATL3 (20°W–0°E, 3°S–3°N) SSTA and Western Atlantic (40°W–20°W, 3°S–3°N) zonal wind stress anomalies, (**b**) contemporaneous October–November Niño3 SSTA and Niño4 zonal wind stress, (**c**) May western Atlantic zonal wind anomalies and following June Equatorial Atlantic thermocline slope anomalies, (**d**) October–November Niño4 zonal wind stress anomalies and following November–December Equatorial Pacific thermocline slope anomalies, (**e**) contemporaneous June ATL3 SSTA and ATL3 thermocline depth anomalies, and (**f**) contemporaneous November–December Niño3 SSTA and Niño3 thermocline depth anomalies. The thermocline depth is defined by the depth of the 23 °C isotherm (z23). In (**b, d, f**), the red and blue lines show the linear regression fit for all data points (dots), while the magenta and pink lines show the fit for the range of Niño3 SSTA limited to those found in ATL3 (crosses). The regression slopes (s) are indicated in the panels for warm (red or pink) and cold events (blue or magenta) separately, and those with bold font indicate that the slopes are asymmetric. (Reproduced from Lubbecke and McPhaden 2017)

and the Tropical Indian Ocean that weakens this influence of Atlantic Niño I over the West African Monsoon (Losada et al. 2012; Mohino et al. 2011).

Northeast Brazil or Nordeste is strongly impacted by Atlantic Niño I (Nobre and Shukla 1996; Ruiz-Barradas et al. 2000). As mentioned earlier (cf. 8.3.1), the warming of the Equatorial Atlantic Ocean during Atlantic Niño I leads to a more southern latitude location of the ITCZ and consequent seasonal delay in the northward migration of the ITCZ. This leads to wetter anomalies over Nordeste during their rainy season (which is in the boreal spring season). Although Atlantic Niño I peaks following the Nordeste rainy season, it is argued that the anomalies in the Equatorial Atlantic from the evolution in the preceding months to its peak are sufficient to trigger the response over Nordeste (Mohino et al. 2011). There is also some evidence to suggest that Atlantic Niño I can lead to the modulation of the ISM by downstream amplification of equatorial Kelvin waves generated by anomalous convection in the Equatorial Atlantic (Kucharski et al. 2009). Atlantic Niño I can also have influence over the extra-tropics, for example, over Europe (Garcia-Serrano et al. 2011), the Mediterranean region (Cassou et al. 2004), and on the NAO (Drevillon et al. 2003) through the interaction of the mean flow with the generated extra-tropical Rossby waves.

8.3.5   Comparing Atlantic Niño I with ENSO Variations

The similarities between El Niño and Atlantic Niño I are not far too many as their name would suggest. Yes, the ENSO model of Zebiak and Cane (1987) was successfully adapted to simulate Atlantic Niño I (Zebiak 1993), and both the Niños occur over the eastern equatorial oceans, with a visible fingerprint of the Bjerknes feedback. The associated wind stress anomalies in the western part of the equatorial oceans, the triggering of the equatorial waves, and the associated thermocline displacements are quite similar between the two Niños. However, the contrast between them is also quite apparent as outlined in Table 8.1. Needless to say, the community has paid far more attention to predicting ENSO variations because of its relatively stronger impact on global climate anomalies than Atlantic Niño I anomalies. However, there is a growing interest to dive deeper into Atlantic Niño I and understand their more subtle impacts on the global climate.

The relationship between ENSO and Atlantic Niño I is somewhat ambiguous and inconsistent. Some studies suggest that there is a robust connection between the two phenomena (e.g., Latif and Barnett 1995; Saravanan and Chang 2000), whereas other studies claim that there is no significant impact of ENSO on Atlantic Niño I (e.g., Zebiak 1993; Ruiz-Barradas et al. 2000). Chang et al. (2006) examined this fragile relationship between ENSO and Atlantic Niño I variations and found that warm ENSO events of 1982/1983 and 1997/1998 were associated with corresponding strong cold and weak warm Atlantic Niño I events, respectively. They suggest that this fragile relationship is due to the conflicting influences of the

Table 8.1 A comparison between Atlantic Niño I and Eastern Pacific ENSO events

	Atlantic Niño I	Eastern Pacific ENSO
1	There is no Southern Oscillation (atmospheric pressure) counterpart	There is a robust atmospheric response as discerned by the Southern Oscillation
2	The SSTA is relatively small ~|1°C		The SSTA is comparatively large and ~ range between −2 °C and 3 °C
3	The variability (SSTA anomalies in ATL3-SST averaged between 3°S and 3°N and 20°W–0°) peak is in boreal summer season	The variability (SSTA anomalies in Niño3.4) peak is in late boreal fall early boreal winter season
4	The seasonal phase locking is stronger	The seasonal phase locking is comparatively weaker
5	The symmetry between Niño I and Niña I is relatively strong; the skewness of SSTA in ATL3 is ~0 °C (Lübbecke and McPhaden 2017)	There is significant asymmetry between El Niño and La Niña; skewness of Niño3 SSTA is ~ + 1 °C (Lübbecke and McPhaden 2017)
6	The periodicity is short and relatively sharp around of 2 years	The periodicity has a broad range of ~ 2–7 years
7	The impact on local and remote seasonal climate anomalies is more subtle; it is yet to be clearly established as an exclusive basis for seasonal climate prediction for any particular region	The impact on local and remote seasonal climate anomalies is comparatively strong and, therefore in many of these regions, forms the basis for seasonal climate prediction
8	Climate models have significant bias over Equatorial Atlantic (Lübbecke et al. 2018)	Seasonal prediction models are more skillful in predicting ENSO (Smith et al. 2015)

tropospheric temperature mechanism of the ENSO teleconnection (cf. Sect. 7.3.4) in the Atlantic Basin and the local air-sea (Bjerknes) feedback mechanism.

This fragile relationship between ENSO and Atlantic Niño I is explained as follows: By way of the tropospheric temperature mechanism, El Niño warms the tropospheric temperature in boreal winter and spring seasons over the Tropical Atlantic (more in northern Tropical Atlantic where SSTs are warmer than in southern Tropical Atlantic Ocean), stabilizing the environment and therefore reducing the moist convection. The moist convection has the role of redistributing the boundary layer energy to the free atmosphere. But because of reduced convection over the Tropical Atlantic due to El Niño, the energy in the atmospheric boundary layer, which is primarily sourced from the underlying ocean through evaporation, accumulates. Eventually, this accumulation of energy in the boundary produces a 'back pressure' that reduces the evaporation, resulting in the warming of the Tropical Atlantic Ocean, which further reduces the zonal SST gradient. This consequently leads to the weakening of the trade winds that feedback in reducing the evaporation further and warming the SST in the Equatorial Atlantic Ocean. But as Chang et al. (2006) noticed, the observed response in the Equatorial and South Tropical Atlantic Oceans is rather opposite to this feedback. First, they found a strong negative correlation between wind stress in the Western Equatorial Atlantic Ocean (60°W–20°W and the 5°S–5°N) and the Niño3 SST index, which suggests that warm ENSO events produce strong easterly surface wind stress anomalies. This easterly wind stress anomaly then causes cooling in the Eastern Equatorial Atlantic by way of Bjerknes feedback, wherein the thermocline shoals through equatorial wave dynamics and local air-sea feedback. The atmosphere responds to this cooling by further strengthening the easterly wind anomaly. It is this tug-of-war between the tropospheric temperature mechanism and the Bjerknes feedback mechanism that Chang et al. (2006) attribute to the fragile relationship between Atlantic Niño I and El Niño. In some El Niño years (like 1997/1998 El Niño), Chang et al. (2006) found a weaker response of the wind stress anomaly over the Western Equatorial Atlantic Ocean that could explain for the weak warm Atlantic Niño I event from the tropospheric temperature mechanism. In other years, like the 1982/1983 El Niño event, the Bjerknes feedback dominates over the tropospheric temperature mechanism. It is, however, intriguing to note the varied response of the wind stress anomaly over the Western Equatorial Atlantic Ocean to ENSO.

8.4 Atlantic Niño II

The Atlantic Niño II or Benguela Niño is the SSTA variation centered around 15°S along the coasts of Angola and Namibia (Fig. 8.9). They occur in a region called the ABF zone in the southeast Atlantic Ocean (Fig. 8.10). This frontal zone separates warm and saline tropical Angola Current waters (with SST >28 °C in austral summer and early fall season) from the cool Benguela Current upwelled waters off Namibia (where SST <20 °C). This frontal zone at around 15°S also demarcates the tropical ecosystem that is nutrient poor to the north and upwelling-driven, a nutrient-rich ecosystem to the south. The Angola Current, which is the southeast branch of the South Equatorial Counter Current (Fig. 8.10), forms the eastern edge of the cyclonic gyre over the Gulf of Guinea in the upper 100 m that has a distinct seasonal cycle with a southeast advance in the austral summer and fall seasons and a northwest retreat during austral winter and spring seasons. On the other hand, the Benguela Current (Fig. 8.10) is the eastern boundary current of the South Atlantic subtropical gyre. The Benguela upwelling system extends from the southern tip of Africa to the Angola-Benguela frontal system at ~15°S. The Benguela upwelling system shows a distinct seasonality that differs between its southern and northern tips. The upwelling north of 25°S is roughly between March and November with a peak at around August. However, the upwelling season south of 30°S is between September and March. This phase lag has a consequence on the seasonal cycle of SST along the southeastern Atlantic Ocean. In the northern part of the Benguela Current, the seasonal cycle of the upwelling is in phase with the seasonal cycle of solar insolation, and therefore, the SST exhibits a very strong seasonal cycle with warm SSTs appearing in austral summer and cold SSTs in austral winter. But in the southern part of the Benguela Current, the seasonal cycle of the upwelling is out of phase with the seasonal cycle of the solar insolation. Therefore, the SST,

Fig. 8.9 The observed SSTA during the peak of Benguela Niño (or Atlantic Niño II) of March 1995. (Reproduced from Florenchie et al. 2003)

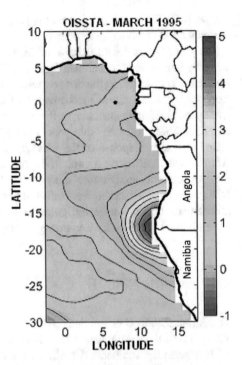

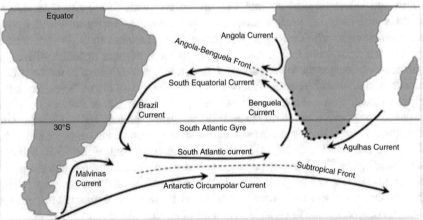

Fig. 8.10 A schematic of the surface ocean current system in South Atlantic Ocean. (Reproduced from Ksepka and Thomas 2012)

for example, near Cape Town in South Africa shows a very weak seasonal cycle (Shannon et al. 1986).

As early as 1983, Hirst and Hastenrath (1983) recognized that seasonal variation of the wind stress in the remote Western Equatorial Atlantic accounted for nearly

23% as opposed to the local wind stress variation that accounted for only 9% of the variance of the SST off the coast of Angola. This led Hirst and Hastenrath (1983) to suggest that this variability in the southeast coast of the Atlantic Ocean is an obvious corollary to El Niño in the Pacific. Similarly, Florenchie et al. (2004) indicated that local air-sea fluxes do not seem to precondition the Angola-Benguela region prior to the development of an Atlantic Niño II event. Florenchie et al. (2003) showed that in the case of Benguela Niños of 1984 and 1995, the perturbations in the thermocline generated in the Western Equatorial Atlantic Ocean propagate along the Equatorial Atlantic as a downwelling Kelvin wave and then propagate further south along the African Coastline before they outcrop at the surface when they reach the upward sloping isotherms of the Benguela upwelling system (Fig. 8.11). The top four panels of Fig. 8.11 show that at the peak of Atlantic Niño II, the SSTA off the coast of Angola-Namibia can be traced out to the ocean temperature anomalies in the Equatorial Atlantic Ocean at a depth of 45 and 75 m below the ocean surface. The time evolution of these ocean temperature anomalies in the bottom three panels of Fig. 8.11 further provides evidence of these anomalies propagating in months prior to the surface manifestation of the Atlantic Niño II SSTA from the equatorial latitudes. These ocean temperature anomalies in Fig. 8.11 also show that unlike Fig. 8.9, the Benguela Niños are not confined to the Angola-Benguela frontal zone but are large-scale events extending from the equator to a latitude of around 20°S along the African west coast. Florenchie et al. (2004) indicate that the surface manifestation of Atlantic Niño II is limited to Angola-Benguela frontal zone region in the March/April time frame. But they can extend further northward in austral winter (in some Atlantic Nino II cases), coinciding with the seasonal peak of the northern Benguela upwelling system, when the thermocline is further closer to the surface.

The variability of the SSTA associated with Benguela Niño has a periodicity of around 10 years (Shannon et al. 1986). Florenchie et al. (2004) introduced an SST index averaged over the Angola-Benguela frontal zone area (19.5°S to 10.5°S and 8.5°E to 15.5°E) and defined a Benguela Niño or Niña when SSTA exceeded |1 °C|. In this way, they were able to identify SSTA in the Angola-Benguela frontal zone area at interannual scales, which were smaller in amplitude than some of the major ones (e.g., Fig. 8.9). Florenchie et al. (2004) note that many of the Atlantic Niño II events seasonally peak in March/April, while nearly equal or more events peak between May and July or later in the year. The Atlantic Niño II has a duration of a few months to 6 months or more, which is on average slightly shorter than the Atlantic Niña II events that have a duration of ~5–8 months (Florenchie et al. 2004). The impact of Benguela Niño is rather varied (Rouault et al. 2003). If the Atlantic Niño II peaks in March/April, then it elevates the ocean temperature beyond the viable temperature range of marine organisms, causing a significant impact on fisheries. Similarly, January–April happen to be the rainy season over southern Angola and northern Namibia and draw significant moisture flux from the southeastern Atlantic Ocean. Therefore, when Benguela Niño reaches maturity in March/April, then it is more impactful during the rainy season over these land masses.

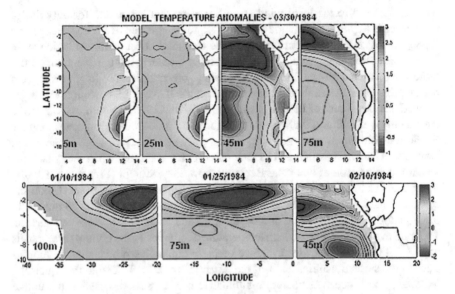

Fig. 8.11 The top four panels show the ocean temperature anomalies on a latitude-longitude plane at four different depths (5 m, 25 m, 45 m, and 75 m) for March 30, 1984, when the SSTA of this Benguela Niño (or Atlantic Niño II) event was at its peak. In order to show that the source of this Benguela Niño event has its origins in the Western Equatorial Atlantic Ocean, the bottom three panels show the time evolution of ocean temperature anomalies at 100 m on January 10, 1984, at 75 m on January 25, 1984, and at 45 m on February 10, 1984. (Reproduced from Florenchie et al. 2003)

From the discussion in Sect. 8.3 and in this section, it should be abundantly clear that Atlantic Niño I and Atlantic Niño II are interrelated. Atlantic Niño II is a manifestation of the interaction between the Benguela upwelling system and the propagating subsurface thermal anomalies from the Equatorial Atlantic Ocean. Therefore, depending on the strength of these interacting components, the Benguela Niño has a far more varied and therefore weaker seasonal phase locking than its Atlantic Niño I counterpart.

8.5   Atlantic Meridional Mode

8.5.1   The Structure of Atlantic Meridional Mode

The early interest in this mode of Tropical Atlantic variability was spurred by its impact on the rainy season rainfall in northeast Brazil (or Nordeste; roughly 11°S–1°S and 46°W–36°W), which is during boreal spring or austral fall time period (Markham and McLain 1977; Hastenrath and Heller 1977; Moura and Shukla 1981). These studies found that when there is a positive or a negative meridional gradient in SSTA across the Equatorial Atlantic Ocean, it results in a wetter or drier rainy season over Nordeste, respectively. Such rainfall variability at interannual scales greatly affects the local agricultural productivity in Nordeste. The AMM was isolated by Ruiz-Barradas et al. (2000) akin to Fig. 8.5 from the same five-variables (viz., ocean heat content, the two components of the wind stress, SSTA, and diabatic heating at 500 hPa) combined rotated EOF analysis, which is shown in Fig. 8.12. This mode of variability (Fig. 8.12) is explained similarly 4.9% of the total Tropical Atlantic Ocean variance. The variations in the surface wind stress anomaly associated with the positive meridional gradient of SSTA of AMM in Fig. 8.12 show a cross-equatorial wind stress anomaly that produces downwelling from the imposed counterclockwise circulation in the northern hemisphere subtropics. This wind stress anomaly not only implies the weakening of the northeasterly trades and strengthening of the southeasterly trades on either side of the equator but also implies Ekman downwelling in the north (and along the equator) and upwelling in the south of the equator. This Ekman downwelling/upwelling feature is reflected in the upper ocean heat content anomalies shown in Fig. 8.12. However, it may be noted that the anomalies of all fields shown in Fig. 8.12 are asymmetrical about the equator.

The diabatic heating anomalies at 500 hPa show that the AMM essentially captures the anomalous meridional migration of the Atlantic ITCZ. The Nordeste rainy season in March through May coincides with the southernmost position of the seasonal cycle of the Atlantic ITCZ. Therefore, when the Northern Tropical Atlantic Ocean warms anomalously, then the ITCZ dislocates further north than usual, producing a drought-like situation in Nordeste and over parts of West Africa (Folland et al. 1986). The time series of the principal component of AMM exhibits a 12-year peak (Fig. 8.12). But there are robust interannual variations as well, which is exploited for seasonal prediction of the Nordeste rainy season (Misra 2006).

8.5.2   Theory for the Atlantic Meridional Mode

It is now clear that the AMM is a coupled ocean-atmosphere mode. Carton et al. (1996) through a series of model experiments showed that the anomalous cross-equatorial SST gradient in the Tropical Atlantic Ocean is dominated by wind-induced changes to the atmospheric heat flux. They showed from separate ocean model integrations that when the interannual variations in the wind stress is ignored but not in the wind effect of the latent heat flux calculations, the model reproduces the AMM manifesting in the cross-equatorial SST gradient. This suggests that the mechanical effects of wind stress (or the momentum flux) are not as important to AMM. Other studies have also confirmed the importance of the wind-induced surface evaporation feedback on AMM (Xie and Tanimoto 1998; Kushnir et al. 2002). This contrasts with Atlantic Niño I variations, where the ocean dynamics forced by the atmospheric momentum flux is important.

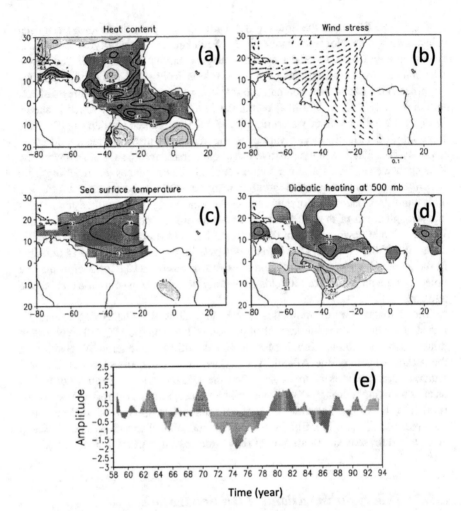

Fig. 8.12 One of the two dominant modes of observed Tropical Atlantic variability isolated from combined five-variables rotated principal component analysis is associated with the AMM. The spatial EOF pattern associated with this AMM is shown for (**a**) ocean heat content ($\times 10^8$ Jm^{-2}), (**b**) wind stress (dynes/cm^2), (**c**) SST (°C), and (**d**) diabatic heating at 500 hPa (°C/day) along with (**e**) the time series of the corresponding principal component. (Reproduced from Ruiz-Barradas et al. 2000; © American Meteorological Society. Used with permission

The importance of the cross-equatorial SST gradient in maintaining the AMM is also recognized by other studies (Moura and Shukla 1981; Chang et al. 1997). For example, a positive gradient of the SSTA associated with AMM sets up a southward pressure gradient in the atmospheric boundary layer, which induces cross-equatorial winds that weaken the northeast trades north of the equator and strengthen the southeasterly trades south of the equator. In turn, these modulated trades reduce or increase the surface evaporation off the ocean surface north or south of the equator, respectively, thereby further strengthening the initial anomalous SST meridional gradient.

In addition to this positive feedback, the interaction between SST and clouds is also considered to be another plausible mechanism for AMM. The observational study of Tanimoto and Xie (2002) indicates that in association with the positive cross-equatorial SST gradient of the AMM, there is an increase and a decrease in cloudiness at ~5°N and ~ 5°S, respectively. These changes in near equatorial cloudiness are a consequence of the northward shift of the ITCZ, which also manifests in the anomalous convergence and divergence of the surface winds at ~5°N and ~ 5°S, respectively. This increased cloudiness at ~5°N is a result of a corresponding increase in high clouds (deep cumulonimbus and cirrus) from enhanced convective activity over the warmer SST. Likewise, the reduction in cloudiness at ~5°S is a result of suppressed convective activity with a corresponding decrease in high clouds over relatively cooler SST. However, Tanimoto and Xie (2002) argue that this SST-cloud feedback in the equatorial region of the Atlantic is not significant in damping the meridional SST gradient of the AMM. They claim that the anomalous SST gradient of the AMM is associated with SSTA centered around 15–20°N/S. They also find cloud anomalies in the subtropics poleward of 10° latitude in both hemispheres, which in contrast to the equatorial cloudiness is a result of low cloud (stratus and stratocumulus) anomalies. These low cloud anomalies are not associated with surface wind convergence and are negatively correlated with in situ SSTA. In other words, higher and lower amounts of low clouds form over regions of relatively colder and warmer SSTA in the subtropics south and north of the equator, respectively. The relatively warmer atmospheric boundary layer in the subtropical regions of warmer SSTA results in reducing the atmospheric static stability, which is a condition that disfavors the formation of stratus clouds. In contrast, in the southern subtropical Atlantic Ocean, the cooler atmospheric boundary layers favor the formation of the stratus clouds. The low cloud-SST feedback is positive unlike the high cloud-SST feedback that is negative. Tanimoto and Xie (2002) find that the shortwave forcing of the low clouds is the dominating influence on the SSTA. Thus, in the south subtropical Atlantic Ocean, more low clouds form over negative SSTA, reducing net radiation into the ocean's mixed layer, which causes more cooling and a further increase in low clouds.

8.5.3   The Influence of ENSO

The diagnosis of the influence of ENSO on Tropical Atlantic variability has led some to suggest that the variability of the northern and the southern branches of the AMM is independent of each other (Nobre and Shukla 1996; Enfield and Mayer 1997; Saravanan and Chang 2000; Misra 2006). Enfield and Mayer (1997) showed that the ENSO variability is strongly, and positively correlated with SSTA in the Tropical North Atlantic (~10–20°N) with a lag of about a season. This correlation with ENSO in the Tropical North Atlantic is most intense and extensive in the area during the April–May-June season. The mechanism for this interannual teleconnection between ENSO variability and the Tropical North Atlantic Ocean following Enfield and Mayer (1997) is illustrated in Fig. 8.13 and is as follows:

- In the boreal winter season, when ENSO peaks, the northeast trades in the Atlantic accelerate, SSTA cools in the Tropical North Atlantic, and there is a corresponding rise in sea level pressure (Fig. 8.13a).
- The resulting anomalous pressure gradient drives the southeasterlies south of the equator to cross into the NH and develop into southwesterlies (Fig. 8.13b–f).
- These anomalous winds are particularly strong in the western part of the Northern Tropical Atlantic, weakening the prevailing northeasterly winds and consequently reducing the latent and sensible heat flux from the ocean. This reduction in heat flux from the ocean causes warming of the SST including the ocean mixed layer with a delay that peaks in April–May-June period between ~10°N and 20°N (Fig. 8.13c–f). Saravanan and Chang (2000) also indicate that in addition to wind-induced changes to surface heat flux, changes in the air-sea difference in humidity and temperature also become important in the warming of the Tropical North Atlantic SSTA following an ENSO event.
- The southeasterly winds south of the equator, which are influenced by ENSO, further enhance the prevailing southeasterly trades and cause further cooling of the local SSTA in conjunction with increased surface evaporation (Fig. 8.13d–f). This in turn enhances the cross-equatorial gradient of the Atlantic SSTA in the boreal spring following an ENSO event. But the correlations of the SSTA in the Southeast Atlantic Ocean with ENSO are relatively much weaker than those in Northern Tropical Atlantic Ocean.
- In the subsequent months after the peak in this teleconnection, the remote ENSO forcing wanes and the correlations with Northern Tropical Atlantic Ocean diminish and even become weakly positive.

As noted earlier, the interest in the interannual variations of the AMM was evoked from the discovery of its influence on the Nordeste rainy season (Hastenrath and Greischar 1993; Nobre and Shukla 1996; Folland et al. 2001; Giannini et al. 2004; Misra 2006). Giannini et al. (2004) further suggested the preconditioning role of the AMM in modulating the ENSO influence of the Nordeste rainy season. They suggested that when AMM evolved consistent with ENSO, i.e., warm or cold ENSO is associated with positive or negative cross-equatorial SST gradient in the Tropical Atlantic Ocean, then AMM and ENSO forcing adds up to force large rainfall anomalies over Nordeste in the boreal spring season, respectively. On the other hand, if the AMM were to develop with negative or positive cross-equatorial SST gradient in the Tropical Atlantic Ocean with warm or cold ENSO episodes, respectively, then they found the Nordeste rainfall in the boreal spring season was far more unpredictable because of the destructive interference from the two external sources. Saravanan and Chang (2000) and Misra (2006) suggest that the teleconnection pattern of the Nordeste seasonal rainfall in the boreal spring season with dipole-like pattern of the SSTA in the Tropical Atlantic Ocean disappears when the largest ENSO events are removed from the time series, which made them to suggest that, at

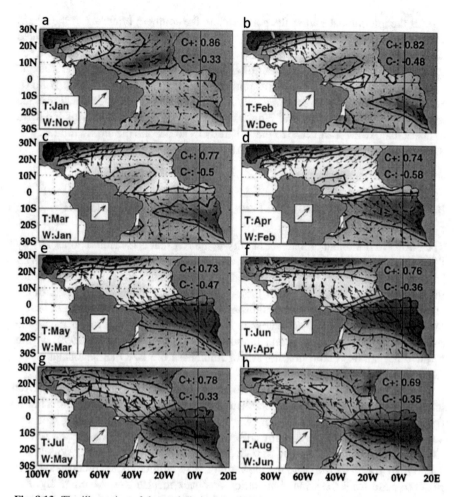

Fig. 8.13 The illustration of the evolution of the ENSOteleconnection with the Atlantic Ocean. The ENSO variations are represented by the principal component time series of the first EOF of the SST over the Pacific basin that clearly isolates the ENSO variations. The Atlantic SSTA is represented by 3-month running averages centered on the months of January through August and lag the ENSO index by 4 months. The lightest and the darkest shadings correspond to the most positive and negative correlations of the SSTA with the ENSO index, respectively. The contour intervals are $-0.25, 0.25$, and 0.5 with the zero contour line dashed. The contemporaneous correlations of the ENSO index with the wind vectors are shown by arrows for the months of November through June, which is 2 months prior to the corresponding SSTA. The inset with 'T' and 'W' represents SSTA and winds followed by the months for which the correlation is shown in each panel. The inset with the wind vector over South America represents the scale for a vector wind correlation of 0.5. The inset statistics show the maximum positive and negative cross-correlation (CC) between Tropical Atlantic and ENSO index. (Reproduced from Enfield and Mayer 1997)

least at the interannual scales, the northern and the southern branches of the AMM vary independently of each other. However, Ruiz-Barradas et al. (2000) repeated their combined EOF analysis on the five variables (namely, ocean heat content, sea surface temperature, zonal and meridional wind stress, and atmospheric diabatic heating anomalies at 500 hPa) after removing all their covariations with Niño3.4 SSTA time series and still recovered the AMM signal, albeit with the reduction in the magnitude of the anomalies and reduced cross-equatorial gradients. This led them to claim that AMM is intrinsic to the Atlantic Ocean despite the strong influence of the interannual variations of ENSO on AMM.

8.5.4   The Relationship with the North Atlantic Oscillation

The NAO is a dominant climate variability in the NH that is measured nominally by the anomalous sea level pressure gradient between the Icelandic Low (in Reykjavik, Iceland) and the NASH (in Ponta Delgada, Azores). There are a number of studies that have tried to link NAO with the AMM with conflicting conclusions. Some studies suggest that there is no relationship between AMM and NAO (Mehta 1998). Other studies claim that NAO can excite forcing in the tropics that forces the dipole pattern of SSTA of AMM (Xie and Tanimoto 1998; Tanimoto and Xie 1999). While a few other studies claim that the AMM influences the NAO (Rajagopalan et al. 1998; Ruiz-Barradas et al. 2000).

In studies where NAO is claimed to affect AMM, it is argued that the NAO affects the northeast trade winds and therefore the underlying SSTA. Using a linear dynamic ocean-atmosphere model, Xie and Tanimoto (1998) produced decadal variations with a pattern suggesting cross-equatorial SSTA gradients like the AMM. They reproduced this result by applying external forcing of weather noise (i.e., random in time) to the model poleward of 20° latitude. They interpreted this result to arise from the wind-evaporation feedback mechanism. Alternatively, Xie and Tanimoto (1998) used observed wind forcing poleward of 20°N/S, while equatorward of this latitude, they did not use any wind forcing in their coupled model integration. This model integration also restored the anomalous cross-equatorial SST gradient in the tropics with its decadal variations that resemble AMM, strongly implying that NAO-like variations could produce AMM. Dong and Sutton (2002), on the other hand, suggest that changes in the deep water formation in the northern Atlantic can modulate the AMM. For example, a sudden weakening of the AMOC in their model integration yielded strong cooling and weak warming in the northern and the southern subtropical Atlantic Oceans that resembled the AMM pattern. This consequently elicited an atmospheric response by shifting the ITCZ further south than usual and thereby further strengthening the dipole structure. Their study is also suggestive of the fact that through these coupled responses, the changes in deep water circulations like the AMOC can be felt in a few years.

Rajagopalan et al. (1998) found a strong coherence between the NAO and the cross-equatorial SST gradient of the AMM in the 8–20-year period, suggesting the potential for significant mid-latitude-tropical interaction. On the other hand, Ruiz-Barradas et al. (2000) found no relationship between NAO and AMM. Some of the reasons for these conflicting results are the period of the analysis is quite different in the two, with the former using a 136-year period from 1865 to 1991, while the latter used a 44-year period from 1950 to 1993. Furthermore, the sources and types of datasets and the methodology used in the analysis were also different in the two studies. Nonetheless, Ruiz-Barradas et al. (2000) when projecting the normalized NAO SLP index and the AMM principal component time series (Fig. 8.12) on the Atlantic SSTA from 20°S to 60°N find that the former has no significant projection in the tropical latitudes while the latter showed its fingerprint very clearly extending into the NH mid-latitudes. This led Ruiz-Barradas et al. (2000) to suggest that AMM has an impact on the NAO variations.

Chapter 9
Large-Scale Aspects of Tropical Weather Extremes

Abstract The current importance of attributing weather extremes to large-scale climate variations like the intraseasonal variability, El Niño, and the Southern Oscillation is discussed in the chapter. A theory for understanding the intensity variations of Atlantic tropical cyclones from large-scale variations attributed to energy transformations between velocity potential and streamfunction fields is discussed. The observed features of tropical thunderstorms from remotely sensed observations are also detailed concluding with the diurnal features of tropics.

Keywords Thunderstorms · Tornadoes · Pacific decadal oscillation · Available potential energy · Kinetic energy · Streamfunction · Velocity potential · Ideal gas law · Irrotational wind · Non-divergent wind · Easterly wave · Gangetic Plains · Diurnal variations · Conditional instability · Lifting condensation level · Semidiurnal variation · Hurricanes · Typhoon · Lightning · Accumulated cyclone energy · Convectively available potential energy

9.1 Attribution Science

Attributing factors to the occurrence of an extreme event is an active field of research. This research activity is stirred by questions that are often asked if the extreme event occurred as a result of climate change. The definition of 'extreme' is quite varied in climate science. In some instances, it could refer simplyto the occurrence of, say, daily temperature or precipitation that exceeds some (e.g., 90th, 95th, 99th) percentile of daily variability as estimated from a climatological base period. In other instances, an extreme event could be based on recurrence interval (e.g., 50 years, 100 years, etc.). Extremes are also contextual. For example, seasonal and annual mean anomalies can be construed as extremes just as short-term weather events. Similarly, TCs and tornadoes are also considered extreme as a class. Field et al. (2012) adopted a somewhat universal definition of extreme to be 'the occurrence of a value of a weather or a climate variable above or below a threshold value near the upper or lower ends of the range of the observed values of the variable'.

© Springer Nature Switzerland AG 2023

221

V. Misra, *An Introduction to Large-Scale Tropical Meteorology*, Springer Atmospheric Sciences, https://doi.org/10.1007/978-3-031-12887-5_9

The current approaches to discerning the attributing factors for an extreme event can be binned into two main categories: (i) using observations to estimate the change in probability of magnitude and frequency of extreme events and (ii) using numerical models to simulate the extreme event in its 'control' environment and to compare that with another simulation done in a hypothetical or counterfactual environment. This latter approach using climate models is a conditional attribution approach (e.g., Lackmann 2015; Fischer and Knutti 2015). Examples of the observational approach to attribution of extreme events could be of the type which explains the long-term variations of, say, observed Atlantic TC activity (e.g., Goldenberg et al. 2001; Emanuel 2005), tornadic activity (e.g., Elsner et al. 2018; Nouri et al. 2021), and extreme rain events in the ISM (Roxy et al. 2017). In their observational study, Kenyon and Hegerl (2008) recognized that modes of climate variability like those associated with ENSO, NAO, and the PDO are important drivers of changes in climate extremes. They suggest that temperature extremes are affected by large-scale anomalous circulation patterns associated with these modes of climate variability. For example, their study finds that positive phases of PDO cause warm extremes in the southeastern United States and eastern Asia. Similarly, they find that NAO exerts a significant influence on temperature extremes in Eurasia.

The basic premise of attribution of extreme events is to calculate the change in the odds of the event between the presence and the absence of the underlying large-scale variability stemming either from natural variations like ENSO or under the anthropogenic influence of climate change. In many instances, very different conclusions on attribution could be arrived at for the same extreme event. These differing conclusions could be a result of the uncertainties in methodology, observational data, and climate models. However, one should also be cautious to focus on the question being asked in attribution studies. For example, Rahmstorf and Coumou (2011) concluded that the Russian heat wave of 2010 was largely due to natural variability in contrast to Dole et al. (2011) who concluded it was due to climate change. These seemingly antithetical conclusions for the same Russian heat wave event from two independent studies actually reflect the different questions being asked: Dole et al. (2011) focused on the magnitude of the heat wave, while Rahmstorf and Coumou (2011) targeted on the likelihood of its occurrence.

9.2   Attribution to the Madden-Julian Oscillation

The World Climate Research Programme established the S2S prediction project in 2013 to explore among many other things the potential to predict the onset, evolution, and decay of some extreme weather events, weeks in advance. The specific lead time of this project ranged from 2 weeks to 2 months (Vitart and Robertson 2018). As discussed in Chap. 5, much of this effort stems from the prospect of the influence of the intraseasonal variations and the ability to predict at least the first cycle of this variation by the current generation of numerical weather prediction and climate models. As the S2S bridges between weather and seasonal timescales, it is considered as both an initial value and a boundary value problem. In some instances, the focus of S2S could be to make a prediction of specific large-scale, long-lived events (e.g., heat waves or droughts), which appear as an initial value problem assisted by the persistence of, say, soil moisture and or SSTA. In other instances, the focus of the S2S could be to make a prediction of event probability that is largely a boundary-forced problem (e.g., the likelihood of active or weak tornado activity over a large area and large temporal window). And in some instances, the S2S objective could be a combination of both (e.g., the onset of wet the season could be an initial value problem, and the likelihood of frequency of extreme rainy days within the season is a boundary-forced issue). The S2S project has created an extensive database of sub-seasonal forecasts and reforecasts from currently around 11 operational centers around the world.

In the case of the extended predictability of the TCs, its main source is the MJO especially in the SH, where the TC season in boreal winter coincides with the seasonal peak in the MJO activity (Vitart and Roberston 2018). The MJO influences the genesis of TC through its strong modulation of the mid-level relative humidity, followed by its influence on low-level cyclonic vorticity that affects the genesis of TCs with minor contributions of the modulation of vertical wind shear (Camargo et al. 2009). The composite tracks of TCs at each of the eight phases of the MJO in Fig. 9.1 shows the gradual eastward shift of the densest TC tracks with the corresponding shift of the convection envelope of the MJO from the Indian Ocean to the Eastern Pacific Ocean. This modulation of the track density of TC in Fig. 9.1 is visible in the Tropical Indian, Western Pacific, and Eastern Pacific Oceans. Vitart and Robertson (2018) show that the S2S models are able to pick this feature of MJO modulation of TC tracks (Fig. 9.2). In Fig. 9.2, the track density anomalies of TC are shown for phases 1 through 8 of the MJO (based on the RMM index discussed in Sect. 6.5). The positive track density anomalies of the TC in S2S models match with the eastward propagation of the convective envelope as the MJO progresses from phases 1 through 8 in Fig. 9.2. Furthermore, it is noted that S2S models display useful skill, on average out to 3 weeks and in the case of ECMWF model out to 4 weeks, in predicting the phases of the MJO (Vitart and Robertson 2018). The similarity in the track density between the observations in the top row of Fig. 9.2 with those from S2S forecast models in Fig. 9.2 is apparent. Therefore, Fig. 9.2 demonstrates that the S2S models can skillfully predict the TC activity in the SH, modulated by the MJO, 3 to 4 weeks in advance.

A classical attribution of an extreme weather event is illustrated in Fig. 9.3, which attributes an MJO event propagating into the Western Pacific Ocean to cause severe cold over western Europe, most of Russia, and parts of North America in March 2013. Vitart and Robertson (2018) isolated one of the S2S models with 16 ensemble members, among which some of the ensemble members produced a verifiable forecast of this cold event at lead times of 26–32 days, while other ensemble members failed to verify (Fig. 9.3). The distinguishing feature of these two sets of forecasts from the same model was that in the verifiable forecasts, the MJO event was successfully propagated from the Indian Ocean (phase 3) at initial time into Western Pacific Ocean (phase 6) by day 26 of the forecast. In contrast,

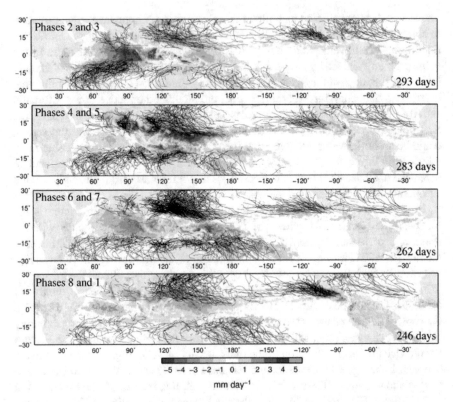

Fig. 9.1 Composite of tracks of TCs (1975–2011) based on the eight phases of the MJO (see Fig. 6.6) when the amplitude of the RMM is greater than one. The total number of TCs in each phase of the MJO is indicated in the bottom right of each panel. The shading corresponds to TRMM rainfall anomalies (1998–2011) corresponding to the eight MJO phases. (Reproduced from Zhang (2013); © American Meteorological Society. Used with permission)

in the set of failed forecasts, the MJO signal diminished over the Western Pacific Ocean.

9.3   Attribution to El Niño and the Southern Oscillation

ENSO is regarded as one of the strongest interannual variations on our planet that has discernible ramifications on global climate and which also influences weather and climate extremes in regions around the world. This global impact of ENSO entails from the large changes observed in the coupled ocean-atmosphere circulation involving the Walker Circulation that also includes the shift of the deep convection to the Central Pacific Ocean during El Niño events (cf. Fig. 6.4). This anomalous equatorial diabatic heat source in the Central Pacific Ocean during a warm ENSO

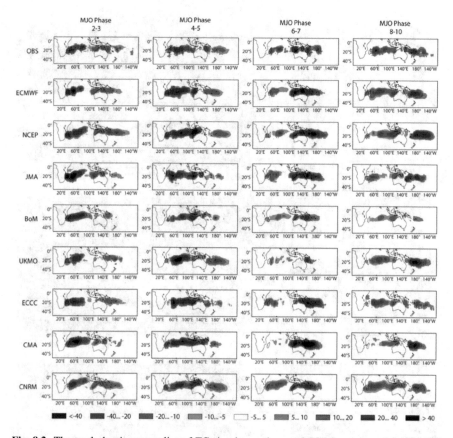

Fig. 9.2 The track density anomalies of TCs in observations and S2S forecast models. The TC track density anomaly is computed as the deviation from climatology computed over the time period of October to March from 1999 to 2010 when MJO is in (left panels) phases 2 and 3, (middle left panels) phases 4 and 5, (middle right panels) phases 6 and 7, and (right panels) phases 8 and 1. The top row is the observations followed by the sub-seasonal (2–3 weeks) forecasts from the eight S2S models for the various phases of the MJO in each of the columns. The track density is computed as the number of TCs passing within 500 km normalized by the total number of TCs of the whole basin. (Reproduced from Vitart and Robertson 2018)

event is known to produce stationary barotropic Rossby wave pattern that radiates to higher latitudes, which is the basis for ENSO teleconnections (Hoskins and Karoly 1981). Lee et al. (2009) further showed that this stationary barotropic Rossby wave is sensitive to the background flow and therefore seasonally dependent. In a set of idealized modeling experiments, they showed that the location of the STJ more than its strength is critical for the atmospheric response to the anomalous diabatic heating of ENSO. In other words, the vertical shear offered by the STJ to the background flow as it shifts equatorward in the boreal winter during a warm ENSO event serves as a waveguide for the barotropic Rossby wave to radiate to higher latitudes in the NH. In contrast, during the boreal summer season when the STJ moves poleward,

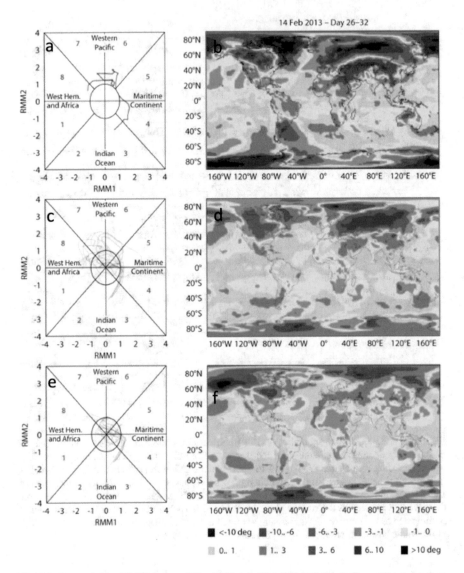

Fig. 9.3 (**a**, **c**, **e**) The RMM index of the MJO (cf. Fig. 6.6) and (**b**, **d**, **f**) weekly mean 2 m air temperature for the verifying week March 7–13, 2013 (lead time of days 26 to 32). (**a**, **b**) Observations from ERA-Interim reanalysis and (**c**, **d**, **e**, **f**) forecasts from one of the S2S models (CFSv2). (**c**, **d**) Successful CFSv2 days 26 to 32 forecasts with strong MJO events predicted over the Western Pacific Ocean and cold anomalies over North America, Russia, and western Europe, which is statistically significant at a 5% significance level. (**e**, **f**) Failed CFSv2 days 26 to 32 forecasts with weak MJO event predicted over the western Pacific Ocean. (Reproduced from Virtart and Roberston 2018)

the ENSO teleconnections become weaker in the NH. These stationary Rossby wave patterns sustain persistent atmospheric circulation anomalies that manifest as anomalies of vertical shear, vorticity, moisture, and heat flux anomalies, which then affect regional climate and weather, remote from the Central Pacific.

Arblaster and Alexander (2012) find that seasonal composites of extreme maximum temperature are significantly influenced by ENSO. They used the maximum of daily maximum temperature in each month (TXx) to represent extreme temperatures of the month. Figure 9.4a shows the climatological distribution of this extreme metric for the DJF season. The largest values of TXx are found in the subtropical belt (Fig. 9.4a). The influence of ENSO on this extreme metric (TXx) is shown in Fig. 9.4b, which displays the DJF mean composite difference for cold (minus)-warm ENSO events. This composite difference illustrates a strong, statistically significant response of TXx to cold, La Niña events from warm El Niño events (Fig. 9.4b). Regions like over South and Southeast Asia, southeast Australia, South Africa, parts of South America, and western Canada exhibit statistically significant cooler maximum temperatures in La Niña events compared to El Niño events (Fig. 9.4b). Similarly, regions like the United States and Siberia display opposite anomalies, i.e., TXx becomes warmer in cold relative to warm ENSO events (Fig. 9.4b). It may be noted that ENSO's impact on worldwide temperature extremes is pronounced in the boreal winter season. The anomalies of TXx due to ENSO are generally much weaker in the boreal summer season (Kenyon and Hegerl 2008).

Similarly, ENSO also exerts a significant influence on daily precipitation extremes. But counterintuitively, the relationship between the oceanic area of extreme dry months and El Niño is strongest despite the surplus heat content over the Tropical Pacific during El Niño (Curtis et al. 2007). This is primarily because much of the vast area of the Tropical Atlantic and water around the Maritime Continent experience extreme dryness during El Niño from the shift of the Walker Circulation relative to La Niña conditions or climatology. Curtis et al. (2007) find that the spatial relationship between extreme daily precipitation frequency and ENSO is like the seasonal rainfall and ENSO relationship (Fig. 9.5). In this instance, extreme precipitation frequency was defined for each grid box ($0.25° \times 0.25°$) within a given season that exceeded the 95th percentile across all seasons from 1998 to 2005. Curtis et al. (2007) find that the relationship between seasonal precipitation and extreme precipitation frequency is significant across the globe, with the strongest relationship occurring over the (tropical and subtropical) oceans. Figure 9.5 represents the areas where extreme daily precipitation frequency is a good indicator of seasonal precipitation and where seasonal precipitation is also strongly related to ENSO. In Fig. 9.5, we observe that this dual relationship, first, between the extreme daily precipitation frequency and seasonal mean precipitation and, second, between seasonal precipitation anomaly and ENSO is strongest over the oceans (e.g., Western Pacific and Central Pacific in Fig. 9.5a, c, Central Pacific and Eastern Atlantic in Fig. 9.5b, the SPCZ and Central Pacific in Fig. 9.5d) and comparatively weaker over land (e.g., southeastern United States and Gulf of Mexico in Fig. 9.5a, northeast Brazil in Fig. 9.5a, b, Southeast Asia and Maritime region in Fig. 9.5b).

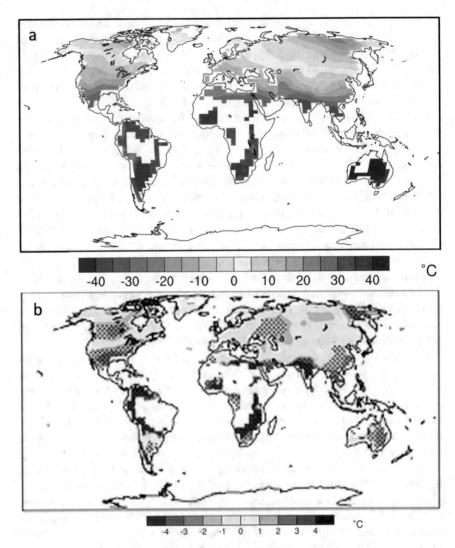

Fig. 9.4 (**a**) The observed DJF climatological composite of maximum monthly daily maximum temperature. (**b**) The observed DJF composites of maximum monthly daily maximum temperature for cold minus warm ENSO events. All values that are significant at a 5% significance level according to the Kolmogorov-Smirnov test are shaded in (**a**). The hatched regions in (**b**) are statistically significant at 5% significant level according to the t-test. The white areas in (**b**) denote grid points with insufficient spatial coverage to calculate gridded values. (Reproduced from Arblaster and Alexander 2012)

ENSO has also a significant bearing on TCs in all tropical ocean basins. Lin et al. (2020) reviewed this impact and conclude the following for the various TC basins:

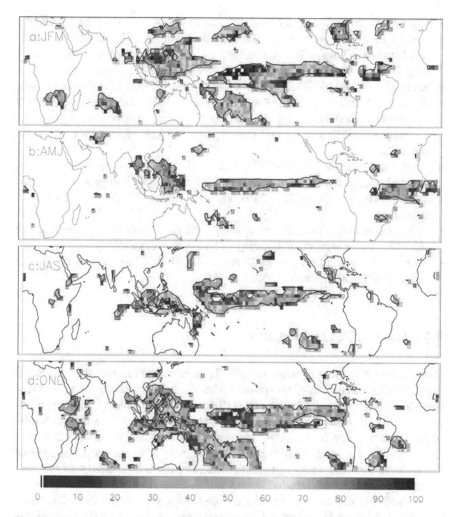

Fig. 9.5 The percentage area of $2.5° \times 2.5°$ grids (shown by the color bar) where extreme precipitation frequency is significantly related (using Kendall's τ nonparametric test) to mean seasonal rain rate and overlaid with significant, contemporaneous seasonal rainfall-Niño3.4 SST index correlations (contoured) for (**a**) January–February–March, (**b**) April–May–June, (**c**) July–August–September, and (**d**) October–November–December seasons. The dark red and blue contours denote areas of positive and negative correlation at the 1% significance level between seasonal rainfall and the corresponding Niño3.4 SST index. (Reproduced from Curtis et al. (2007); © American Meteorological Society. Used with permission)

- Western North Pacific (~110°E–180°E and 0°–60°N): During developing warm ENSO years, TCs form closer to the date line and consequently are longer-lived, larger in size, and more intense and generate more *accumulated cyclone energy* despite a significant reduction in the ocean heat content. The TCs also tend to track away from the Asian coast during El Niño years.

- Central Eastern Pacific (~150°W–60°W and 0°–35°N, not including the Atlantic Ocean): El Niño years increase the ocean heat content, which realizes a significant increase in intense TCs (Category 3 and above); but the tracks tend to shift away from the Mexican coast and toward international date line.
- North Atlantic (~110°W–20°W and 0°–60°N, not including the Pacific Ocean): Atlantic TC activity is significantly suppressed during El Niño years, with a reduction of landfall along the US coastline. However, El Niño is not a guarantor for suppressed TC activity in North Atlantic as AMO and AMM also exert significant influence.
- North Indian Ocean (~50°E–100°E and 0°–30°N): TCs in this basin occur in pre- and post-monsoon seasons, with ENSO exerting an influence on the post-monsoon season TCs. In El Niño years, fewer intense TCs form in the Bay of Bengal.
- South Indian Ocean (~10°E–135°E and 0°–60°S): TC activity is diminished or enhanced east or west of 75°E during El Niño years, respectively.
- South Pacific Ocean (~135°E–75°W and 0°–60°S): El Niño enhances TC activity east of 170°E and diminishes it west of 170°E.

9.4   The Psi-Chi Framework

One relatively novel aspect of assessing the large-scale controls on extreme weather events like the TCs is through the 'psi-chi' framework. In explaining the evolution of the ISM, Krishnamurti and Ramanathan (1982) used the so-called psi-chi (or stream function (ψ)-velocity potential (χ)) framework that illustrated the time evolution of the transformation of the kinetic energy from irrotational to the non-divergent component of the wind. They showed the progression of the transformation of the available potential energy (APE) to eventually the kinetic energy of the non-divergent component of the wind, which was consistent with the evolution of the kinematic field from onset to maturity and then the demise of the ISM. They even used this framework to understand the intraseasonal variations of the ISM. In this section, it will be shown that the psi-chi framework can be adapted to understand the differences in the simulation of TCs that vary in their intensity over the period of their model simulation.

The $\psi - \chi$ interaction equation is obtained from the familiar vorticity and divergence equations with pressure as the vertical coordinate system. The vorticity equation is given by

$$\frac{\partial \zeta}{\partial t} + \vec{V}_H . \nabla (\zeta + f) + \omega \frac{\partial \zeta}{\partial p} + (\zeta + f) \nabla . \vec{V}_H + \vec{k} . \nabla \omega \times \frac{\partial \zeta}{\partial p} = F_\zeta \quad (9.1)$$

and the divergence equation is given by

$$\frac{\partial D}{\partial t} + \nabla \cdot \left(\vec{V}_H . \nabla \vec{V}_H \right) + \nabla \cdot \left(\omega \frac{\partial \vec{V}_H}{\partial p} \right) + \nabla \cdot \left(f \vec{k} \times \vec{V}_H \right) + \nabla^2 \emptyset = F_D$$

$$(9.2)$$

Now, separating the horizontal wind vector, $\vec{V}_H$ in to irrotational ($\vec{V}_\chi$) and non-divergent ($\vec{V}_\psi$) components of the wind, such that $\vec{V}_H = \vec{V}_\chi + \vec{V}_\psi$ so that $\vec{V}_\chi = -\nabla\chi$ and $\vec{V}_\psi = \vec{k} \times \nabla\psi$, where χ and ψ are velocity potential and stream function, respectively. The relative vorticity (ζ) then is

$$\zeta = \vec{k} . \nabla \times \vec{V}_H = \nabla^2 \psi$$

$$(9.3)$$

and divergence (D) is

$$D = \nabla . \vec{V}_H = -\nabla^2 \chi$$

$$(9.4)$$

Furthermore, the total kinetic energy of the flow (K) is given by

$$K = \frac{1}{2} \vec{V}_H . \vec{V}_H = \frac{1}{2} |\nabla\psi|^2 + \frac{1}{2} |\nabla\chi|^2 - J(\psi, \chi) = K_\psi + K_\chi - J(\psi, \chi)$$

$$(9.5)$$

where the Jacobian (J) term in Eq. 9.5 represents the energy interaction term between the non-divergent and irrotational components of the flow. The Jacobian term can be expanded as

$$J(\psi, \chi) = \left| \frac{\partial (\psi, \chi)}{\partial (x, y)} \right| = \left| \begin{matrix} \frac{\partial \psi}{\partial x} & \frac{\partial \chi}{\partial x} \\ \frac{\partial \psi}{\partial y} & \frac{\partial \chi}{\partial y} \end{matrix} \right| = \frac{\partial \psi}{\partial x} \frac{\partial \chi}{\partial y} - \frac{\partial \chi}{\partial x} \frac{\partial \psi}{\partial y}$$

$$(9.6)$$

Substituting Eqs. 9.3 and 9.4 in 9.1 and 9.2, we get for vorticity equation

$$\frac{\partial \nabla^2 \psi}{\partial t} = - J\left(\psi, \nabla^2 \psi + f \right) + \nabla\chi . \nabla \left(\nabla^2 \psi + f \right) - \omega \frac{\partial \nabla^2 \psi}{\partial p} + \left(\nabla^2 \psi + f \right) \nabla^2 \chi$$

$$- \nabla\omega . \nabla \frac{\partial \psi}{\partial p} + J\left(\omega, \frac{\partial \chi}{\partial p} \right) + F_\zeta$$

$$(9.7)$$

Similarly, the divergence equation can be cast in terms of χ and ψ as

$$\frac{\partial \nabla^2 \chi}{\partial t} = \nabla^2 \left[\frac{1}{2} (\nabla\psi)^2 + \frac{1}{2} (\nabla\chi)^2 - J(\psi, \chi) \right] - \left(\nabla^2 \psi \right)^2 - \nabla\psi . \nabla \left(\nabla^2 \psi \right) - \omega \frac{\partial \nabla^2 \chi}{\partial p}$$

$$- J\left(\omega, \frac{\partial \psi}{\partial p} \right) - \nabla\omega . \nabla \frac{\partial \chi}{\partial p} - \nabla f . \nabla\psi + J(f, \psi) + J\left(\nabla^2 \psi, \chi \right)$$

$$+ f\nabla^2 \chi + \nabla^2 \emptyset + F_D$$

$$(9.8)$$

where F_ζ and F_D represent friction in the vorticity and divergence equations, respectively.

Note that

$$\psi \frac{\partial \nabla^2 \psi}{\partial t} = \nabla \cdot \left(\psi \nabla \frac{\partial \psi}{\partial t} \right) - \frac{\partial}{\partial t} \left(\frac{|\nabla \psi|^2}{2} \right) \tag{9.9}$$

or Eq. 9.9 can be rearranged as

$$\frac{\partial K_\psi}{\partial t} = -\psi \frac{\partial \nabla^2 \psi}{\partial t} + \nabla \cdot \left(\psi \nabla \frac{\partial \psi}{\partial t} \right) \tag{9.10}$$

and similarly note that

$$\chi \frac{\partial \nabla^2 \chi}{\partial t} = \nabla \cdot \left(\chi \nabla \frac{\partial \chi}{\partial t} \right) - \frac{\partial}{\partial t} \left(\frac{|\nabla \chi|^2}{2} \right) \tag{9.11}$$

or Eq. 9.11 can be rewritten as

$$\frac{\partial K_\chi}{\partial t} = -\chi \frac{\partial \nabla^2 \chi}{\partial t} + \nabla \cdot \left(\chi \nabla \frac{\partial \chi}{\partial t} \right) \tag{9.12}$$

Multiplying Eq. 9.7 by ψ and Eq. 9.8 by χ and subsequently substituting them with Eqs. 9.10 and 9.12, respectively, followed by integrating over a closed domain (represented by one of the overbars) and also vertically integrating (represented by the other overbar), we can obtain

$$\frac{\partial \overline{\overline{K}}_\psi}{\partial t} = \underbrace{f \nabla \overline{\overline{\psi}} . \nabla \chi}_{\text{Term 1}} + \underbrace{\nabla^2 \psi \overline{\overline{\nabla}} \psi . \nabla \chi}_{\text{Term 2}} + \underbrace{\nabla^2 \chi \left(\frac{\overline{\overline{\nabla}} \psi}{2} \right)^2}_{\text{Term 3}} + \underbrace{\omega J \left(\overline{\overline{\psi}}, \frac{\partial \chi}{\partial p} \right)}_{\text{Term 4}} + \overline{\overline{F}}_\psi + B_\psi \tag{9.13}$$

$$\frac{\partial \overline{\overline{K}}_\chi}{\partial t} = - \underbrace{f \nabla \overline{\overline{\psi}} . \nabla \chi}_{\text{Term 1}} - \underbrace{\nabla^2 \psi \overline{\overline{\nabla}} \psi . \nabla \chi}_{\text{Term 2}} - \underbrace{\nabla^2 \chi \left(\frac{\overline{\overline{\nabla}} \psi}{2} \right)^2}_{\text{Term 3}} - \underbrace{\omega J \left(\psi, \frac{\overline{\overline{\partial \chi}}}{\partial p} \right)}_{\text{Term 4}}$$

$$- \underbrace{\chi \overline{\overline{\nabla}}^2 \emptyset}_{\text{Term 5}} + \overline{\overline{F}}_\chi + B_\chi \tag{9.14}$$

where J, K_ψ, K_χ, ϕ, B_y are Jacobian operator, the kinetic energy of irrotational and non-divergent components of the winds, geopotential, and corresponding boundary flux terms, respectively. The wind fields are converted to ψ and χ fields using the Poisson's Eqs. 9.3 and 9.4 in the limited domain using a double Fourier transform method (Krishnamurti and Bounoua 1996). This method assumes periodic boundary conditions for the wind field and allows for a correction to account for nonvanishing domain-averaged winds.

The Terms 1, 2, 3, and 4 of Eqs. 9.13 and 9.14 are the $\psi - \chi$ energy interaction terms, which represent the contributions from the orientation of the gradients of the ψ and χ, the orientation of the gradients of ψ and χ in the vicinity of the vorticity, the covariance of the horizontal divergence, and the kinetic energy of the non-divergent component, respectively. All of these Terms (1 through 4) appear as a source and as a sink (with an opposite sign) in Eqs. 9.13 and 9.14.

Term 1 is a dot product of two vectors. Term 1 represents the interaction of the large-scale ψ and χ fields. Term 1 can be expanded as

$$Term\ 1 \equiv \overline{\overline{f\nabla\psi.\nabla\chi}} = \overline{\overline{f|\nabla\psi||\nabla\chi|cos\theta}},$$

where θ is the angle between $\nabla\psi$ and $\nabla\chi$. Therefore, if the two vectors are oriented in parallel, then the dot product in Term 1 will be maximized. On the other hand, if $\nabla\psi$ and $\nabla\chi$ are orthogonal to each other, then Term 1 will be minimized.

Term 2 can be written as $Term2 \equiv \overline{\overline{\nabla^2\psi\nabla\psi.\nabla\chi}} = \overline{\overline{\xi|\nabla\psi||\nabla\chi|cos\theta}}$. Term 2 represents the same interaction as Term 1 but in the vicinity of the cyclonic vorticity (e.g., TCs). Therefore, Term 2 can be regarded as small-scale interaction between ψ and χ fields.

Term 3 can be written as $Term3 \equiv \overline{\overline{Term3 \equiv \nabla^2\chi\left(\frac{\nabla\psi}{2}\right)^2}} = \overline{\overline{D.K_\psi}}.$ If we approximate Eq. 9.13 for argument's sake to the following:

$$\frac{\partial \overline{\overline{K}}_\psi}{\partial t} \asymp \underbrace{\nabla^2\chi\left(\frac{\nabla\psi}{2}\right)^2}_{Term\ 3} = D.K_\psi \tag{9.15}$$

Equation 9.15, can be rewritten as

$$\frac{\partial \ln\left(\overline{\overline{K}}_\psi\right)}{\partial t} \asymp D \tag{9.16}$$

It is evident from Eq. 9.16 that Term 3 can lead to exponential growth or decay of non-divergent kinetic energy in the presence of strong convergence or divergence, respectively.

Term 4 can be expanded as $Term4 \equiv \overline{\omega J\left(\psi, \dfrac{\partial \chi}{\partial p}\right)} = \omega \dfrac{\partial \psi}{\partial x}\dfrac{\partial^2 \chi}{\partial y \partial p} - \omega \dfrac{\partial \psi}{\partial y}\dfrac{\partial^2 \chi}{\partial x \partial p} =$ $\omega \left(\dfrac{\partial \psi}{\partial x}\dfrac{\partial D_y}{\partial p} - \dfrac{\partial \psi}{\partial y}\dfrac{\partial D_x}{\partial p}\right)$, where D_x and D_y represent divergence and convergence in the zonal and meridional directions. Therefore, Term 4 represents the interaction between K_ψ and K_χ in the presence of strong vertical gradients of convergence and horizontal gradients of ψ.

Term 5 is unique to Eq. 9.14, which does not appear in Eq. 9.13. It represents APE and serves as the source for the interaction between K_ψ and K_χ. APE is typically the covariance between vertical motion (a proxy for diabatic heating or precipitation in the tropics) and temperature, which can be shown as follows:

For a closed domain (a large enough domain where the lateral boundary flux is negligible),

$$\nabla^2 \emptyset \chi = 0 \tag{9.17}$$

Equation 9.14 can be expanded as

$$\chi \nabla^2 \emptyset + \emptyset \nabla^2 \chi = 0 \tag{9.18}$$

or

$$\chi \nabla^2 \emptyset = -\emptyset \nabla^2 \chi \tag{9.19}$$

Using Eq. 9.4 and the mass continuity equation in Eq. 9.17, we get

$$\chi \nabla^2 \emptyset = -\emptyset \dfrac{\partial \omega}{\partial p} \tag{9.20}$$

For an inviscid bottom and a rigid lid at the top,

$$\int_{P_{top}}^{P_{bot}} \dfrac{\partial \overline{\omega \emptyset}}{\partial p} = 0 \tag{9.21}$$

Note that in Eq. 9.21, we have used only one overbar because the vertical integral is explicitly indicated. The single overbar denotes the area average in Eq. 9.21. Equation 9.21 can be rewritten as

$$\int_{P_{top}}^{P_{bot}} \overline{\emptyset} \dfrac{\partial \overline{\omega}}{\partial p} = -\int_{P_{top}}^{P_{bot}} \overline{\omega} \dfrac{\partial \overline{\emptyset}}{\partial p} \tag{9.22}$$

or

$$\phi \dfrac{\partial \omega}{\partial p} = -\omega \dfrac{\partial \phi}{\partial p} \tag{9.23}$$

Substituting Eq. 9.23 with Eq. 9.20, we get

$$\text{APE} \equiv \chi \nabla^2 \emptyset = \omega \frac{\partial \phi}{\partial p} \tag{9.24}$$

Using the hydrostatic equation and ideal gas law in Eq. 9.24 we can arrive at

$$\text{APE} = -\omega \alpha = -\frac{R}{P} \omega T \tag{9.25}$$

From Eq. 9.25, it follows that APE can be interpreted as a covariance of the vertical motion with temperature. This covariance denotes the ascent of warm air and the descent of cold air within the closed domain. Alternatively, Eq. 9.25 may be interpreted as APE is generated when heating occurs where the air is warmer and cooling occurs where the air is cooler. Krishnamurti and Ramanathan (1982), from their diagnostic analysis, show that mean ISM is maintained by this covariance of heating and temperature that produces the APE, which is converted to the kinetic energy of the irrotational component of the wind (Eq. 9.14) and then to the non-divergent component of the wind (Eq. 9.13). It should be noted that the conversion of the kinetic energy of the irrotational component of the wind to non-divergent component of the wind is instantaneous. Of course, at every stage of this conversion, some of this energy is lost to dissipation through friction and the boundary flux term.

Figure 9.6 shows the time series of the various terms of the $\psi - \chi$ interactions (Eq. 9.13) during the period of the ISM of 1979. The Term 1 of Eq. 9.13 ($= \overline{f \nabla \psi . \nabla \chi}$) depends on the orientation of the gradients of ψ and χ and is found to be the leading term in the interaction of $\psi - \chi$. If the gradients of ψ and χ are parallel, then the energy goes efficiently from the irrotational to the non-divergent wind through Term 1. On the other hand, if the gradients are orthogonal to each other, then the energy transformation from the irrotational to the non-divergent wind is zero or negligible. The period of the rapid increase in Term 1 in the first week of June 1979 (Fig. 9.6a) coincides with the onset of the ISM. A period of a relative decrease of this term (around mid-July 1979) from the period of onset is related to the intraseasonal break period of the monsoon. Comparatively, all the other three terms shown in Fig. 9.6a, b, and c are smaller in magnitude relative to Term 1. Term 2 of Eq. 9.13 ($= \overline{\nabla^2 \psi \nabla \psi . \nabla \chi}$) signifies the same transformation as Term 1 but in the vicinity of cyclonic vorticity (when $\nabla^2 \psi > 0$). Therefore, this term will tend to increase when a monsoon low forms during the season. This term is largely small compared to Term 1 and is usually negative because the geometry of the low-level summer monsoon flow is anticyclonic north of the equator (i.e., $\nabla^2 \psi < 0$). Term 3 of Eq. 9.13 ($= \overline{\nabla^2 \chi \left(\frac{\nabla \psi}{2} \right)^2}$) suggests the covariance of divergence with the kinetic energy of the non-divergent component of the wind, which exponentially increases when this covariance is maximized because (cf. Equation 9.16). Figure 9.6c suggests that Term 3 increases episodically and can reach the magnitudes of Term 1 in such

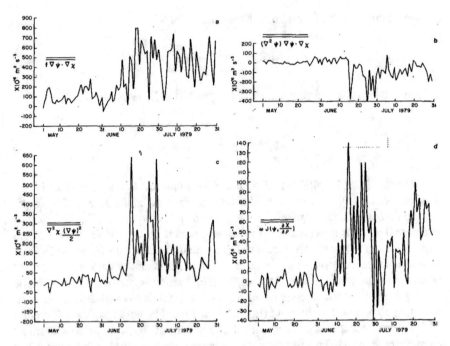

Fig. 9.6 Daily time variations of the various terms of the $\psi - \chi$ interactions (Eq. 9.12) domain averaged over 50°E–70°E and 4°S–20°N over the period of the ISM of 1979. (Reproduced from Krishnamurti and Ramanathan (1982); © American Meteorological Society. Used with permission)

moments when the covariance is maximized. Term 4 of Eq. 9.13 ($=\overline{\nabla^2 \chi \left(\frac{\nabla \psi}{2}\right)^2}$) is large when the vertical gradient of convergence is large. However, they largely remain much smaller than Term 1 during the evolution of the ISM (Fig. 9.6d).

The same $\psi - \chi$ framework can also be applied to understand the evolution of TCs. Even in the context of the TCs, the choice of the domain has to be carefully chosen to compute the domain average of the forcing terms of Eqs. 9.13 and 9.14. Otherwise, the boundary flux terms can become significant and make the interpretation of the terms in the $\psi - \chi$ interaction equations difficult in terms of their contribution to the redistribution of the kinetic energy. Furthermore, we have to differentiate between the inflow and outflow layers of the TC, which limits the vertical integration within these relatively shallow layers, or examine these terms at a single level. At lower levels, it is advisable to mark the lateral boundaries of the domain at the radius where the tangential wind varies asymptotically with the radius. At the outflow level of the TC, the lateral boundary can be marked where the radial wind varies asymptotically with the radius.

To demonstrate the application of this $\psi - \chi$ framework on understanding the evolution of TCs, we take the example of three Atlantic TCs simulated using a

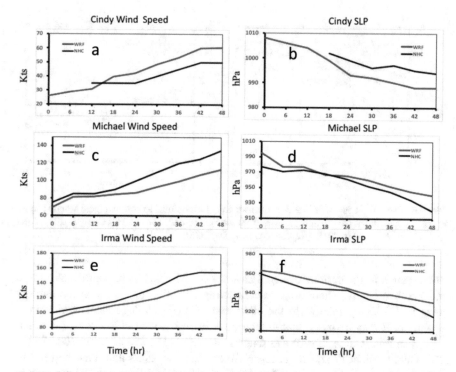

Fig. 9.7 The observed (HURDAT2; Landsea and Franklin 2013) and the simulated intensity of the three cases for Atlantic TCs (**a, b**) Cindy in 2017, (**c, d**) Michael in 2018, and (**e, f**) Irma in 2017. The intensity is indicated both in terms of (**a, c, e**) V_{max} and (**b, d, f**) minimum SLP. (Reproduced from Das 2020)

regional atmospheric model, WRF3.2 (Fig. 9.7). We note that the model simulation overestimates the wind speed and minimum SLP in the case of TC Cindy (of 2017), throughout its evolution (Fig. 9.7a, b). In the remaining two cases (Atlantic TCs Michael of 2018 [Fig. 9.7c, d] and Irma 2017 [Fig. 9.8e, f]), the model simulation underestimates the maximum wind speed and minimum SLP, which contrasts with the simulation of TC Cindy. In Fig. 9.8a, we show the time series of the fractional conversion of $\overline{K}_\chi$ to $\overline{K}_\psi (\equiv \overline{C}_{\chi \to \psi} = \left(\frac{\overline{K}_\psi - \overline{K}_\chi}{\overline{K}_\psi} \right) \times 100)$ at 850 hPa for all three TCs computed at 3-h interval from the corresponding WRF simulations. It is seen clearly from Fig. 9.8 that consistent with the evolution of the intensity of the TC (Fig. 9.7), this fractional conversion ($\overline{C}_{\chi \to \psi}$) at 850 hPa is highest for Irma followed by that for Michael and then Cindy (Fig. 9.8a). In addition, we also see from the figure that $\overline{C}_{\chi \to \psi}$ at 850 hPa evolves with the intensity of the TC: $\overline{C}_{\chi \to \psi}$ is small at the beginning of the simulation and grows to its peak value toward the end of the simulation, when the TC attained its maximum intensity (Fig. 9.8a).

In contrast, the evolution of $\overline{C}_{\chi \to \psi}$ in the outflow level in Fig. 9.8b shows its magnitude decreasing through the simulation in the case of Irma and Michael but

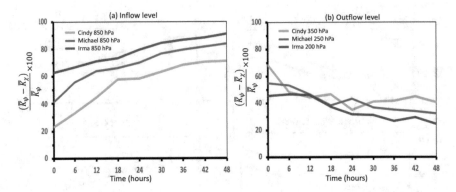

Fig. 9.8 The time series of the fractional conversion of irrotational to non-divergent kinetic energy at (**a**) 850 hPa and at (**b**) the outflow level from the WRF simulations of TCs Cindy in 2017, Irma in 2017, and Michael in 2018. (Reproduced from Das 2020)

not as much in the case of Cindy. TCs Irma and Michael were significantly more intense than Cindy. Therefore, the contrasting evolution of $\overline{C}_{\chi\to\psi}$ at the outflow level in TC Cindy relative to the other two TCs (Fig. 9.8b) reflects their varied evolution. It may be noted that at their peak intensity, the outflow level (at which $\overline{C}_{\chi\to\psi}$ in Fig. 9.8b is shown) was diagnosed as 350 hPa, 250 hPa, and 200 hPa for TCs Cindy, Michael, and Irma, respectively. It is suggested here that the relative reduction in $\overline{C}_{\chi\to\psi}$ in the outflow level with time in Michael and Irma causes an increased likelihood for the intensification of the TC as it retains more kinetic energy for the irrotational component of the flow and hence for the radial outflow. This follows from Rappin et al. (2009), who showed that the energy sink of the secondary circulation is minimized by the weak inertial stability in the outflow layer, which leads to the rapid intensification of the TC. They suggest that the preferred outflow channels in rapidly intensifying TCs form in regions of least resistance that is characterized by weak environmental inertial stability. In a similar vein, Merrill (1988) documented that in intensifying TCs, the outflow is influenced by upper-level synoptic features in contrast to non-intensifying TCs. Therefore, the relative reduction of the growth in $\overline{K}_{\psi}$ in the outflow layer at the expense of the growth of $\overline{K}_{\chi}$ leads to a favorable environment for the intensification of the TC.

In Fig. 9.9, we show the time series of the first three terms of Eqs. 9.12 and 9.13 and Term 5 of Eq. 9.13 for all three TCs at 850 hPa, which indicate that Term 1 is dominant followed by Term 2, and then Term 3. The generated APE is proportionately higher in the stronger TCs (Michael and Irma) relative to the weaker TC (Cindy). The comparative magnitudes of all these terms between the three TC cases are also reflective of their relative differences in intensity (Fig. 9.9). The magnitude of Term 1 $\left(= \overline{|\nabla\psi|\,|\nabla\chi|\,cos\theta}\right)$ is not only dictated by the magnitude of the gradients of ψ and χ but also by the angle (θ) between the gradients. In Fig. 9.10a–c, we show the product of the gradients of ψ and χ as a function of θ at the time of peak intensity of all three TCs. Systematically we see in Fig. 9.10a–c that

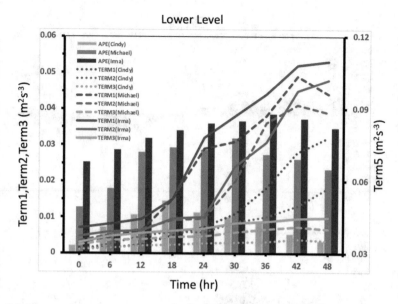

Fig. 9.9 The transformation terms in 10^{-2} m^2s^{-3} (Terms 1, 2, and 3, of Eq. 9.13 and Term 5 of Eq. 9.14 in text) at 850 hPa for Atlantic TCs Cindy in 2017, Irma in 2017, and Michael in 2018. (Reproduced from Das 2020)

Table 9.1 The tally of the points with the positive and negative orientation of the gradients of ψ and χ in Figs. 9.8 and 9.10

TC	Cindy	Michael	Irma
Figure 9.8 (at 850 hPa)			
Angles (θ) $\leq 90°$	28,443	34,289	37,291
Angles (θ) $> 90°$	25,812	2071	17,709
Figure 9.10 (at outflow level)			
Angles (θ) $\leq 90°$	39,818	30,941	25,407
Angles (θ) $> 90°$	40,244	48,827	57,775

the dot products of the gradients of ψ and χ are lowest in the weakest TC (Cindy) and highest in the strongest TC (Irma), a reflection of relative differences in the magnitudes of their gradients. We also consistently observe in Fig. 9.10a–c that there is a greater density of points with gradients of ψ and χ to be aligned with negative values of $\cos\theta$ (i.e., the angles [θ] are between 90° and 180°) for weaker storms (e.g., Cindy; Fig. 9.10a) relative to the stronger storms (e.g., Michael [Fig. 9.10b] and Irma [Fig. 9.10c]; see also Table 9.1). Likewise, between Michael (Fig. 9.10b) and Irma (Fig. 9.10c), one observes that Michael displays a larger density of points with angles between 90° and 180° than Irma at their peak intensity (Table 9.1). In other words, gradients of ψ and χ at 850 hPa in the large-scale TC environment are relatively less parallel in Michael than in Irma.

In the outflow level, these three forcing terms of Eq. 9.12 have opposite signs relative to those at 850 hPa (Fig. 9.11). At the initial time of the evolution of TCs Michael and Irma, these terms are marginally positive and then eventually they grow

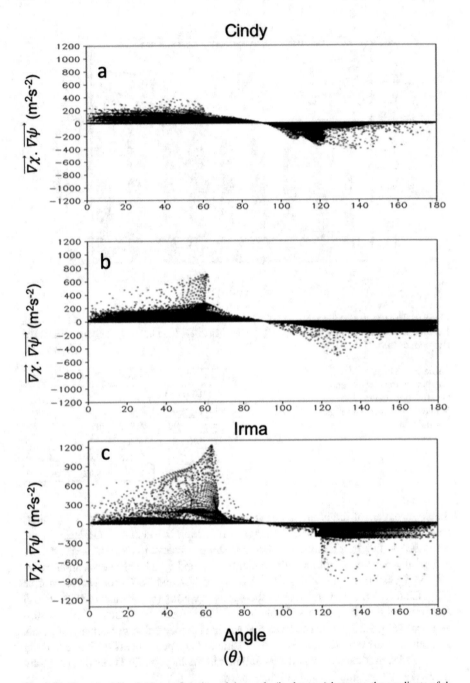

Fig. 9.10 Term 1 of Eq. 9.13 as a function of the angle (in degrees) between the gradients of ψ and χ at 850 hPa for Atlantic TCs (**a**) Cindy in 2017, (**b**) Irma in 2017, and (**c**) Michael in 2018, when they reach their peak intensity. (Reproduced from Das 2020)

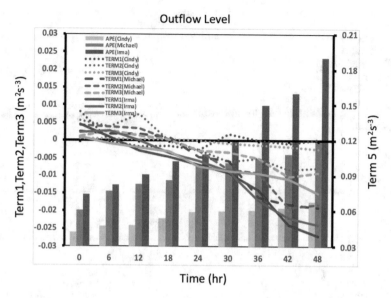

Fig. 9.11 The transformation terms in 10^{-2} m^{2}s^{-3} (Terms 1, 2, and 3 of Eq. 9.13 and Term 5 of Eq. 9.14 in text) at the outflow level for Atlantic TCs Cindy in 2017 at 350 hPa, Irma in 2017 at 250 hPa, and Michael in 2018 at 200 hPa. (Reproduced from Das 2020)

to large negative values toward the end of the simulations. These negative values suggest they are a sink to $\overline{K}_\psi$ and a source for $\overline{K}_\chi$ at the outflow level within the domain wherein the radial velocity is relatively large. Again, the relative magnitude of these terms in Fig. 9.11 suggests that the larger negative values at the outflow level serve to intensify the TC, while smaller negative values tend to weaken the TC. It is interesting to note that APE at this outflow level is higher than at 850 hPa (Fig. 9.9) for all three TCs, and yet the conversion of $\overline{K}_\chi$ to $\overline{K}_\psi$ is far less than at 850 hPa. This is because the orientation of the gradients of ψ and χ is not conducive to this conversion (Fig. 9.12). Unlike at 850 hPa (Fig. 9.10a–c), the gradients of ψ and χ at the outflow level across the three TCs at their peak intensity are far more comparable (Fig. 9.12a–c). As Fig. 9.12a–c shows in comparison to Fig. 9.10a–c, most of the points in the environment of these TCs through their evolution have an angle between 90° and 180° (Table 9.1), which renders their dot product to be overall negative. In Fig. 9.12, we observe that the density of points with the gradients of ψ and χ having an angle between 90° and 180° is lower for weaker TCs like Cindy than stronger TCs like Michael and Irma (Table 9.1). In addition, between Michael and Irma, it is apparent that the density of points with angles between 90° and 180° is higher for the latter than the former (Fig. 9.12b, c; Table 9.1). Therefore, the conversion $\overline{C}_{\chi \to \psi}$ at the outflow level of the stronger TCs is less efficient than the weaker TCs, thereby reducing the inertial stability and the resistance of the outflow.

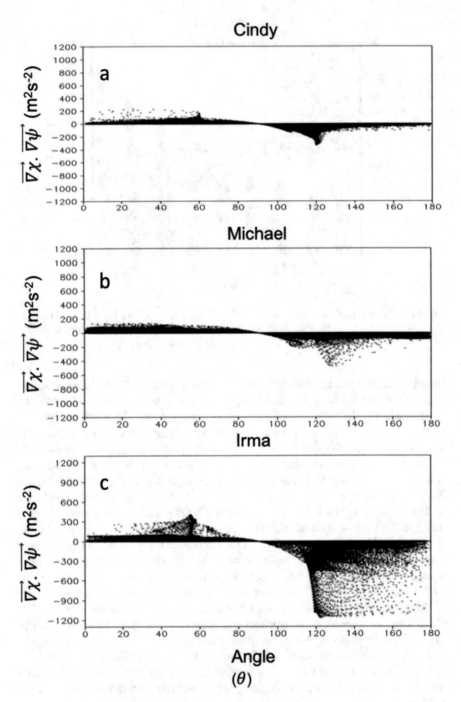

Fig. 9.12 Term 1 of Eq. 9.13 as a function of the angle (in degrees) between the gradients of ψ and χ at the outflow level of Atlantic TCs (**a**) Cindy in 2017 at 350 hPa, (**b**) Irma in 2017 at 250 hPa, and (**c**) Michael in 2018 at 200 hPa, when they reach their peak intensity. (Reproduced from Das 2020))

Therefore, when the gradients of the stream function and velocity potential are large and oriented parallel to each other both in the large-scale TC environment and in the region of the primary circulation of the TC, then the TC is favored to intensify with the robust conversion of $\overline{K}_\chi$ to $\overline{K}_\psi$ at 850 hPa. In contrast, however in the outflow layer, we require a slower conversion of $\overline{K}_\chi$ to $\overline{K}_\psi$ for the TC to intensify; otherwise, it leads to increased inertial instability and weakening of the TC.

9.5 The Easterly Wave

The tropical easterly wave is a ubiquitous tropic-wide phenomenon that produces synoptic disturbances ranging from hurricanes in the Atlantic and Eastern Pacific Ocean basins to typhoons in the Western Pacific Ocean and monsoon lows and depressions in the Indian Ocean. They also produce other organized convective systems like mesoscale convective complexes or squall lines. The easterly waves are off-equatorial Rossby waves, whose characteristics vary around the tropics as a result of the varying background winds around the tropics, which imply varying horizontal and vertical wind shear (Fig. 9.13). The easterly waves are observed year-round but have their peak activity from July to September and have maximum intensity near the West African coast (0–10°W).

The AEWs have the following characteristics:

- Mean wavelength of 2500 km.
- Period of 3–5 days. Avila and Pasch (1992) indicate on an average there are 59 AEWs a year with a standard deviation of only 4.4.
- The phase speed of 6–8 ms^{-1} or 6 to 7 longitude degrees per day.
- They account for nearly 60% of Atlantic tropical storms and nonmajor hurricanes, while they account for 85% of major hurricanes (Landsea 1993). However, the fraction of AEWs that strengthen to Atlantic TCs is quite small (~6%; Hopsch et al. 2010).
- The Eastern Pacific Ocean, which is regarded as the most active TC formation region on Earth in terms of genesis events per unit area per unit time, has the majority of its TCs forming from AEWs that have propagated across the Atlantic, and the Caribbean into Eastern Pacific (Avila and Pasch 1992).

The AEW features a trough or cyclonic flow that develops on a low-level zonal jet (at around 600 or 700 hPa) called the AEJ. The surface thermal contrast between the semiarid region of the Sahara and the rain-belt region of the sub-Saharan region produces the AEJ from the thermal wind balance, which occurs between 8°N and 17°N with speeds of 10–25 ms^{-1}. The mechanism for the source of these waves follows the combined barotropic and baroclinic instability discussed in Sect. 4.3.

The AEWs develop along two axes, just north and south of the AEJ (Thorncroft and Hodges 2000). The northern location of the AEW (poleward of 15°N) grows from relatively dry baroclinic instability near the southern margin of the Sahara and

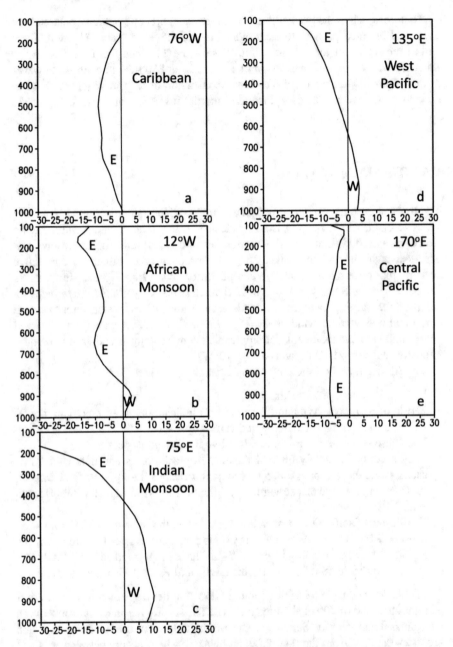

Fig. 9.13 Climatological zonal wind (from CFSR) for August at 10°N and at (**a**) 76°W (Caribbean region), (**b**) 12°W (West African monsoon region), (**c**) 75°E (ISM region), (**d**) 135°E (Western Pacific), and (**e**) 170°E (Central Pacific), with E and W indicating easterlies and westerlies. The x-axis is zonal wind in ms^{-1}, and the y-axis is pressure in hPa

maximizes around 850 hPa. The southern track of the AEW grows from barotropic instability and interaction with deep convection in the rainy zone (equatorward of 15°N) and maximizes at around 600 hPa. The southern and the northern tracks of the AEW merge in the main development region of the Atlantic (~ 10°N–20°N and 17.5°W–65°W; Goldenberg and Shapiro 1996). AEWs are generally cold core up to about 600 hPa with a weak warm core above. The AEWs are operationally monitored across four stations: from interior Africa in Niamey (Niger; 13.5°N and 2°E), Bamako (Mali; 12.6°N and 8°W), on the west coast of Africa in Dakar (Senegal; 15°N and 18°W), to over Eastern Atlantic Ocean in Praia (Cape Verde; 14.9°N and 23.5°W). The AEWs are usually tracked by tracking the propagation of a reference point like the trough axis. The trough axis is identified by the location where the meridional wind vanishes (i.e., where the wind shifts, say, from northerlies to southerlies at around 700 hPa). The AEWs are more comprehensively identified by not only examining the meridional winds but also looking for the low-level cold core in the temperature cross section and peak in relative humidity through deep layers of the troposphere or high equivalent potential temperature that is indicative of conditional instability.

9.6 Thunderstorms

TCs grab our attention in terms of the destruction they can cause, especially, with intense convective thunderstorms that can cause harm to life and property. However, at any given instant of time, there are far more convective thunderstorms than TCs. It should be noted there are roughly about 16 million thunderstorms that occur each year worldwide, and there are roughly 2000 thunderstorms at any instant (https://www.nssl.noaa.gov/education/svrwx101/thunderstorms/). These thunderstorms can cause destruction from associated flooding, strong winds, and lightning. Therefore, the sheer volume of events of convective thunderstorms at any given instant of time across the globe poses a potential threat. These convective thunderstorms also play an important role in redistributing heat, moisture, and momentum in the vertical in the tropics, where they occur frequently. It is generally believed that the intensity of the thunderstorm is proportional to the intensity of the vertical velocity—the greater the vertical velocity, the more intense is the thunderstorm (Dosewell 2001; Zipser et al. 2006).

TRMM had some unique attributes at the time of its launch in late 1997, which included a space-borne precipitation radar, a passive microwave imager (TMI), a VIRS, and a LIS. This overlap of powerful observing instruments in addition to its low (35°)-inclination, low-altitude, and non-synchronous orbit permitted frequent sampling over the tropics throughout the day. Using these TRMM datasets, Zipser et al. (2006) cataloged thunderstorms (they defined them as precipitation features that are contiguous regions of precipitation with an area greater than 75 km^2) across the global tropics. However, Zipser et al. (2006) had to develop proxies for ascertaining the intensities of the thunderstorms as measurements of convective updraft speed were difficult to obtain. Their proxies included four criteria: (1) the height of the 40 dBZ reflectivity level—the higher the level, the more intense is the thunderstorm; (2) the brightness temperature at 37 GHz microwave channel, (3) the brightness temperature at 85 GHz microwave channel, with lower temperatures at these channels suggesting a more intense thunderstorm; and (4) the lightning flash rate, with higher rates indicating a more intense precipitation feature. It should be however noted that although the thunderstorms isolated from each of these four criteria have significant overlap (Fig. 9.14), their relationship is not really one-to-one. Zipser et al. (2006) indicated that 40 dBZ echo top is a function of the sixth power of the diameter of the scattering particles, while the low brightness temperatures are related to the third power and that too only in an integrated ice water content sense. These variations in the methodology generate subtle differences in the diagnosis of thunderstorms like over Northern Australia, Equatorial Africa, and the foothills of the Himalayas in India (Fig. 9.14).

Zipser et al. (2006) using 7 years of this TRMM data compiled a sort of a climatology of these thunderstorms between the latitudes of 36°N and 36°S (latitudes over which TRMM operated), which is captured in Fig. 9.14. There are many remarkable features one can discern from this figure, like:

- All four proxies to isolate intense thunderstorms qualitatively produce similar results. They all show that southern/central United States (Great Plains regions), southeastern South America, parts of India, and the sub-Saharan region have some of the most intense thunderstorms (marked in black color in Fig. 9.14).
- There is a strong preference for the intense convective events to occur over land. It is well known that thunderstorms over oceans are weaker (Lucas et al. 1994; Williams and Stanfill 2002). The oceanic ITCZ produces large rainfall totals largely from moderate strength thunderstorms (marked in green in Fig. 9.14) and contains very few thunderstorms from the three most intense categories.
- The most intense convective events over the ocean occur in the coastal oceans, adjacent to land where the storm motion is from land to ocean (e.g., tropical oceans west of Central America and West Africa and subtropical oceans east of southeastern United States, South America, Australia, and Africa).
- Interestingly, equatorial Amazonia and Indonesia have a high number of moderate-intensity thunderstorms and fewer intense thunderstorms than Equatorial Africa.

There are other important features of the tropical thunderstorms that are high-lighted in Zipser et al. (2006) like:

- In most continental regions, there is a strong seasonal cycle for the most intense thunderstorms. For example, the Sahel region (10°N–18°N in West Africa) and northern India have their peak in JJA, south-central United States in boreal spring and summer, pre-monsoon storms in the Gangetic Plains of India and Bangladesh in boreal spring, and northern Australia in boreal winter, while southeastern South America and Equatorial Africa have intense thunderstorms in all seasons.

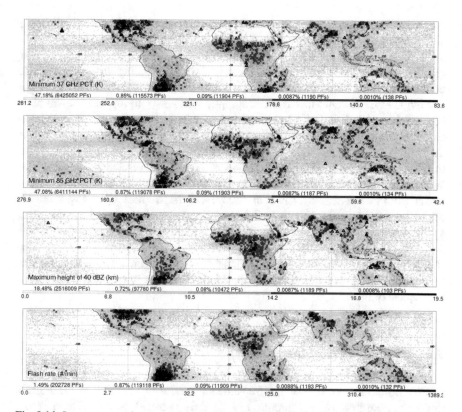

Fig. 9.14 Intense convective events (or precipitation features) identified by minimum brightness temperature in (**a**) 37 GHz and (**b**) 85 GHz microwave channels, (**c**) maximum height of the 40 dBZ echo top, and d) flash rate. The color code indicates the frequency and intensity of the event. For example, in panel (**d**), locations marked in black occur at 0.001% of the 12.8 million precipitation features = 128 precipitation features that have more than 314.7 lightning flashes per minute. This figure is compiled over 7 years of TRMM sub-daily data from January 1, 1998, through December 31, 2004. (Reproduced from Zipser et al. (2006); © American Meteorological Society. Used with permission)

- The diurnal cycle of intense thunderstorms shows a preference for a strong afternoon maximum over land and a nocturnal maximum over tropical oceans. The peak in the diurnal activity of thunderstorms over land in the afternoon (~0017–0018 h) is sharper than that of the diurnal peak in rainfall. In contrast, over the oceans, the amplitude of the broad nocturnal peak in the diurnal activity of thunderstorms is comparable to that of the diurnal amplitude of the oceanic rainfall, but the latter has a peak around sunrise, while the former can occur throughout the night.

There is some clarity on the favorable environmental conditions for intense thunderstorms to develop, which include the presence of *conditional instability*, high CAPE, abundant moisture in the lower troposphere, dynamical lifting, and strong shear in the lower troposphere. Using these criteria, Brooks et al. (2003)

found that severe thunderstorms are usually concentrated downstream of high terrain and poleward of moisture sources. They suggest that the high mid-tropospheric lapse rates are advected off the high terrain over the underlying boundary layer, creating a high CAPE environment. In addition, the subsiding air on the lee side of the mountains helps to cap (or create convective inhibition above) the layer of high CAPE. In some instances, orography provides a lifting mechanism as in convergence associated with low-level jets, which assist in breaking through the cap over the boundary layer. Furthermore, in these regions with the upper-level flow being from the direction of the high terrain and poleward flow in the lower troposphere, it creates naturally a region of significant vertical wind shear that is important for influencing storm severity and longevity (Weisman and Klemp 1982). But in a recent study, Liu et al. (2020) find that tropical regions with intense thunderstorms (except for the Sahel) have far lower vertical wind shear than their subtropical counterparts like the south-central United States and southeastern South America. They conclude that higher CAPE, lower CIN, weaker low-level wind shear, and helicity are found in thunderstorm environments of tropical regions (except for Sahel which displays high CAPE and high wind shear) relative to subtropical regions.

9.7 Diurnal Variations

Diurnal variations are one of the fundamental modes of variation in the Earth's climate system that was first reported in the early twentieth century (Hann 1901). Diurnal variations refer to variability that occurs on the sub-daily scale, typically reaching a diurnal maximum (or zenith) and a diurnal minimum (or nadir) once a day. Diurnal variations are reported to occur in many variables including precipitation, thunderstorm frequency and intensity, outgoing longwave radiation, surface pressure, land surface temperatures, soil moisture, surface heat fluxes, sea surface temperature, surface salinity, upper air geopotential height and temperature, horizontal winds and boundary layer circulations, large-scale horizontal moisture fluxes, vertical motion, and mass divergence.

The analysis of 3 hourly weather reports in Dai et al. (2007) over a couple of decades (1975–1997) revealed the following about diurnal variations:

- Thunderstorms over land, characterized by deep convection, occur more often in the late afternoon than at other times, owing to diurnal variations.
- The diurnal harmonic explains about 40–80% of the total daily variance of precipitation over land, especially in the summer.
- The diurnal harmonic explains about 40% of the total daily variance of precipitation over oceans.
- The *semidiurnal variation* (twice daily) of precipitation accounts for 15–25% of the total daily variance over land, while over the oceans, it accounts for 20–40% of the total variance.

- Over the oceans, the amplitudes of the semidiurnal cycle and diurnal cycle of precipitation is comparable. However, the amplitude is less than half the diurnal amplitude of the land areas in the summer.

The diurnal cycle is regarded as a form of manifestation of the Earth's climate system response to solar radiation. However, hidden in this rather generalized statement about the diurnal cycle is a very complex set of interactions that is best illustrated by the fact that Mars (farther from the Sun than the Earth) has a very large amplitude of diurnal variations, while Venus (closer to the Sun than the Earth) displays very small diurnal amplitude. Ruppert (2016), through novel experiments using a cloud resolving model, showed that the diurnal cycle can impact climate in profound ways. He examined the impact of modulating the diurnal cycle by changing the duration of the day from 12 to 48 h and even making the diurnal cycle completely absent. In these experiments, it was found that humidity in the atmospheric column (as measured by precipitable water) remained the driest with no diurnal cycle, while the experiment with the longest diurnal cycle was the most moist. Furthermore, the greater accumulation of humidity in the atmospheric column with longer diurnal cycles results in deeper clouds and increased convection, resulting in a very different climate evolution than the more extreme case of no diurnal cycle or short diurnal cycle..

The diurnal variations are typically diagnosed by harmonic analysis. This is equivalent to taking, for example, hourly data of the variable in interest ($P(t)$, where t is time) and conducting a Fourier analysis on it, from which we isolate the first harmonic component diurnal (24-h) period. Mathematically, this may be represented as

$$P(t) = P_o + \sum_{i=1}^{N/2} A_i \cos (i\theta - \Phi_i) \tag{9.26}$$

where N, P_o, A_i, $\theta = \frac{2\pi t}{N}$, Φ_i is the number of intervals ($=24$ if it is hourly), the daily mean value of the variable, amplitude, and phase angle of the ith harmonic (with $i = 1$ corresponding to diurnal harmonic), respectively.

The broad observations of diurnal variation of convective activity in the tropics following Yang and Slingo (2001) are as follows: (1) the amplitude of the diurnal cycle over land is much larger than that over the oceans, (2) the summer hemisphere has a larger diurnal amplitude than that of the winter hemisphere, (3) the maximum continental precipitation occurs at late afternoon/early evening and in the early morning over oceans.

This distinction in the diurnal phasing of convection between land and ocean in the tropics led Yang and Slingo (2001) to suggest that mechanisms for diurnal variations over land and ocean are different. Over land, the diurnal variation of convection is a thermodynamical response to variations of solar radiation. During the day, the solar radiation raises the instability by raising the lower tropospheric temperature and moisture in the atmospheric boundary layer through more evaporation, and turbulent eddies that gradually lead to convective precipitation maximizing in the late afternoon/evening. However, this simple mechanism is modified by local orography, propagation, and decay of transient weather systems, proximity to the ocean, and land surface type to name a few. Such influences lead to exceptions with some tropical continental regions displaying continental diurnal precipitation maximizing at nighttime (Yang and Slingo 2001).

Over the oceans, the mechanism for diurnal convection is varied and complex. Randall et al. (1991) explain the early morning maximum in oceanic precipitation from direct radiation-convection interaction, which suggests that in the daytime, solar absorption at cloud top leads to an increase in stability and reduces convection, while the higher longwave cooling at cloud top relative to cloud base at nighttime results in destabilization of the upper troposphere, leading to an eventual maximum in precipitation in the early morning. Alternatively, Gray and Jacobson (1977) suggest that the spatial gradients in the radiative heating/cooling between the cloudy and cloud free regions lead to variations in the horizontal divergence that give rise to diurnal variation of convective activity over the oceans. Chen and Houze (1997) believe that cloud-radiation interaction is insufficient to explain diurnal variations of convection over the ocean and instead suggests complex surface-cloud-radiation interaction. They make this argument because large-scale organized oceanic convective systems that exhibit strong diurnal variation initiate in the afternoon when it is unfavorable in terms of atmospheric radiative processes. They indicate that skin SST reach its maximum in the afternoon, which leads to the initiation of the convection, which under optimal environmental conditions through cloud-radiation interaction grows and reaches a maximum in the early morning. They also refer to something called 'diurnal dancing', wherein diurnal convection pulsates spatially. Over the life cycle of these large oceanic convective systems that can last up to a day, they can leave the underlying boundary layer with low entropy air and cloud canopy that shades the ocean surface. As a result, the next day, convection tends to occur in neighboring regions unaffected by the previous convective system since the surface conditions under the current convective system are unfavorable for another round of convection. But as Yang and Slingo (2001) point out, there are again exceptions when oceanic precipitation in some regions maximizes in the afternoon.

Dai et al. (2007) indicates that CAPE has small diurnal variations over the oceans and therefore does not serve as a driving force to diurnal convection. However, the diurnal variation in the land-ocean circulation that manifests in the surface divergence field is evident with the diurnal variation of rainfall in tropical oceans. This variation of circulation is a consequence of the air being warmest over the continents in the summer afternoon, which drives the rising motion and air mass convergence while inducing subsidence and surface divergence over the neighboring ocean. But from midnight to early morning, the air over the ocean becomes warmer than over the adjacent land, thereby reversing the land-ocean circulation from that in the daytime. Dai et al. (2007) found evidence of this reversal in the low-level vertical velocity in the NCEP-NCAR reanalysis to be in phase with diurnal variations of convection.

Yang and Slingo (2001) also observed that spatially coherent diurnal convective variations over coastal areas propagate out into the ocean for several hundreds of kilometers. For example, diurnal convection propagates away from the coasts of Indonesian islands, southward from the Indo-Gangetic plains to the Bay of Bengal, westward off the West African coast, and southwestward from the Mexican coast. Similarly, there are instances when diurnal convection propagates inland from the coastal oceans. For example, the precipitation diurnal peak phase propagates from the coastline to inland of Sumatra (Mori et al. 2004) and over Brazil (Garreaud and Wallace 1997). Kikuchi and Wang (2008) from their analysis of remotely sensed precipitation principally identified three diurnal regimes of tropical precipitation as oceanic, continental, and coastal, based on their amplitude, peak time, and phase propagation characteristics that basically fit with the description provided earlier (Fig. 9.15). Within the coastal regime, Kikuchi and Wang (2008) identify two other sub-regimes: The seaside coastal regime (characterized by large amplitude, offshore phase propagation with diurnal zenith from late evening to noon of the next day) and landside coastal regime (featuring large amplitude, inland phase propagation with a peak from noon to evening). The propagation speed of both these sub-regimes is found to be around 10 ms^{-1}. Kikuchi and Wang (2008) indicate that the propagation of the landside coastal regime is sustained by sea breeze, while its interactions with local topography could also play a significant role in the propagation. The propagation of the seaside coastal regime cannot be just explained by the land breeze or the concavity of the land or the mean flow primarily because the land breeze is not as strong as sea breeze (Mapes et al. 2003), the morphology of the coastline cannot explain propagation, and the mean flow is not necessarily sufficient to explain the propagation speed of the diurnal convection in the seaside coastal regime. Yang and Slingo (2001) and Mapes et al. (2003) suggest that gravity waves emanating from land convection could explain the maintenance of the propagating seaside coastal regime diurnal convection.

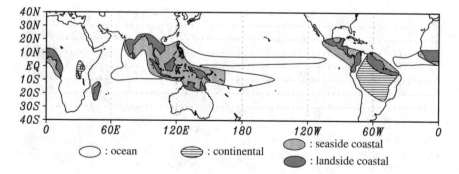

Fig. 9.15 A schematic of the three diurnal regimes of tropical precipitation (oceanic, continental, and coastal) based on amplitude, diurnal zenith time, and phase propagation characteristics. (Reproduced from Kikuchi and Wang (2008); © American Meteorological Society. Used with permission)

Chapter 10
Climate Change

Abstract The alternating ways to define tropics meteorologically are discussed in this chapter in the context of a warming world. The impact of global climate change on tropical climate and its variations are presented with particular emphasis on the tropical oceans.

Keywords Upped ante scheme · Walker circulation · Climate change · Global warming · ENSO · Atlantic tropical cyclones · Clausius Clapeyron equation

10.1 The Widening Tropics

Often when we refer to tropics, we simply think of the latitude range of 23.5°S to 23.5°N or sometimes more broadly ranging from 30°S to 30°N. However, meteorologically, the tropics can be defined more dynamically in the sense that they could vary temporally and spatially rather than remaining fixed at a specific latitude. Waugh et al. (2018) have identified at least nine different metrics (Fig. 10.1) to define the tropical latitudinal range, which includes:

1. The latitude at which the zonal mean OLR acquires a value of 250 Wm^{-2} (OLR in Fig. 10.1a)
2. The latitude at which the zonal mean OLR is 20 Wm^{-2} less than the subtropical maximum and is poleward of the latitude of the subtropical maximum (ΔOLR in Fig. 10.1a)
3. The latitude defining the edge of the Hadley Circulation by identifying the latitude where the zonal mean meridional stream function at 500 hPa goes to zero (PSI in Fig. 10.1b)
4. The latitude at which the EDJ appears by identifying the latitude at which the maximum of the zonal mean of zonal wind at 850 hPa occurs (EDJ in Fig. 10.1b)
5. The latitude at which the STJ appears by identifying the latitude at which the maximum in the average of the zonal mean of the zonal wind between 100 and 400 hPa appears after removing the zonal mean of the zonal wind at 850 hPa is removed (STJ in Fig. 10.1b)

© Springer Nature Switzerland AG 2023
V. Misra, *An Introduction to Large-Scale Tropical Meteorology*, Springer
Atmospheric Sciences, https://doi.org/10.1007/978-3-031-12887-5_10

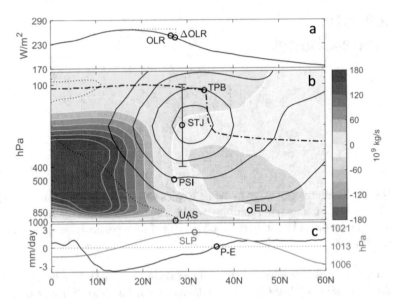

Fig. 10.1 A schematic illustration of the definitions for defining the latitude of the tropics in the NH based on (**a**) OLR, (**b**) zonal winds (black contour), tropopause height (dot-dashed), stream function (shaded), and (**c**) SLP (blue) and P-E (black). The acronyms are explained in the text (Sect. 10.2). (Reproduced from Waugh et al. 2018; © American Meteorological Society. Used with permission)

6. The latitude where the largest meridional gradient of zonal mean tropopause pressure occurs (TPB in Fig. 10.1b)
7. The latitude at which the zonal mean of the surface zonal wind becomes zero (UAS in Fig. 10.1b)
8. The latitude at which the difference of the zonal mean precipitation (P) and zonal mean evaporation (E) goes to zero (P-E in Fig. 10.1c)
9. The latitude where the subtropical high is situated by identifying the latitude at which the maximum of the zonal mean of the mean sea level pressure occurs (SLP in Fig. 10.1c)

Some of these studies have found that the conclusions on the impact of global warming on the trends of the width of the tropics could vary by the definition of the tropics one chooses. For example, Waugh et al. (2018) find that PSI, UAS, P-E, EDJ, and SLP (in SH only) are closely related to the width of the Hadley Circulation. Therefore, any of these metrics will provide similar measures of changes to tropical width. On the other hand, measures like SLP (in NH only), STJ, OLR, ∆OLR, and TPB are not good proxies for the width of the Hadley Circulation because they do not covary strongly with the remaining measures of the tropical width or respond differently to increases in greenhouse gas concentration. Many studies based on metrics that measure the width of the Hadley Circulation have however shown evidence of the broadening or widening trend of the tropics both based on observations (~0.2–0.4° latitude per decade) and model projections of the future climate under high emission scenarios.

Although the observed expansion of the tropical latitudes in the annual mean in both hemispheres shows similar magnitudes, the factors driving them are quite different. In the SH, increasing GHGs and stratospheric ozone depletion are found to be the main drivers of the expansion in the current climate, while in the future climate, it is bound to expand poleward despite the recovery of the stratospheric ozone owing to increasing concentrations of GHGs (Grise et al. 2019). In the NH, the contribution of the GHGs to tropical expansion is much weaker and is unlikely to dominate over the natural variability even in the late twenty-first century under the highest emission scenario (Grise et al. 2019). The negative phase of the PDO is found to contribute significantly to the expansion of the Hadley Circulation in the NH. It is also observed that the largest poleward expansion of the Hadley Circulation is observed in the summer and the fall seasons in both hemispheres.

10.2 The Upped Ante Scheme

During warm ENSO events, the tropospheric temperature (weighted average temperature in the column from 1000 hPa to 200 hPa) exhibits uniform positive temperature anomalies (~0.5–1°C) across the tropical strip (20°S to 20°N; Yulaeva and Wallace 1994). This uniform warming of the tropospheric temperature during warm ENSO events is a result of the fact that the tropical troposphere cannot maintain horizontal pressure gradients. Chiang and Sobel (2002) show that the Niño3 SST anomalies lead the tropical mean SST and the tropical mean tropospheric temperature anomalies by 2–6 months, suggesting that ENSO variations control the interannual variability of the tropical tropospheric temperature anomalies. Chiang and Sobel (2002) further show that this remote, tropics-wide warming of the tropospheric temperature has important implications on the remote SST and precipitation responses. For example, from the quasi-equilibrium theory (cf. 3.2), we understand that the temperature profile in the convective layer is tied to the boundary layer entropy (θ_{eb}). Therefore, it follows that warming of the tropospheric temperature anomalies leads to an instantaneous adjustment of θ_{eb}. But because of this instantaneous change in θ_{eb}, a thermodynamic disequilibrium between the atmosphere and the underlying ocean surface arises. This results in changes in the surface turbulent fluxes, inducing gradual warming of the remote SST like over the Tropical Atlantic and the Indian Oceans during El Niño years.

Chiang and Sobel (2002) show that the response of the remote tropical SST and precipitation to the tropospheric temperature anomalies is a function of the mixed layer depth or surface thermal inertia. They indicate that the response of the remote SSTA decreases and the phase lag with the imposed tropospheric temperature anomalies increases as the mixed layer depth increases. On the other hand, the amplitude of the remote precipitation anomalies increases and so does the phase lag with the imposed tropospheric temperature anomalies as the mixed layer depth of the ocean increases. This is because as you raise the heat storage capacity, for example, in an ocean with a deep mixed layer, the SST takes a longer time to come to equilibrium with the imposed tropospheric temperature forcing. The more the SST and the overlying atmospheric boundary layer entropy (or θ_{eb}) are in disequilibrium, the larger the fluxes between them, implying more evaporation and convection. Their modeling study further reveals that the faster the rate of change of tropospheric temperature anomalies (initiated by the amplitude of the convection in the Central Pacific in the case of ENSO), the larger the precipitation response. Furthermore, they find that the tropospheric temperature signal propagates to the surface only in regions of either deep and or shallow convection.

The above discussion serves as a prelude to introducing the upped ante scheme. There is a growing consensus from global climate model projections under high emission scenarios, and observations of the past few decades suggest that the tropical tropospheric temperature is getting warmer (Chou and Neelin 2004; Thorne et al. 2011). But the remote tropical precipitation response is far less uniform with both excess and deficit trends observed in various parts of the tropics from the different global climate models (Fig. 10.2). The upped ante scheme is illustrated schematically in Fig. 10.3. As the greenhouse gas increases, the troposphere warms, resulting in high θ_{eb}, which then leads to higher evaporation and convection following our discussion in the previous paragraph. In order to maintain stronger convection, CAPE must increase, which implies that the atmospheric boundary layer moisture would have to increase, largely by stronger surface evaporation. However, in the non-convective regions of the tropics, the atmospheric boundary layer moisture cannot increase so easily in response to changes in θ_{eb}. This leads to spatial humidity gradients in the atmospheric boundary layer, which is further accentuated in the margins of tropical convective regions (or subtropics) where it imports drier air from non-convective regions by way of advection. As a result, the margins of convective regions are unable to sustain the original convection in the face of newly imposed warmer tropospheric temperature anomalies from global warming. Therefore, convection reduces further at these margins. However, in the convective regions, the evaporation becomes stronger, and precipitation also becomes stronger. So in essence, the rich (wet regions) get richer (wetter), and the poor (dry regions) get poorer (drier). Chou and Neelin (2004) claim that this process is analogous to a poker game, where the player must meet the upped ante (increased demand of atmospheric boundary layer moisture) to sustain convection in a warm troposphere. The ante here is applicable to the margins of the convective region where it is unable to supply sufficient moisture in the boundary layer to sustain convection.

10.3   ENSO Variability

The response of the Tropical Pacific to global warming has been investigated with great interest as it has a bearing on the variations of remote climate via

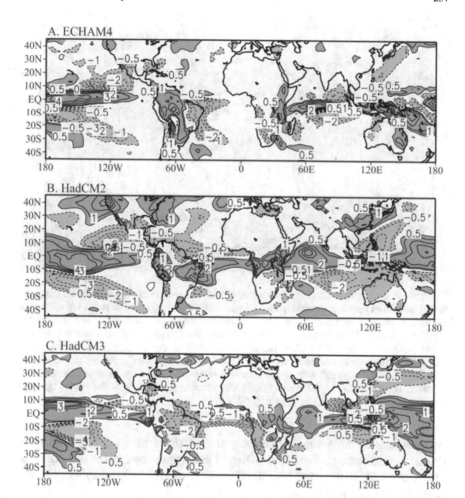

Fig. 10.2 The projection of precipitation anomalies obtained as the climatological annual mean difference between the periods of 2070–2090 and 1971 to 1990 from three different CMIP3 model runs (A: ECHAM4; B: HadCM2; C: HadCM3) with SRESA2 emission scenario. (Reproduced from Neelin et al. 2003)

teleconnections. However, despite extensive studies, the conclusions seem elusive. Some studies claim that in a future warm climate, the Tropical Pacific could be in a permanent El Niño state (Knutson and Manabe 1995; Meehl and Washington 1996; Vecchi and Soden 2007), while other studies claim to the contrary that the climate of the Tropical Pacific could go into a permanent La Niña state (Clement et al. 1996; Seager and Murtugudde 1997). There is also the aspect of the amplitude of the change with some studies claiming weak and others suggesting large changes in the magnitude of the SST and heat content of the Equatorial Pacific Ocean in a future warm climate (e.g., Park et al. 2009; Kohyama et al. 2017). In addition, there

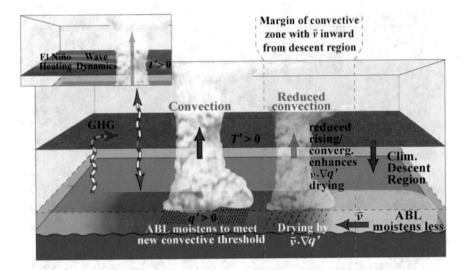

Fig. 10.3 A schematic illustration of the upped-ante mechanism to explain the tropical precipitation anomalies in response to tropospheric temperature warming (shaded in red) from increasing concentrations of greenhouse gases. This mechanism shows that convection increases in the convective regions (like the ITCZ) and reduces near the margins of these convective regions (subtropics) as a result of moisture increase and deficit in the atmospheric boundary layer, respectively. The moisture deficit in the subtropics is further accentuated by the implied subsidence from the increased convection in the convective regions and reduction in the convergence of moisture. For the warm ENSO events (inset), the warming spreads across the tropics by wave dynamics. (Reproduced from Neelin et al. 2003)

is also debate on the changing frequency of the CP and EP types of El Niño in a future warm climate (Collins et al. 2010; Timmermann et al. 2018). We will now try to parse some of these uncertainties in projections of ENSO variability under the global warming scenario.

There are two different mechanisms to explain the El Niño and La Niña types of mean states in the projected climate (Vecchi et al. 2008). In the projections that suggest a La Niña type of mean state, Clement et al. (1996) claim an 'ocean thermostat' mechanism is in play. In this mechanism, under global warming, the Eastern Equatorial Pacific will tend to warm less than the Western Pacific raising the zonal temperature gradient. This is because the upwelling in the Eastern Pacific will oppose the surface heating (from increased atmospheric heat fluxes due to global warming) in contrast to Western Pacific where the thermocline is deep and upwelling is much weaker. The increased zonal gradient of the equatorial SST in the Pacific will then lead to stronger easterlies, promoting further cooling by upwelling and further enhancing the SST gradient leading to a La Niña-type mean state. On the other hand, the projections of the El Niño-type mean state claim that the anticipated weakening of the Walker Circulation in a warming climate will cause a weakening of the trade wind easterlies and, by way of the Bjerknes feedback mechanism, resulting in the weakening of the zonal SST gradient in the Equatorial Pacific.

These mechanisms were tested with numerical models (Fig. 10.4). Clement et al. (2006) used the Cane-Zebiak model in which the atmospheric processes are simplified and do not represent the weakening of the Walker Circulation in response to global warming. In this model, the ocean thermostat mechanism tends to dominate, resulting in a La Niña type of mean state in the projected climate (Fig. 10.4a). In contrast, in GCMs coupled to simplified (mixed layer or slab) ocean models, the weakening Walker Circulation dominates the Tropical Pacific response, leading to an El Niño type of mean state (Fig. 10.4b). In these types of models, the ocean is thermodynamically coupled to the atmosphere (with exchange of heat fluxes), but ocean dynamics is fixed (with no ocean currents or upwelling), which eliminates the ocean thermostat mechanism. In the coupled GCMs (coupled between AGCM and OGCM), where both ocean thermostat and weakening Walker Circulation mechanisms are operating, the Tropical Pacific response shows a more uniform warming across the basin (Fig. 10.4c). This suggests that the diminished El Niño type of mean state in the coupled GCM relative to that of the AGCM coupled to mixed layer ocean models is a result of the opposing forcing between the ocean thermostat and the weakening Walker Circulation mechanisms.

In comparing two observational datasets of SST over 100 years long, Vecchi et al. (2008) found very contrasting temperature trends in the Equatorial Pacific (Fig. 10.4d, e). The SST trends from the HadISST dataset (Rayner et al. 2003), suggest a La Niña type of pattern with an increase in the zonal gradient of the Equatorial Pacific SST over time (Fig. 10.4d). But in the case of NOAA ERSST (Smith and Reynolds 2004), the long-term SST trends show an El Niño type of pattern, suggesting a decreased zonal gradient of the Equatorial Pacific SST over time (Fig. 10.4e). This rather ironic contrast in the SST trends from two different reconstructions of historical SST has been however resolved (Vecchi et al. 2008). It is shown that over two periods of time, around the 1930s and 1980s, the two datasets showed very different trends in the zonal gradient of equatorial SST (Fig. 10.5). These periods coincided with periods of greatest change in the observations of SST. In the 1930s, the switch was made from measuring using buckets lowered into the water to using the ship engine room water intake. The correction applied to address this change in the observation of the SST is different in HadISST and ERSST. Furthermore, the 1980s marked the beginning of the satellite infrared SST retrievals, which are used in HadISST and not in ERSST. However, the structure of the SST trends in ERSST is confirmed with SLP data that confirm the weakening of the Walker Circulation (Vecchi et al. 2006; Bunge and Clarke 2009).

Paleoclimate records and paleoclimate modeling studies indicate that ENSO variability continues to be robust in both warmer (Huber and Caballero 2003; von der Heydt et al. 2011) and cooler (Zheng et al. 2008) climates than the present. During the warm, early Pliocene period (~4.5 to 3 million years ago), the zonal gradient of the Equatorial Pacific SST was about 2 °C, which is like the gradient prevalent during modern El Niño events (Wara et al. 2005). In addition, Pliocene records suggest that ENSO frequency and amplitude were not significantly different from today (Watanabe et al. 2011; Scroxton et al. 2011).

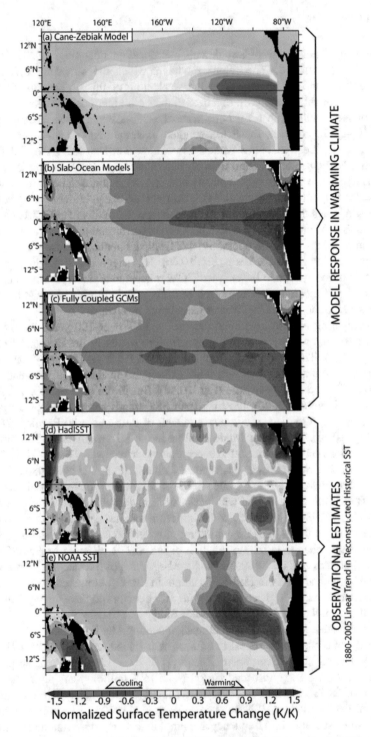

Fig. 10.4 The response of the Tropical Pacific SST to a warming climate of varying complexity of the climate model from (**a**) the Cane-Zebiak model (that has a simple atmosphere and

There is, however, evidence emerging that there is an increasing number of CP El Niño events during the most recent decades (since the 1980s) that is unusual in a multi-century context (Freund et al. 2019). However, the mechanisms leading to the increased frequency of the CP El Niño events remain elusive. The future changes in El Niño diversity from coupled GCMs also are inconclusive with model biases and decadal variability providing further challenges. Freund et al. (2020) suggest that future changes in El Niño are stronger for CP than for EP events. Furthermore, this study claims that GCMs with La Niña-like projected warming in the Tropical Pacific tend to produce more frequent EP than CP events compared to models that project El Niño-like warming patterns. In addition, it is found that models that project El Niño-like warming produce more and intense El Niño events during the positive phase of the PDO, suggesting that future changes in ENSO variability depend on mean state warming and decadal scale natural variability.

Alternatively, Cai et al. (2020) claim that the chaotic variability of the climate system has a significant role to play in the future projections of ENSO variability. Through numerical climate model experiments, they show that infinitesimal random perturbation to identical initial conditions (the so-called butterfly effect) can lead to drastically different trajectories of ENSO variations in a future climate. For example, when the butterfly effect leads to higher initial (ENSO) variability, a greater cumulative Equatorial Pacific oceanic heat loss occurs, which reduces stratification of the upper Equatorial Pacific Ocean, leading to a smaller increase in ENSO variability toward the end of the century under greenhouse warming. By similar arguments, it is seen that there is an increase in ENSO variability toward the end of the twenty-first century in model simulations with lower initial ENSO variability. In other words, ENSO stores memory of its past variability and organizes its future, accordingly, touting a self-modulation ENSO mechanism. This may explain some of the diversity in the ENSO projections from the climate models. Many studies have pointed to the fact that sharper thermocline (or strong vertical gradients of potential temperature or strong stratification) in the mean state produces stronger ENSOs. Meehl et al. (2001) suggest that for a given surface wind variation, it is easier for equatorial upwelling in a more stratified ocean to draw the cold water below the thermocline to produce greater SST variability. This has a positive feedback on the surface winds, which further accentuates the zonal gradient of SST and the zonal slope of the thermocline that feeds back to ENSO variability.

Fig. 10.4 (continued) relatively complex ocean model that includes ocean dynamics) and from the ensemble average of 13 different models from the CMIP3 suite to doubling of CO_2 concentration in (**b**) AGCM coupled to slab ocean models (representing simple atmosphere and complex ocean) and (**c**) the same AGCM as in (**b**) but now coupled to OGCM. The trends of the observed SST in the Tropical Pacific were computed over the period of 1880–2005 from (**d**) HadISST and (**e**) ERSST. (Reproduced from Vecchi et al. 2008)

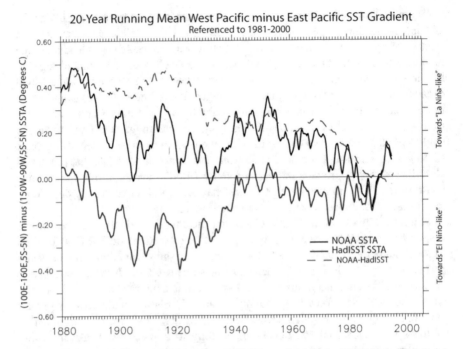

Fig. 10.5 The time series of the 20-year mean of the zonal gradient of SST in the Equatorial Pacific from two historical, reconstructed SST datasets: ERSST (NOAA SSTA; Smith and Reynolds (2004)) and HadISST (HadISST SSTA; Rayner et al. 2003). (Reproduced from Vecchi et al. 2008)

10.4   The Tropical Atmospheric Circulation

Amid all the uncertainty of the climate change projections, it is quite certain that lower-tropospheric water vapor content will rise as the climate warms. This is a direct result of the Clausius-Clapeyron equation:

$$\frac{de_s}{dT} \approx \frac{L_v e_s}{R_v T^2} \tag{10.1}$$

where e_s, T, L_v, R_v are saturation vapor pressure, temperature, latent heat of vaporization, and gas constant for water vapor. This equation for the rate of change of water vapor with respect to temperature implies that warmer temperatures can hold more water vapor than colder temperatures. Linearizing Eq. 10.1 around 0 °C and using $L_v = 2.5 \times 10^6 \text{JKg} - 1$ and $R_v = 461.5 \text{ JKg}^{-1}\text{K}^{-1}$, we get

$$\frac{de_s}{dT} \cong 0.073\,(1 - 0.007T) \tag{10.2}$$

where T in Eq. 10.2 is in Celsius. So at the surface for T=0 °C, we find from Eq. 10.2 that saturation vapor pressure increases by approximately 7% per degree Celsius and by around 6% per degree at 24 °C. The column-integrated water vapor content is dominated by water vapor in the lower troposphere. Held and Soden (2006) show that the changes in column-integrated water vapor both in future climate projections with CO_2 doubling by the end of the twenty-first century (Fig. 10.6a) and in the current climate simulations with CO_2 concentration following the observed trajectory (Fig. 10.6c) follow the changes as dictated by the Clausius-Clapeyron equation. But the changes in the global mean precipitation scale comparatively at a much lower rate of ~2%K^{-1} both in the future (Fig. 10.6b) and in the current (Fig. 10.6d) climate. In fact, in the twentieth-century simulations, the models show a reduction in global mean precipitation rather uniformly below the linear fit for

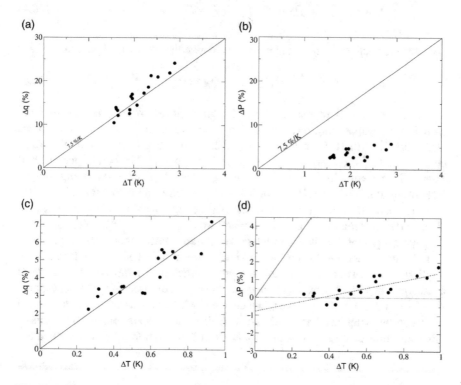

Fig. 10.6 The scatter plot of the percentage change in the global mean (**a, c**) column-integrated water vapor and (**b, d**) precipitation with global mean change in surface air temperature averaged over 20 CMIP3 models for (**a, b**) A1B emission scenario projections and (**c, d**) twentieth-century simulations. The changes are computed as the difference between the first-year and the last 20-year means that correspond for (**a, b**) as the difference between the means of 2080–2100 and 2000–2020 and (**c, d**) 1980–2000 and 1860–1880. The solid lines are the rate of increase in column-integrated water vapor with temperature (7.5%K^{-1}), and the dashed line is the linear fit for the scatter between ΔP and ΔT, whose slope is 2.2%K^{-1}. (Reproduced from Held and Soden 2006; © American Meteorological Society. Used with permission)

the twenty-first-century projections by about 1%, which is likely the result of the increasing concentration of aerosols (Ramanathan et al. 2001).

It is expected that the change in precipitation with temperature should be smaller than that of the integrated column water vapor as implied by the Clausius-Clapeyron equation. This is because the changes in the global mean precipitation or evaporation stem either from a change in Bowen's ratio (=sensible heat flux/latent heat flux) and or net change in radiative flux at the surface. This follows from globally and vertically integrating Eqs. (4.1) and (4.2), which give rise to

$$Q_1 \equiv \oint \int_{p_s}^{P_{top}} q1 = Q_R + P - E \qquad (10.3)$$

$$Q_2 \equiv \oint \int_{p_s}^{P_{top}} q2 = P - E \qquad (10.4)$$

and then subtracting the resulting equations to give the following equation:

$$Q_1 - Q_2 - Q_R = 0 \qquad (10.5)$$

where $\oint$, P, and E represent global integral, global mean precipitation, and global mean evaporation, respectively. Therefore, in order to maintain the balance represented in Eq. 10.5, an increase in radiative flux and or evaporation must be compensated by a corresponding change in precipitation. Held and Soden (2006) show that the combination of the radiative effect of uniform warming and an increase in albedo can explain at best $1\%K^{-1}$ increase in precipitation. They also indicate that a decrease in Bowen's ratio plays a significant role in generating the slope between ΔP and ΔT but is still much weaker than that implied by Clausius-Clapeyron equation for change in integrated column water vapor with temperature.

Held and Soden (2006) propose a simple conceptual model to understand the implication of this scaling of moisture and rainfall with temperature on tropical circulation. In a broad sense, one can conceive air parcels ascending the troposphere from the moist boundary layer, which eventually condenses, precipitates, and returns to the boundary layer with much smaller water vapor content. If we ignore this return flow of water vapor, the global mean precipitation (P) can be written as

$$P = Mq \qquad (10.6)$$

where M is the mass flux ($\equiv \rho w \equiv \frac{kg}{m^2 s}$; where ρ, w are the density of air and vertical velocity, respectively) and q is boundary layer mixing ratio. Since q in Eq. 10.6 scales with temperature as per the Clausius-Clapeyron equation and P increases much more slowly, then by Eq. 10.6, M must decrease at a rate less than the Clausius-Clapeyron rate. In other words, the vertical velocity must decrease as the climate warms. Since the bulk of precipitation and evaporation occurs in the tropics, it is suggested that the mass flux will decrease in the precipitating towers as the temperature increases. Held and Soden (2006) show that in all climate models of CMIP3, indeed the mass flux decreases. In the non-convecting regions of the tropics, the radiative cooling balances the adiabatic warming $\left(\equiv \omega \frac{\partial \theta}{\partial p} \right)$, but with reduced upward convective mass flux in the precipitating regions, the implication is that the corresponding subsidence in the non-precipitating region decreases. Since radiative cooling cannot change as rapidly as the mass flux, then the only way to maintain the energy balance in the precipitating region is for the static stability ($\equiv \frac{\partial \theta}{\partial p}$) to increase. Even in the non-convective regions of the tropics, the temperature in the free troposphere follows the moist adiabat, which averaged over the column is proportional to the boundary layer mixing ratio. Therefore, in a warming climate, as the boundary layer moisture increases, the static stability increases proportionately. Held and Soden (2006) argue that redistribution of the mass flux can affect the strength of the circulation independent of the radiatively driven subsidence. They measure the former by examining the spatial variance of the mass flux. They divide the total mass flux into its zonal mean component and the deviations about it (or the stationary eddy component). They find that the reduction in the total mass flux is dominated by the reduction of the stationary eddy component, while the change in the zonal mean component is much smaller (Fig. 10.7a and b). In other words, the implication here is that the zonal mean meridional (Hadley) circulation does not decrease in strength as fast as the zonal overturning Walker Circulation. He and Soden (2015) conducted extensive model experiments with AGCM that tested the impact of direct CO_2 forcing, mean SST warming, and the pattern of SST warming on the anthropogenic weakening of the tropical circulation. Their conclusions were as follows:

- The mean SST warming is the largest contributor to tropical mean circulation through increased tropospheric stratification from its influence on raising tropical mean moisture and latent heat release.
- The direct CO_2 forcing contributes to the moderate weakening of the tropical mean circulation through radiative warming of the atmosphere but with moderate influence on the stratification of the lower troposphere.
- The pattern of SST warming has very little impact on the circulation.
- The direct CO_2 forcing produces the greatest weakening of the tropical circulation over the convective zones in the oceans, largely as a result of larger land-ocean thermal contrast and strengthening convection over land.
- The mean warming of the SST weakens convection both over the ocean and land; the weakening of the Walker Circulation is primarily caused by the mean SST warming.

Unlike the conclusions drawn on the 1–3% increase in global mean precipitation with every degree increase in global mean surface temperature, for example, in Held and Soden (2006), the annual maximum daily extreme rainfall is estimated to increase between 5.9% and 7.7% per degree Celsius rise of globally averaged near surface temperature (Westra et al. 2013). In fact, there is emerging evidence that tropical extreme rainfall intensity increases with temperature at rates of about double the Clausius-Clapeyron scaling rate—the so-called 'super-Clausius-Clapeyron' rate

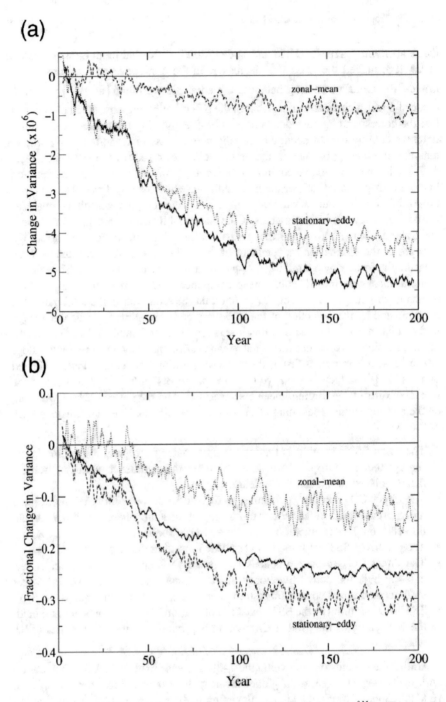

Fig. 10.7 The time series of the **(a)** absolute (M) and **(b)** fractional change ($\frac{\delta M}{M}$) of the monthly mean mass flux at 500 hPa averaged over the tropics (30°N–30°S) from the GFDL CM2.1 model using A1B emission scenario. The total mass flux, zonal mean, and the stationary eddy component of the mass flux are shown in solid, dashed, and dotted lines, respectively. The results are smoothed using a 5-year running mean. (Reproduced from Held and Soden 2006; © American Meteorological Society. Used with permission)

(Allan et al. 2010; Lau and Wu 2011). This enhancement of extreme precipitation is partly attributable to the increased contribution of convective over the stratiform type of precipitation (Berg and Haerter 2013). Convective rainfall by nature is more intense and has shorter durations, and intensity tends to increase with the increase in temperature as the humidity content increases. However, regions where the daily mean temperature exceeds ~24 °C display a decreasing extreme rainfall intensity with increasing temperature as a result of a reduction in relative humidity, suggesting the decrease of moisture availability as the temperature rises above this threshold (Hardwick-Jones et al. 2010; Utsumi et al. 2011). The mechanism for the associated decrease in moisture availability beyond the temperature threshold of 24 °C, which leads to a decrease in extreme rainfall intensity, however, remains elusive (Westra et al. 2014). van Oldenborgh et al. (2017) in examining the impact of climate change on Hurricane Harvey which made landfall in Houston, Texas, in 2017 suggested that the extreme rainfall intensity increase at the super-Clausius-Clapeyron rate is likely a result of an increase in moisture flux both from higher moisture content and stronger winds or updrafts driven by the heat of condensation of the moisture.

10.5   The Tropical Atlantic Ocean

One of the most critical aspects of the impact of global warming on the Tropical Atlantic Ocean is its impact on the fate of the Atlantic TCs. How will they change in a future warm climate? Knutson et al. (2010) concluded from their study that owing to observational uncertainties, the historical trends in Atlantic TC intensity and frequency are comparatively small relative to natural variations. The uncertainties of observing Atlantic TCs have been highlighted in Landsea et al. (2010). They show, for example, that short-lived TCs in the Atlantic are observed more frequently in the recent decades from remotely sensed observations compared to the late nineteenth century and early parts of the twentieth century. Similarly, even landfalling Atlantic TCs in the United States have failed to show long-term trends. However, the SSTs in the main development region of the TCs in the Tropical Atlantic Ocean (~10°N–20°N and 17.5°W–65°W) have increased rapidly in the recent decades, which is partly a result of increasing GHG emissions and partly a result of the natural variations (e.g., the Atlantic Multidecadal Oscillation; Murakami et al. 2018). But this warming in the Tropical Atlantic has remained typical and has not outpaced the warming in other tropical ocean basins, which has a bearing on the projected trends of the Atlantic TCs (Knutson et al. 2010). The emerging projections from high-resolution climate models are that the overall frequency of the Atlantic TCs will decrease from anywhere between 6% and 34%, but the frequency of the strongest hurricanes (Categories 3–5) may increase between 2% and 11% (Knutson et al. 2010) in a future warm climate. But even at very high levels of CO_2 concentration (e.g., around 9000 ppm), Korty et al. (2017) find that there is a tendency for decreased frequency of the low-latitude TCs owing to increased vertical shear, drying in the middle troposphere from a reduced mass flux in deep convection and broadening of the TC activity to higher latitudes (subtropics). It should be noted that these are estimates based on model projections, which unfortunately display grave errors in the Tropical Atlantic Ocean (Richter et al. 2014). So over time, as models improve, the projections could vastly differ from prior estimates.

10.6   The Tropical Indian Ocean

Among all the tropical ocean basins, the Tropical Indian Ocean is the smallest and the warmest. It also exhibits some of the largest warming trends (~1 °C/century), which are basin-wide, manifesting at the surface and in the subsurface ocean (Hoerling et al. 2004; Alory et al. 2007; Gnanaseelan et al. 2017; Fig. 10.8a). The peak surface warming in the basin is observed over the Equatorial Indian Ocean with the warming rate exceeding by almost two to three times over that in the Tropical Pacific Ocean. It may be noted that in Fig. 10.8, the trend is computed at each grid point after removing the corresponding global mean SST so that the warming trend can be interpreted as SST warming in excess of global warming. This warming in the Tropical Indian Ocean is significantly stronger during the ISM season than in the remaining part of the year. This warming is intriguing given that the long-term trend of the net heat flux over the Indian Ocean are decreasing (Rahul and Gnanaseelan 2016). Furthermore, the warming rate is exceptional to the Tropical Indian Ocean among other tropical basins, which confounds the theory of increasing GHG concentration as the obvious cause of this warming. Swapna et al. (2014) suggest that this warming in the Tropical Indian Ocean coincides

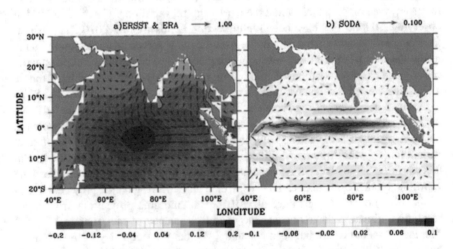

Fig. 10.8 (a) The linear trend of the annual mean SST (shaded; °C/decade) overlaid with linear trend 10 m surface wind (vectors; ms^{-1}/decade) computed over the period of 1958–2015. (b) Linear trend in the annual zonal current (shaded; ms^{-1}/decade) overlaid with linear trend of the surface current (vectors; ms^{-1}/decade). (Reproduced from Gnanaseelan et al. 2017)

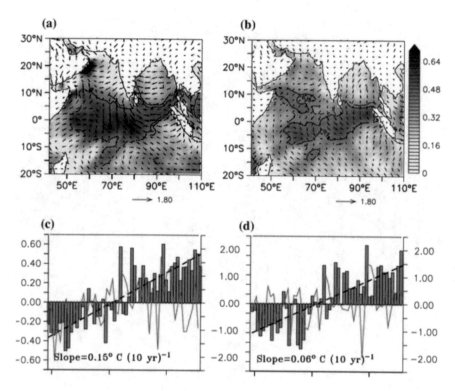

Fig. 10.9 Linear trends in SST (shaded; °C per 62 years) and 10 m surface winds (vectors; ms^{-1} per 54 years) for (**a**) the summer monsoon season (June to September) and (**b**) the remaining calendar months (October to May). The contours in panels (**a**) and (**b**) suggest that the trends exceed the 99% confidence interval. The time series of SST (°C; bars) and zonal wind (ms^{-1}; red line) anomalies averaged over the Equatorial Indian Ocean (50°E–100°E and 5°S–5°N) for (**c**) the ISM season and the (**d**) the remaining calendar months. The linear trends indicated in panels (**c**) and (**d**) exceed the 95% confidence interval. The trends have been computed after removing the corresponding global mean SST so that the warming trends can be interpreted as warming in excess of the global mean SST warming rates. (Reproduced from Swapna et al. 2014)

with the corresponding weakening trend of the cross-equatorial surface winds and strengthening trend of the westerly flow along the Equatorial Indian Ocean (Fig. 10.9). These enhanced westerlies promote downwelling and thermocline deepening in the Equatorial Indian Ocean. The validity of the weakening trend of the monsoon circulation was also verified by the weakening trend of the meridional gradient of the mean sea level pressure between the monsoon trough and the subtropical high over the Southern Indian Ocean (Krishnan et al. 2012). This warming of the Indian Ocean has also coincided with the reorientation of the northward-propagating intraseasonal modes of the ISM. For example, many studies find that in recent decades, stationary convection over the Tropical Indian Ocean has stalled the northward propagation, leading to more frequent and extended monsoon breaks (Krishnan et al. 2006; Ramesh Kumar et al. 2009; Sabeerali et al. 2014).

Glossary

Accumulated Cyclone Energy This is a metric to measure global tropical cyclone activity or tropical cyclone activity in a region/ocean basin. It is obtained by taking the cumulative sum of the square of the maximum sustained wind speed measured every 6 h over the lifetime of the named tropical cyclone.

Adiabatic Cloud Water Content This is the difference between the air parcel's initial and final water vapor mixing ratio. It is also referred to as the condensed water mixing ratio. It can be easily ascertained from a thermodynamic diagram as shown in Fig. G1.

However, adiabatic cloud water content is rarely observed because the air parcel in the rising updraft is often diluted by entrainment of the environment air. Furthermore, the precipitation or rain out of excess condensed water also reduces the adiabatic cloud water content.

Altimetry The measurement of altitude is called altimetry. It is usually used in the context of measuring ocean surface topography using a satellite altimetry. The ocean surface topography is the deviation of the height of the ocean surface with respect to a reference, which is the geoid (center of the Earth where the Earth's gravity field is uniform). The ocean surface topography is caused by the loading of the overlying atmospheric pressure, waves, tides, and currents. Satellite altimetry combines precise orbit determination with accurate ranging of the satellite from the ocean surface by a microwave altimeter that bounces microwave pulses off the ocean surface and measures the time it takes for the pulses to return to the spacecraft.

Angular Momentum The angular momentum about a point O of a body with a linear momentum (L) is its cross product with the position vector (r) of the body relative to O. Mathematically, it can be written as $\vec{M} = \vec{L} \times \vec{r}$. However, in a more common application of angular momentum in meteorology, we consider absolute angular momentum per unit mass (γ), given by

© Springer Nature Switzerland AG 2023

V. Misra, *An Introduction to Large-Scale Tropical Meteorology*, Springer Atmospheric Sciences, https://doi.org/10.1007/978-3-031-12887-5

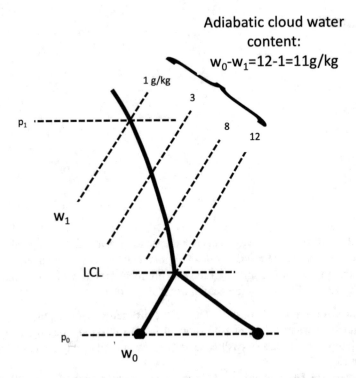

Fig. G1 An illustration of adiabatic cloud water content on a skew-T log p diagram is the difference between the initial and final mixing ratio of the air parcel, which in this case is 11 g/kg. The slanted dashed lines are mixing ratio lines, p_0 and p_1 refer to the initial and the final pressure of the air parcel, and w_0 and w_1 refer to the initial and final mixing ratio of the air parcel, respectively. The LCL is the Lifting Condensation Level

$$\gamma = \omega R \cos \phi = \overbrace{(\Omega R \cos \phi + u)}^{\text{Absolute velocity}}. \quad \overbrace{R \cos \phi}^{\text{radius of rotation}} \qquad \text{(A1)}$$

The absolute velocity in Eq. A1 is the sum of the angular velocity of the Earth and the relative zonal wind speed (u), and R is the radius of the Earth ($=6370$ km). The angular velocity of Earth (Ω) is $\frac{2\pi}{86164.1} = 7.292 \times 10^{-5}$ rad s^{-1}. In the absence of external forces or torques, angular momentum is conserved. An enlightening consequence of this conservation is illustrated by the following example. Let us suppose an air parcel moves north from latitude ϕ_1 to latitude ϕ_2. Initially, its radius of rotation, r_1, is

$$r_1 = R \cos \phi_1 \qquad \text{(A2)}$$

Its angular momentum per unit mass is

$$\gamma_1 = \omega_1 \left(R \cos \phi_1 \right) \tag{A3a}$$

At latitude ϕ_2, the new angular momentum is

$$\gamma_2 = \omega_2 \left(R \cos \phi_2 \right) \tag{A3b}$$

But by conservation of angular momentum, $\gamma_1 = \gamma_2$, which implies

$$\omega_2 = \omega_1 \left(\frac{\cos \phi_1}{\cos \phi_2} \right) \tag{A4}$$

Eq. A4 then suggests that $\omega_2 > \omega_1$, which means that the air parcel at latitude ϕ_2 rotates around the axis of the Earth more quickly than at latitude ϕ_1. Since Ω is constant, the relative zonal speed will change when the distance from the axis of rotation changes as the parcel moves from latitude ϕ_1 to ϕ_2. So an observer in NH will see the air parcel as having a component to the east, which is a deviation to the right from its original path in this case when the air parcel moves from latitude ϕ_1 to ϕ_2. In the SH, the same movement of the air parcel moving south will result in the departure from its original path to the west.

Buoyancy (Chap. 1) From Archimedes' principle, the upward force (buoyancy) exerted on a body immersed in a fluid is equal to the weight of the fluid it displaces. In the context of an air parcel, the net upward force (F) experienced by it is

$$F = \left(\rho' V - M \right) g \tag{A5}$$

where ρ' is the density of the surrounding fluid (environment), V is the total volume of the displaced fluid, and M is the mass of the air parcel. Substituting $M = \rho V$, where ρ is the density of air parcel, and dividing Eq. A5 all through by M, we get

$$f = \frac{F}{M} = \frac{\left(\rho' - \rho \right) V g}{\rho V} = \left[\frac{\left(\rho' - \rho \right)}{\rho} \right] g \tag{A6}$$

Using the ideal gas law, Eq. A6 may also be written as

$$f = \left[\frac{\left(T_v - T_v' \right)}{T_v'} \right] g \tag{A7}$$

where T_v and T_v' are the virtual temperature of the air parcel and the surrounding air (environment), respectively.

Clausius-Clapeyron Equation This is an important relationship that relates saturation vapor pressure to temperature. It is named after Rudolf Clausius and Benoit Paul Emile Clapeyron. This equation comes as a consequence of the first law of thermodynamics when it is applied at the point of phase transition of water (which

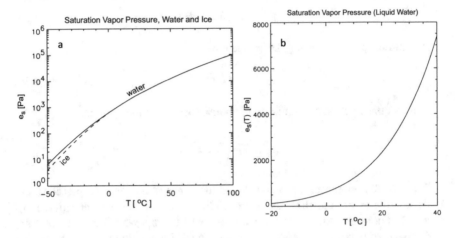

Fig. G2 Plot of saturation vapor pressure with temperature (**a**) for both water and ice with ordinate in log scale and (**b**) for water with ordinate in linear scale. Both the panels are identical and describe the Clausius-Clapeyron equation but plotted on different scales for saturation vapor pressure on the y-axis and a different temperature range on the x-axis

is isobaric and isothermal). This point allows for the Gibbs free energy to be the same in either phase of water to provide the following complete relationship of the Clausius-Clapeyron equation:

$$\frac{de_s}{dT} = \frac{L_v}{T(\alpha_2 - \alpha_1)} \tag{A8}$$

where e_s, T, L_v, α_1, and α_2 are saturation vapor pressure, temperature, latent heat of vaporization, the specific volume of liquid water, and specific volume of water vapor, respectively. However, under typical atmospheric conditions, $\alpha_2 \gg \alpha_1$, and further assuming water vapor behaves as an ideal gas, then the above equation can be written in the more familiar form of Clausius-Clapeyron equation as

$$\frac{de_s}{dT} \approx \frac{L_v}{T\alpha_2} = \frac{L_v}{R_v T^2} \tag{A9}$$

where R_v is the gas constant for water vapor. The Clausius-Clapeyron equation describes the slope of the tangent to the curves plotted between temperature and saturation vapor pressure in Fig. G2.

Conditional Instability Conditional instability refers to atmospheric stability of a parcel of air becoming unstable, conditioned on the premise that the air parcel becomes moist enough so that its lapse rate ($\Gamma = \frac{\partial \theta}{\partial z}$) exceeds the saturated lapse rate ($\Gamma_m = \frac{\partial \theta_e}{\partial z}$) but is less than that of the dry adiabatic lapse rate ($\Gamma_d = \frac{g}{c_p} =$

9.8 °C/km). In relation to conditional instability, the following relationships are useful to know:

$\Gamma = \Gamma_d$ air parcel is dry neutral
$\Gamma = \Gamma_m$ air parcel is saturated neutral
$\Gamma_m < \Gamma < \Gamma_d$ air parcel is conditionally unstable
$\Gamma > \Gamma_d$ air parcel is absolutely unstable

Equilibrium Level The equilibrium level is also referred to as the level of neutral buoyancy. This is the height or level at which the temperature of the buoyant air parcel becomes equal to the temperature of the environment.

Equivalent Potential Temperature The equivalent potential temperature is the temperature an air parcel would have after all its moisture was condensed out by a pseudo-adiabatic process (i.e., latent heat of condensation being used to heat the air parcel), and the air parcel brought dry adiabatically back to 1000 hPa. Mathematically, it is given by

$$\theta_e = \theta^{\frac{L_v w}{c_p T_{LCL}}} \tag{A10}$$

where θ_e, θ, w, L_v, c_p, and T_{LCL} are equivalent potential temperature, potential temperature, mixing ratio, latent heat of condensation, the specific heat capacity of dry air, and temperature at LCL, respectively. It may be noted that while θ is conserved only for the dry adiabatic process, θ_e is conserved for both dry and moist adiabatic processes. Since θ_e is conserved during the saturated adiabatic process, moist adiabats in a thermodynamic diagram (e. g., skew $-$ T log p diagram) may be regarded as lines of constant θ_e.

Expendable Bathythermograph An expendable bathythermograph (XBT) is a probe that is dropped at a known rate from a ship, which measures temperature as it falls through the depth of the water. A resistance in the head of the probe connected by a thin twin wire to the equipment on the ship measures the temperature of the water. XBTs represented the largest fraction of temperature profile observations in the oceans since the 1970s up until 2007, when Argo profiling floats were fully implemented.

Geostrophy Geostrophy refers to the winds resulting from the balance between the pressure gradient force and the Coriolis force. The Coriolis force occurs when the pressure gradient force sets air into motion. In the NH, the Coriolis force pulls to the right until it completely opposes the pressure gradient force, and the net force becomes zero. At this stage of the equilibrium, winds blow uniformly, parallel to isobars. Mathematically, the geostrophic balance is given by

$$f u_g = -\frac{\partial \phi}{\partial y} \tag{A11}$$

and

$$f v_g = \frac{\partial \phi}{\partial x} \tag{A12}$$

where u_g, v_g, and ϕ are zonal and meridional components of the wind and geopotential height, respectively. Under this geostrophic balance, the winds are non-divergent.

In the quasi-geostrophic approximation, the total wind is divided into geostrophic ($\vec{V}_g$) and ageostrophic ($\vec{V}_{ag}$) components of the wind so that the Eulerian form of the momentum equation may be written as

$$\frac{\partial \left(\vec{V}_g + \vec{V}_{ag} \right)}{\partial t} + \left(\vec{V}_g + \vec{V}_{ag} \right) \nabla \left(\vec{V}_g + \vec{V}_{ag} \right) + \omega \frac{\partial \left(\vec{V}_g + \vec{V}_{ag} \right)}{\partial p}$$

$$= -f\hat{k} \times \left(\vec{V}_g + \vec{V}_{ag} \right) + \hat{k} \times \nabla \phi \tag{A13}$$

It turns out from scale analysis that all the ageostrophic components of the winds on the LHS (only) of Eq. A13 are very small relative to the geostrophic component and therefore neglected. Furthermore, the vertical advection term is also small and neglected. On the other hand, the ageostrophic components of the winds on the RHS of Eq. A13 are of the same order of magnitude as the retained geostrophic terms on the LHS ($\sim 10^{-4}$). Therefore, the resulting equation from scale analysis of Eq. A13 yields the quasi-geostrophic momentum equation:

$$\frac{\partial \vec{V}_g}{\partial t} + \vec{V}_g \nabla \vec{V}_g == -f\hat{k} \times \left(\vec{V}_g + \vec{V}_{ag} \right) + \hat{k} \times \nabla \phi \tag{A14}$$

Hypsometric Equation The hypsometric equation is an equation relating the thickness between two pressure levels to the mean virtual temperature of the layer. It is given by

$$\Delta z = z_2 - z_1 = \frac{R_d \overline{T}_v}{g} \ln \left(\frac{p_1}{p_2} \right) \tag{A15}$$

where Δz is the thickness between two pressure levels, p_1 and p_2, at geometric heights of z_1 and z_2, respectively. R_d is the gas constant for dry air, and $\overline{T}_v$ is the mean virtual temperature of the layer.

Inertial Stability) Inertial stability refers to the resistance of the transverse horizontal displacement. Mathematically,

$$\frac{\partial M}{\partial r} > 0 : \text{inertially stable} \tag{A16}$$

$$\frac{\partial M}{\partial r} = 0 : \text{inertially neutral} \tag{A17}$$

$$\frac{\partial M}{\partial r} < 0 : \text{inertially unstable} \tag{A18}$$

where M (in natural coordinate with coordinate r to the right of the wind and coordinate n in the direction of the geostrophic wind) is absolute momentum$\equiv f_0 r + v_g$ and $\frac{\partial M}{\partial r} = \eta \equiv$ absolute vorticity, v_g is the geostrophic wind speed on an f-plane. In many studies, absolute momentum, M is defined for purely zonal geostrophic flow ($\equiv f_0 y - u_g$) in which case the derivatives in Eqs. A16, A17, and A18 are taken with respect to y instead of r. Therefore, inertial instability occurs when absolute vorticity is negative in the NH or positive in the SH. Thus, negative absolute vorticity in NH or positive absolute vorticity in SH is rarely seen at synoptic scales because if it does occur, then inertial instability would occur. The concept of inertial stability is also applicable to curved flow as in a vortex, with the resistance of the vortex to maintain its structure in the presence of forcing from other weather circulation systems termed inertial stability. In this instance, inertial stability (I^2) is defined as

$$I^2 = \eta = (\xi + f_o)\left(f_o + \frac{v}{r}\right) \tag{A19}$$

In the case of the curved flow, the background flow is assumed to be in gradient wind balance, and the radial gradient of the angular momentum ($M_a \equiv vr + \frac{f_o r^2}{2}$) is considered instead of the gradient of the absolute momentum for transverse flow. From Eq. A19, inertial stability increases proportionally with relative vorticity (ξ), local Coriolis parameter (f_o), and relative angular momentum ($\frac{v}{r}$).

Level of Free Convection The level of free convection refers to a level wherein the air parcel will ascend freely further without any further input of energy because it is positively buoyant (i.e., the surrounding air is colder and denser than the air parcel). As illustrated in Fig. G3, on a thermodynamic diagram, the level of free convection occurs when the moist adiabat followed by the air parcel crosses from the cold side of the environmental profile to the warm side.

Mass Continuity Equation The mass continuity equation is another version of the principle of conservation of mass, which states that mass can neither be created nor destroyed but can be redistributed. The mass continuity equation or equation of continuity states that the amount of air or density of air can change for a fixed volume of air (or air column extending from the surface to the top of the atmosphere) either by air mass transformation (or density advection) or by divergence (i.e., removing or adding air to the column). Mathematically, this may be written as

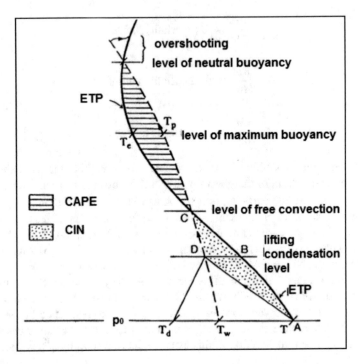

Fig. G3 Thermodynamic diagram showing surface air parcel ascent along A-D-C to the level of neutral buoyancy or EL. The LFC is also depicted. (Reproduced from https://www.weather.gov/source/zhu/ZHU_Training_Page/Miscellaneous/inversion/inversion.html)

$$\frac{\partial \rho}{\partial t} + \vec{V}.\nabla \rho + \rho \nabla.\vec{V} = 0 \tag{A20}$$

At synoptic (≥ 1000 km) scales, density advection is negligible compared to the three-dimensional divergence term, and so the mass continuity equation is more simply in x, y, z coordinate system is written as

$$\nabla.\vec{V} \equiv \frac{\partial u}{\partial x} + \frac{\partial v}{\partial y} + \frac{\partial w}{\partial z} \tag{A21}$$

Mixed Layer Depth The mixed layer depth features nearly uniform properties of water (e.g., temperature, density). There are a couple of ways for defining the mixed layer depth. One such definition is based on temperature, which defines the mixed layer depth as the depth at which the sea surface temperature cools by 0.2 °C. Alternatively, mixed layer depth can also be based on density, which is the depth at which density is equal to surface density plus the density difference brought about by a surface temperature decrease of 0.2 °C. Interestingly, the barrier layer (a layer of freshwater that causes a strong stratification in the upper ocean) is diagnosed as the difference between the mixed layer depth calculated from the temperature threshold and the mixed layer depth based on the density criterion.

Mixed Layer Ocean Model A mixed layer ocean model is a one-dimensional ocean model that is used to represent the variations of the mixed layer from the variations in the atmospheric flux. There are no ocean current and dynamics included in this model. The equation for a mixed layer ocean model is given by

$$\frac{dT}{dh} = \overbrace{\frac{Q}{\rho h C_p}}^{\text{Term 1}} - \overbrace{\frac{\Delta T}{h} w_{\text{entrain}}}^{\text{Term 2}} \tag{A22}$$

where T, ρ, h, C_p, and w_{entrain} are mixed layer ocean temperature, density of ocean water in the mixed layer, depth of the mixed layer, specific heat capacity of seawater, and entrainment velocity. In Eq. A22, Term 1 corresponds to change in mixed layer temperature due to surface heat flux, while Term 2 corresponds to change in mixed layer temperature due to entrainment of cold deep water.

Ocean Heat Content The ocean heat content (HC) is defined by the complete ocean volume integration of temperature:

$$HC = \int_x \int_y \int_z C_{SW} \rho_{SW} T \, dx \, dy \, dz \tag{A23}$$

where C_{SW} is the specific heat capacity of seawater and ρ_{SW} is the density of seawater. However, a common simplification adopted in the tropics to measure upper ocean heat content and its variations in place of computing ocean heat content from Eq. A23 is to measure the depth of the 20 °C isotherm. This choice of simplification reduces the three-dimensional thermal field to a two-dimensional variable that can be mapped and viewed easily. The choice of the 20 °C isotherm stems from initial work done in the Tropical Pacific, which may be approximated as a two-layer system divided by a sharp pycnocline (a layer in the ocean with strong vertical gradients in density), for which 20 °C isotherm is a proxy (Kessler 1990).

Reduced Gravity Two-Layer Ocean Model The reduced gravity two-layer ocean model is a simplified substitute for an OGCM, which recognizes that there is a prominent main thermocline in the oceans that separates the warm and less dense upper ocean from the relatively colder and denser deeper ocean. The reduced gravity model uses the thermocline as a step function in density to stratify the ocean as a two-layer fluid. It further assumes as a good approximation that the fluid below the thermocline is stagnant with the fluid layer above the thermocline being the only active layer that is in motion. The advantage of a reduced gravity model is its ability to capture the first baroclinic mode of the circulation and the structure of the main thermocline. If only the upper layer is in motion, it is called reduced gravity or

$1\frac{1}{2}$ layer ocean model. If both the first and second layers are in motion, but the layer below is held stagnant, then it is termed $2\frac{1}{2}$ layer model.

Rigid Lid The rigid-lid approximation is a commonly used simplification in the study of density-stratified fluids. Under this approximation, it is assumed that the vertical motion is nonexistent across this lid or level, and therefore, the flow is horizontally non-divergent. Because of the strong stability of the tropopause and stratosphere above it, the tropopause is often thought of as a rigid lid.

Sawyer-Eliassen Equation This equation describes a stream function solution for the cross-frontal circulation due to ageostrophic motion. It takes the following form:

$$\underbrace{\left(-\gamma\frac{\partial\theta}{\partial p}\right)}_{\text{Static stability}}\frac{\partial^2\psi}{\partial y^2} + \underbrace{\left(2\frac{\partial M}{\partial p}\right)}_{\text{Baroclinicity}}\frac{\partial^2\psi}{\partial p\partial y} + \underbrace{\left(-\frac{\partial M}{\partial y}\right)}_{\text{Inertial stability}}\frac{\partial^2\psi}{\partial p^2}$$

$$= \underbrace{Q_g}_{\text{Geostrophic forcing}} - \underbrace{\gamma\frac{\partial}{\partial y}\left(\frac{d\theta}{dt}\right)}_{\text{Diabatic forcing}} \tag{A24}$$

where γ, θ, ψ, and M are thermodynamic constant, potential temperature, stream function, and absolute geostrophic momentum ($= U_g - fy$), respectively. The coordinate system is such that the x-axis is along front and the y-axis is across the front into the cold air. The geostrophic forcing term is

$$Q_g = \underbrace{2\gamma\left(\frac{\partial U_g}{\partial y}\right)\left(\frac{\partial\theta}{\partial x}\right)}_{\text{Shearing deformation}} + \underbrace{2\gamma\left(\frac{\partial V_g}{\partial x}\right)\left(\frac{\partial\theta}{\partial y}\right)}_{\text{Stretching deformation}} \tag{A25}$$

Equation A17 is akin to the QG omega equation so that if the RHS of the equation is positive or negative, then the LHS has to be also positive or negative, resulting in stream function being a minimum or maximum (remember Laplacian operator!), respectively. Once the stream function is determined, then the winds associated with the Sawyer-Eliassen Circulation (in the y-z plane) can be determined directly from the stream function within a cross-sectional plane perpendicular to the frontal boundary as

$$v_{ag} = -\frac{\partial\psi}{\partial p} \text{ and } \omega = \frac{dp}{dt} = \frac{\partial\psi}{\partial y} \tag{A26}$$

The major difference between the Sawyer-Eliassen Equation from the QG theory is that the former accounts for horizontal and vertical variations of stability in the vicinity of the front and accounts for the local variation of absolute vorticity. As a result, the Sawyer-Eliassen equation provides a transverse frontal ageostrophic circulation characteristic of frontal tilting, a tighter temperature gradient at the front, and a much shorter time to generate a front than with the QG theory.

Scatterometry Scatterometry refers to deriving surface winds over oceans from scatterometers. Scatterometers are active microwave sensors that send out a signal and measure how much of the signal returns after interacting with the target. The fraction of energy returned to the satellite (backscatter) is a function of wind speed and wind direction. The wind speed is determined by the amplitude of the backscatter signal, while the wind direction is found by determining the angle that is most likely consistent with the backscatter observed from multiple angles (the polar orbiting satellite scans a given point on water from angles spanning 90°).

Semidiurnal Variation The tropics are well known to show a semidiurnal peak of sea level pressure with an amplitude of ~1 hPa at ~0945 and 2145 LT. Therefore, semidiurnal variation, as the name suggests, peaks twice daily. This semidiurnal variation can be explained on the basis of atmospheric tidal theory: atmospheric thermal tides are caused by heating from the absorption of solar radiation by ozone in the stratosphere and, to a lesser extent, absorption of solar radiation by water vapor and clouds in the troposphere. This heating excites vertically propagating waves that travel from the stratosphere through the troposphere to the surface. The surface tidal pressure signal is essentially semidiurnal because a large fraction (~80%) of the forcing driving diurnal variability goes into modes whose energy is trapped near the level of forcing (or the level of major absorption of solar radiation). It is theorized that the surface pressure tides over the tropical oceans are large enough to induce diurnal variations in low-level convergence that can trigger diurnal variations of convection. It is observed that the global mean amplitude of the diurnal cycle of precipitation and thunderstorm frequencies is larger than those of the semidiurnal oscillation. This is consistent with corresponding variations in surface winds and divergence. However, the global mean amplitude of the semidiurnal variation of surface pressure is greater than that of the diurnal oscillation.

Subcloud Layer The subcloud layer refers to the layer extending from the surface to the base of the cloud, which includes (a portion of the) atmospheric boundary layer.

Thermal Wind The thermal wind is not actual wind but a useful relationship that allows us to relate the kinematic (or wind) field to mass (or temperature field). It is defined as the vector difference between geostrophic winds (u_T, v_T) at two pressure levels (p_1 and p_0 with $p_0 > p_1$), which is directly related to horizontal layer mean (between p_1 and p_0) virtual temperature gradient ($\overline{T}_v$). Mathematically, this is expressed as

$$v_T = v_g\,(p_1) - v_g\,(p_0) = \frac{R_d}{f}\frac{\partial \overline{T}_v}{\partial x}\ln\left(\frac{p_0}{p_1}\right) \tag{A27}$$

$$u_T = u_g\left(p_1\right) - u_g\left(p_0\right) = -\frac{R_d}{f}\frac{\partial \overline{T}_v}{\partial y}\ln\left(\frac{p_0}{p_1}\right) \tag{A28}$$

Virtual Temperature The virtual temperature is the temperature a dry parcel of air would have at the same pressure and density as the moist air parcel. The origin of virtual temperature comes from the use of the ideal gas law for moist air. Just as for dry air, we use gas constant for dry air (R_d) and for water vapor, we use gas constant for water vapor (R_v), we do not have an equivalent gas constant for moist air because the water vapor content in the moist air parcel can vary. Therefore, we define virtual temperature ($\overline{T}_v$), which makes it convenient to use the ideal gas law for moist air as

$$p = \rho R_d \overline{T}_v \tag{A29}$$

Wet Bulb Temperature The wet bulb temperature(Tw) is the temperature at which a given volume of air attains saturation upon evaporating water into the volume keeping pressure constant. This is in contrast to dew point temperature (Td) which is the temperature at which saturation occurs at fixed vapor pressure. Typically $Td \leq Tw \leq T$, where T is the dry air temperature. However at saturation, $Td=Tw=T$.

References

Adames, A. F. and J. M. Wallace, 2015: Three-dimensional structure and evolution of the moisture field in the MJO. J Atmos Sci., 72(10):3733–3754.

Adames, A. F. and Y. Ming, 2018: Interactions between water vapor and potential vorticity in synoptic-scale monsoonal disturbances: moisture vortex instability. J Atmos Sci., 75(6):2083–2106.

Adames, Á.F., D. Kim, S.K. Clark, Y. Ming, and K. Inoue, 2019: Scale Analysis of Moist Thermodynamics in a Simple Model and the Relationship between Moisture Modes and Gravity Waves J. Atmos. Sci., 76, 3863–3881.

Adames, A. F. and E. D. Maloney, 2021: Moisture mode theory's contribution to advances in our understanding of the Madden-Julian Oscillation and other tropical disturbances. Curr. Clim. Change reports, 7, 72–85. https://doi.org/10.1007/s40641-021-00172-4.

Adames, A. F., S. W. Powell, F. Ahmed, V. C. Mayta, and J. D. Neelin, 2021: Tropical precipitation evolution in a buoyancy-budget framework. J Atmos Sci., 78(2), 509–528.

Alaka, G. J. and E. D. Maloney, 2014: The intraseasonal variability of African easterly wave energetics. J. Climate, 27, 6559–6580, https://doi.org/10.1175/JCLI-D-14-00146.1.

Allan R. J., N. Nicholls, P. D. Jones, I. J. Butterworth, 1991: A further extension of the Tahiti-Darwin SOI, early SOI results and Darwin pressure. J. Climate, 4, 743–749.

Allan, R. P., B. J. Soden, V. O. John, W. Ingram, and P. Good, 2010: Current changes in tropical precipitation, Environ. Res. Lett., 5, 025205.

Alory, G., S. Wijffels, and G. Meyers, 2007: Observed temperature trends in the Indian Ocean over 1960–1999 and associated mechanisms. Geophysical Research Letters 34, L02606, https://doi.org/10.1029/2006GL028044.

Annamalai, H., R. Murtugudde, J. Potemra, S.-P. Xie, and B. Wang, 2003: Coupled dynamics over the Indian Ocean: spring initiation of the Zonal Mode, Deep Sea Res., PartII, 50, 2305–2330.

Ananthakrishnan, R. and M. K. Soman, 1988: The onset of the southwest monsoon over Kerala: 1901–1980. Int. J. Climatol., 8, 283–296.

Arakawa, A. and W. H. Schubert, 1974: Interaction of cumulus cloud ensemble with the large-scale environment. Part I. J. Atmos. Sci., 31, 674–701.

Arakawa, A., 2003: The cumulus parameterization problem: past, present, and future. J. Clim., 17, 2493-2525.

Arakawa, A., 1969: Parameterization of cumulus clouds. Proc. Symp. on Numerical Weather Prediction, Tokyo, Japan, WMO/International Union of Geodesy and Geophysics, 1–6.

Arblaster, J. M. and L. V. Alexander, 2012: The impact of the El Niño-Southern Oscillation on maximum temperature extremes. Geophys. Res. Lett., https://doi.org/10.1029/2012GL053409.

© Springer Nature Switzerland AG 2023

V. Misra, *An Introduction to Large-Scale Tropical Meteorology*, Springer Atmospheric Sciences, https://doi.org/10.1007/978-3-031-12887-5

Ashok, K., S. K. Behera, S. A. Rao, H. Weng, and T. Yamagata, 2007: El Niño Modoki and its possible teleconnections. J. Geophys. Res., 112, C11007, https://doi.org/10.1029/2006JC003798.

Avila, L. A., 1991: Eastern North Pacific hurricane season of 1990. Mon. Wea. Rev., 119, 2034–2046.

Avila, L. A., and R. J. Pasch, 1992: Atlantic tropical systems of 1991. Mon. Wea. Rev., 120, 2688–2696.

Barnston AG, Chelliah M, Goldenberg SB. 1997. Documentation of a highly ENSO-related SST region in the Equatorial Pacific. Atmosphere-Ocean 35: 367–383.

Basu, S., S. D. Meyers, and J. J. O'Brien, 2000: Annual and interannual sea level variations in the Indian Ocean from TOPEX/Poseidon observations and ocean model simulations. Journal of Geophysical Research, 105, 975–994.

Battisti, D. S. and A. C. Hirst, 1989: Interannual variability in a tropical atmosphere–ocean model: Influence of the basic state, ocean geometry and nonlinearity. J. Atmos. Sci., 46, 1687–1712.

Berg, P. and J. O. Haerter, 2013: Unexpected increase in precipitation intensity with temperature—A result of mixing precipitation types? Atmos. Res., 119, 56–61.

Betts, A. K., and W. Ridgway, 1989: Climatic equilibrium of the atmospheric convective boundary layer over a tropical ocean. J. Atmos. Sci., 46 , 2621–2641.

Bjerknes, J., 1966: A possible response of the atmospheric Hadley circulation to equatorial anomalies of ocean temperature. Tellus, 18, 820–828.

Bjerknes, J., 1969: Atmospheric teleconnections from the equatorial Pacific. Mon. Wea. Rev., 97, 163–172.

Bolin, B., 1950: On the influence of the Earth's orography on the westerlies. Tellus, 2, 184–195.

Bombardi, R. J., J. L. KinterIII, and O. W. Frauenfeld, 2019: A global gridded dataset of the characteristics of the rainy and dry seasons. Bull. Amer. Soc., 100(7), 1315–1328, https://doi.org/10.1175/BAMS-D-18-0177.1.

Boos, W. R., and Z. Kuang, 2010: Dominant control of the South Asian monsoon by orographic insulation versus plateau heating, Nature, 463, 218–222.

Boos, W. R. and Z. Kuang, 2013: Sensitivity of the South Asian Monsoon to Elevated and non-elevated heating. Scientific Reports, 3, 1192, https://doi.org/10.1038/srep01192.

Brandt, P., A. Funk, V. Hormann, M. Dengler, R. J. Greatbatch, and J. M. Toole, 2011a: Interannual atmospheric variability forced by the deep equatorial Atlantic Ocean. Nature, 473, 497–500. https://doi.org/10.1038/nature10013.

Brandt, P., G. Caniaux, B. Bourlès, A. Lazar, M. Dengler, A. Funk, V. Hormann, H. Giordani, and F. Marin, 2011b: Equatorial upper-ocean dynamics and their interaction with the West African monsoon. Atmospheric Science Letters, 12, 24–30. https://doi.org/10.1002/asl.287

Bretherton, F. P., 1964: Low frequency oscillations trapped near the equator, Tellus, 16, 181–185.

Broccoli A. J., K. A. Dahl, R. J. Stouffer, 2006: Response of the itcz to northern hemisphere cooling. Geophys Res Lett 33:L01702. https://doi.org/10.1029/2005GL024546.

Brooks, H. E., J. W. Lee, and J. P. Craven. 2003: The spatial distribution of severe thunderstorm and tornado environments from global reanalysis data. Atmos. Res., 67–68: 73–94.

Bruce, J. G., M. Fieux, and J. Gonella, 1981: A note on the continuance of the Somali eddy after the cessation of the Southwest monsoon. Oceanologica Acta, 4, 7–9.

Bunge, L. and A. J. Clarke, 2009: A verified estimate of the El Niño Index Niño3.4 since 1877. J. Clim., 22, 3979–3992.

Burgers, G., and D. B. Stephenson, 1999: The "normality" of El Niño. Geophys. Res. Lett., 26, 1027–1030.

Burpee, R. W., 1972: The origin and structure of easterly waves in the lower troposphere of North Africa. J. Atmos. Sci., 29, 77– 90.

Cai, W., B. Ng, T. Geng, L. Wu, A. Santoso, M. J. McPhaden, 2020: Butterfly effect and self-modulating El Niño response to global warming. Nature, 585, 68–73.

Camargo, S. J., M. C. Wheeler, and A. H. Sobel, 2009: Diagnosis of the MJO modulation of tropical cyclogenesis using an empirical index. J. Atmos. Sci., 66, 3061–3074.

Cane, M. A., S. E. Zebiak, and S. C. Dolan, 1986: Experimental forecasts of El Niño. Nature, 321, 827–832.

Cane, M., M. Munnich, and S. E. Zebiak, 1990: A study of self-excited oscillations of the tropical ocean–atmosphere system. Part I: Linear analysis. J. Atmos. Sci., 47, 1562–1577.

Capotondi, A. and Co-Authors, 2015: Understanding ENSO diversity. Bull. Amer. Meteor. Soc., 96, 921–938, https://doi.org/10.1175/BAMS-D-13-00117.1.

Carlson, T. N., 1969: Some remarks on African disturbances and their progress over the tropical Atlantic. Mon. Wea. Rev., 97, 716–726.

Carton, J. A., X. Cao, B. S. Giese, and A. M. da Silva, 1996: Decadal and interannual SST variability in the tropical Atlantic Ocean, J. Phys. Oceanogr., 26, 1165–1175.

Carvalho, L. M. V., C. Jones, and T. Ambrizzi, 2005: Opposite phases of the Antarctic Oscillation and relationships with intraseasonal to interannual activity in the tropics during the austral summer. J. Climate, 18, 702–718.

Cassou, C., Deser, C., Terray, L., Hurrell, J. W., & Drévillon, M. (2004). Summer sea surface temperature conditions in the North Atlantic and their impact upon the atmospheric circulation in early winter. Journal of Climate, 17(17), 3349–3363.

Chang, C. P., and H. Lim, 1988: Kelvin wave-CISK: A possible mechanism for the 30–50 day oscillations, J. Atmos. Sci., 45,1709–1720.

Chang, P., L. Ji, and H. Li, 1997: A decadal climate variation in the tropical Atlantic ocean from thermodynamic air-sea interactions, Nature,385, 516–518.

Chang, P., Y. Fang, R. Saravanan, L. Ji and H. Seidel, 2006: The cause of the fragile relationship between the Pacific El Niño and the Atlantic Niño. Nature, 443, 324–328. https://doi.org/10.1038/nature05053.

Chao, W. C., 2000: Multiple quasi-equilibria of the ITCZ and the origin of monsoon onset. J Atmos Sci 57: 641–651.

Chao, W. C. and B. Chen, 2004: Single and double ITCZ in an aqua-planet model with constant sea surface temperature and solar angle. Clim. Dyn., 22, 447–459.

Charney, J. G., 1971: Tropical cyclogenesis and the formation of the ITCZ. In: Reid WH (ed) Mathematical problems of geophysical fluid dynamics. Lectures in applied mathematics. Am Math Soc 13: 355–368.

Charney, J. G. and A. Eliassen, 1964: On the growth of the hurricane depression. J. Amos. Sci., 21, 68–75.

Chen, D., N. Smith, W. Kessler, 2018: The evolving ENSO observing system. National Science Review, 5, 805–807, https://doi.org/10.1093/nsr/nwy137.

Chen, H. -C., Y. -H. Tseng, Z. -Z. Hu, and R. Ding, 2020: Enhancing the ENSO predictability beyond the Spring barrier. Sci Rep 10, 984 (2020). https://doi.org/10.1038/s41598-020-57853-7.

Chen, S. S., and R. A. Houze Jr., 1997: Diurnal variation of deep convective systems over the tropical Pacific warm pool, Q. J. R. Meteorol. Soc., 123, 357–388.

Chiang, J. C. H., S. E. Zebiak, and M. A. Cane, 2001: Relative roles of elevated heating and surface temperature gradients in driving anomalous surface winds over tropical oceans. J. Atmos. Sci., 58, 1371–1394.

Chiang, J. C. H. and A. H. Sobel, 2002: Tropical tropospheric temperature variations caused by ENSO and their influence on the remote tropical climate. J. Climate, 15, 2616–2631, https://doi.org/10.1175/1520-0442(2002)015<2616:TTTVCB>2.0.CO;2

Chiodi, A. M., and D. E. Harrison, 2013: El Niño impacts on seasonal U.S. atmospheric circulation, temperature, and precipitation anomalies: The OLR-event perspective. J. Climate, 26, 822–837, https://doi.org/10.1175/JCLI-D-12-00097.1

Chou, C., J. D. Neelin, and H. Su, 2001: Ocean–atmosphere–land feedbacks in an idealized monsoon. Quart. J. Roy. Meteor. Soc., 127, 1869–1891.

Chou, C., and J. D. Neelin, 2004: Mechanisms of global warming impacts on regional tropical precipitation. J. Climate, 17, 2688–2701.

Cho, H. R., and D. Pendlebury, 1997: Wave CISK of equatorial waves and the vertical distribution of cumulus heating. J. Atmos. Sci., 54 , 2429–2440.

Clemens, S., 2006: Extending the historical record by proxy. The Asian Monsoon, B. Wang, Ed., Praxis, 615–629.

Clement, A. C., R. Seager, M. A. Cane, and S. E. Zebiak, 1996: An ocean dynamical thermostat, J. Clim., 9, 2190–2196.

Clement, A. C., R. Seager, and R. Murtugudde, 2006: Why are there tropical warm pools? J. Climate, 18, 5294–5310.

Collins, M., et al., 2010: The impact of global warming on the tropical Pacific Ocean and El Niño. Nat. Geosci., 3, 391–397, https://doi.org/10.1038/ngeo868.

Compo, G. P. and P. D. Sardeshmukh, 2010: Removing ENSO-related variations from the climate record. J. Climate, 23, 1957–1978.

Convoy, J.L., J.T. Overpeck, J.E. Cole, T.M. Shanahan, M. Steinitz-Kannan, 2008:Holocene changes in eastern tropical Pacific climate inferred from a Galápagos lake sediment record. Quaternary Science Reviews, 27, 1166–1180.

Cook, E. R., K. J. Achukaitis, B. M. Buckley, R. D. D. Arrigo, G. C. Jacoby, and W. E. Wright, 2010: Asian monsoon failure and megadrought during the last millennium. Science, 486, 486–490.

Cronin, T. W. and A. A. Wing, 2017: Clouds, circulation, and climate sensitivity in a radiative-convective equilibrium channel model. J Adv Model Earth Syst., 9, 2833–905. https://doi.org/10.1002/2017MS001111.

Curtis, S., A. Salahuddin, R. F. Adler, G. J. Huffman, G. Gu, and Y. Hong, 2007: Precipitation extremes estimated by GPCP and TRMM: ENSO relationships. J. Hydromet., 8, 678–689, https://doi.org/10.1175/JHM601.1.

Dai, A., X. Lin, and K.-L. Hsu, 2007: The frequency, intensity, and diurnal cycle of precipitation in surface and satellite observations over low- and mid-latitudes. Climate Dyn., 29, 727–744, https://doi.org/10.1007/s00382-007-0260-y.

Das, S., 2020: Understanding the Evolution of Tropical Cyclones Through the Psi-Chi Framework. Ph. D. dissertation, Florida State University, 91 pp.

Davis, M., 2001: Late Victorian Holocausts: El Niño Famines and the Making of the Third World (Verso, London).

De Deckker, P., 2016: The Indo-Pacific Warm Pool: critical to world oceanography and world climate. Geosci. Lett. 3, 20 (2016). https://doi.org/10.1186/s40562-016-0054-3.

DeMott, C. A., N. P. Klingaman, and S. J. Woolnough, 2015: Atmosphere-ocean coupled processes in the Madden-Julian oscillation. Rev Geophys. 53(4):1099–1154.

DeMott, C. A., N. P. Klingaman, W. -L. Tseng, M. A. Burt, Y. Gao, and D. A. Randall, 2019: The convection connection: how ocean feedbacks affect tropical mean moisture and MJO propagation. J Geophys Res Atmosph., 124(22):11910–11931.

Dessler, A. E., S. P. Palm, and J. D. Spinhirne, 2006: Tropical cloud-top height distributions revealed by the Ice, Cloud, and Land Elevation Satellite (ICESat)/Geoscience Laser Altimeter System (GLAS), J. Geophys. Res., 111, D12215, https://doi.org/10.1029/2005JD006705.

de Szoek, S. P., S. -P. Xie, 2008: The tropical eastern Pacific seasonal cycle: assessment of errors and mechanisms in IPCC AR4 coupled ocean-atmosphere general circulation models. J. Climate, 21, 2573–2590.

de Szoeke, S. P., J. B. Edson, J. R. Marion, C. W. Fairall, and L. Bariteau, 2015: The MJO and air-sea interaction in toga coare and dynamo. J. Clim., 28(2):597–622.

Diaz, M. and W. R. Boos, 2019: Monsoon depression amplification by moist barotropic instability in a vertically sheared environment. Q J R Meteorol Soc., 145(723):2666–2684.

Ding Y., J. C. L. Chan, 2005: The East Asian summer monsoon: an overview. Meteorology and Atmospheric Physics 89: 117–142.

Dippe, T., R. J. Greatbatch, and H. Ding, 2017: On the relationship between Atlantic Niño variability and ocean dynamics. Climate Dynamics. https://doi.org/10.1007/s00382-017-3943-z.

Dole, R., and Coauthors, 2011: Was there a basis for anticipating the 2010 Russian heat wave? Geophys. Res. Lett., 38, L06702, https://doi.org/10.1029/2010GL046582.

Domeisen, D. I. V., and Coauthors, 2020: The role of the stratosphere in subseasonal to seasonal prediction: 2. Predictability arising from stratosphere-troposphere coupling. J. Geophys. Res. Atmos., 125, e2019JD030923, https://doi.org/10.1029/2019jd030923.

Dommenget, D., T. Bayr, and C. Frauen, 2012: Analysis of the non-linearity in the pattern and time evolution of El Niño southern oscillation, Clim. Dyn., https://doi.org/10.1007/s00382-012-1475-0.

Dong, B.-W., and R. T. Sutton, 2002: Adjustment of the coupled ocean atmosphere system to a sudden change in the thermohaline circulation, Geophys. Res. Lett., 29(15), 1728 https://doi.org/10.1029/2002GL015229.

Doswell, C. A., III, 2001: Severe convective storms—An overview. Severe Convective Storms, Meteor. Monogr., No. 5, Amer. Meteor. Soc., 1–26.

Drévillon, M., C. Cassou, and L. Terray, 2003: Model study of the North Atlantic region atmospheric response to autumn tropical Atlantic sea-surface-temperature anomalies. Quarterly Journal of the Royal Meteorological Society, 129(593), 2591–2611.

Elsner, J. B., T. Fricker, and Z. Schroder, 2018: Increasingly powerful tornadoes in the United States. Geophys. Res. Lett., https://doi.org/10.1029/2018GL080819.

Emanuel, K. A., 1995: The behavior of a simple hurricane model using a convective scheme based on subcloud-layer entropy equilibrium. J Atmos Sci., 52(22):3960–3968.

Emanuel, K. A., 2005: Increasing destructiveness of tropical cyclones over the past 30 years. Nature, 436, 686–688.

Emanuel, K. A., J. D. Neelin, and C. S. Bretherton, 1994: On large-scale circulations in convecting atmospheres. Quart. J. Roy. Meteor. Soc., 120, 1111–1143.

Enfield, D. B., and D. A. Mayer, 1997: Tropical Atlantic SST variability and its relation to El Nino-Southern Oscillation. J. Geophys. Res., 102, 929–945.

Fairall, C. W., E. F. Bradley, D. P. Rogers, J. B. Edson, and G. S. Young, 1996: Bulk parameterization of air–sea fluxes for TOGA-COARE. J. Geophys. Res., 101, 3747–3764.

Fasullo, J., and P. J. Webster, 2003: A hydrological definition of Indian monsoon onset and withdrawal, J. Clim., 16(19), 3200–3211, https://doi.org/10.1175/1520-0442(2003)0162.0.CO;2.

Fedorov, A. V., S. Hu, M. Lengaigne, E. Guilyardi, 2014: The impact of westerly wind bursts and ocean initial state on the development, and diversity of El Niño events. Clim Dyn 44:1381–1401.

Feng, J., T. Li, and W. Zhu, 2015: Propagating and nonpropagating MJO events over Maritime Continent. J Clim., 28(211), 8430–8449.

Field, C. B., and Coauthors, Eds., 2012: Managing the Risks of Extreme Events and Disasters to Advance Climate Change Adaptation: Summary for Policymakers. A Special Report of Working Groups I and II of the Intergovernmental Panel on Climate Change, 19 pp. [Available online at http://ipcc-wg2.gov/SREX/images/uploads/SREX-SPMbrochure_FINAL.pdf.].

Fischer, E. M., and R. Knutti, 2015: Anthropogenic contribution to global occurrence of heavy-precipitation and high-temperature extremes. Nat. Clim. Change, 5, 560–564, https://doi.org/10.1038/nclimate2617.

Ferreira, R. N., and W. H. Schubert, 1997: Barotropic aspects of ITCZ breakdown. J. Atmos. Sci., 54, 261–285.

Findlater, J., 1969: A major low-level air current near the Indian Ocean during the northern summer. Quart. J. Roy. Meteor. Soc., 95, 362–380.

Florenchie, P., J. R. E. Lutjeharms, C. J. C. Reason, S. Mason, and M. Rouault, 2003: The source of Benguela Niños in the South Atlantic Ocean. Geophys. Res. Lett., 30, https://doi.org/10.1029/2003GL017172.

Florenchie, P., C. J. C. Reason, J. R. E. Lutjeharms, M. Rouault, C. Roy, and S. Mason, 2004: Evolution of interannual warm and cold events in the Southeast Atlantic Ocean. J. Climate, 17, 2318–2334.

Foltz, G. R., and M. J. McPhaden, 2010: Interaction between the Atlantic meridional and Niño modes. Geophysical Research Letters, 37, 1–5. https://doi.org/10.1029/2010GL044001.

Folland, C. K., T. N. Palmer, and D. E. Parker, 1986: Sahel rainfall and world-wide sea temperatures, Nature, 320, 602–607.

Folland, C. K., A. W. Colman, D. P. Powell, and M. K. Davey, 2001: Predictability of northeast Brazil rainfall and real-time forecast skill,1987–98. J. Clim., 14, 1937–1958.

Frank, N. L., 1969: The "inverted V" cloud pattern—An easterly wave? Mon. Wea. Rev., 97, 130–140.

Freund, M. B., B. J. Henley, D. J. Karoly, H. V. McGregor, N. J. Abram, and D. Dommenget, 2019: Higher frequency of central Pacific El Niño events in recent decades relative to past centuries. Nat. Geosci., 12, 45–455. https://doi.org/10.1038/s41561-019-0353-3.

Freund, M. B., J. R. Brown, B. J. Henley, D. J. Karoly, and J. N. Brown, 2020: Warming patterns affect El Niño diversity in CMIP5 and CMIP6 MODELS. J. Climate, 33(19), 8237–8260. https://doi.org/10.1175/JCLI-D-19-0890.1.

Frierson, D. M. W., Y. T. Hwang, N. Fuckar, R. S. S. Kang, A. Donohoe, E. Maroon, 2013: Contribution of ocean overturning circulation to tropical rainfall peak in the Northern Hemisphere. Nature Geoscience, 6(11), 940–944.

Fu, R., A. D. Del Genio, W. B. Rossow, and W. T. Liu, 1992: Cirrus-cloud thermostat for tropical sea surface temperatures tested using satellite data. Nature, 358, 394–397.

García-Serrano, J., T. Losada, B. Rodríguez-Fonseca, and I. Polo, 2008: Tropical Atlantic variability modes (1979–2002). Part II: Time-evolving atmospheric circulation related to SST-forced tropical convection. Journal of Climate, 21(24), 6476–6497.

Garcia-Serrano, J., T. Losada, and M. B. Rodriguez-Fonseca, 2011: Extratropical atmospheric response to the Atlantic Nino decaying phase. J. Clim., 24(6), 1613–1625, https://doi.org/10.1175/2010JCLI3640.1.

Garreaud, R. D., and J. M. Wallace, 1997: The diurnal march of convective cloudiness over the Americas. Mon. Wea. Rev., 125, 3157–3171.

Giannini, A., R. Saravanan, and P. Chang, 2004: The preconditioning role of tropical Atlantic variability in the development of the ENSO teleconnection: implications for the prediction of Nordeste rainfall. Clim. Dyn. 22, 839–855. https://doi.org/10.1007/s00382-004-0420-2

Giannini, A., R. Saravanan, and P. Chang, 2005: Dynamics of the boreal summer African monsoon in the NSIPP1 atmospheric model. Climate Dynamics, 25, 517–535.

Gill, A. E., 1980: Some simple solutions for heat-induced tropical circulations. Q. Roy. Met. Soc., 106, 447–462.

Gill, A. E., 1982: Atmosphere Ocean Dynamics. Academic Press.

Glantz, M.H., Katz, R. and Krenz, M. 1987: The societal impacts associated with the 1982–83 worldwide climate anomalies. Report based on the workshop on Economic and Societal impacts Associated with the 1982–83 Worldwide Climate Anomalies, 11–13 November 1985, Lugano , Switzerland: UNEP, NCAR, Boulder, Colorado.

Gnanaseelan, C., M. K. Roxy, and A. Deshpande, 2017: Variability and trends of sea surface temperature and circulation in the Indian Ocean. In Observed Climate Variability and Change over the Indian Region, M. N. Rajeevan and S. Nayak (eds.), Springer Geology, https://doi.org/10.1007/978-981-10-2531-0_10.

Godfrey, J. S., 1996: The effect of the Indonesian throughflow on ocean circulation and heat exchange in the atmosphere: A review. J. Geophys. Res., 101, 12 217–237.

Godfrey, J. S., M. Nunez, E. F. Bradley, P. A. Coppin, and E. J. Lindstrom, 1991: On the net surface heat flux into the western equatorial Pacific, J. Geophys. Res., 96, suppl., 3391–3400.

Godfrey, J. S., R. A. Houze Jr., R. H. Johnson, R. Lukas, J.-L. Redelsperger, A. Sumi, and R. Weller, 1998:The Coupled Ocean Atmosphere Response Experiment (COARE): An interim report, J. Geophys. Res., 103, 14,395–14,450.

Goldenberg, S. B., C. W. Landsea, and A. M. Mestas-Nunez, 2001: The recent increase in Atlantic Hurricane Activity: Causes and Implication. Science, 293, 474-479. https://doi.org/10.1126/science.1060040.

Goswami, B. N., R. N. Keshavamurty, and V. Satyan, 1980: Role of barotropic, baroclinic, and combined barotropic-baroclinic instability for the growth of monsoon depressions and mid-tropospheric cyclones. Proc. Indian Acad. Sci. (Earth Planet Sci.), 89, 77–97.

Goswami, P. and V. Mathew, 1994: A mechanism of scale selection in tropical circulation at observed intraseasonal frequencies. J. Atmos. Sci., 5, 3155–3166.

Goswami, B. N., R. S. Ajayamohan, P. K. Xavier, and D. Sengupta, 2003: Clustering of synoptic activity by Indian summer monsoon intraseasonal oscillations. Geophys. Res. Lett., 30(8), https://doi.org/10.1029/2002GL016734.

Gould, J.,D. Roemmich, S. Wijffels, H. Freeland, M. Ignaszewsky, X. Jianping, S. Pouliquen, Y. Desaubies, U. Send, K. Radhakrishnan, K. Takeuchi, K. Kim, M. Danchenkov, P. Sutton, B. King, B. Owens, S. Riser Argo profiling floats bring new era of in situ ocean observations, Eos, Transactions American Geophysical Union, 85 (2004), pp. 185–191, https://doi.org/10.1029/2004EO190002.

Gray, W. M., 1979: Hurricanes: Their formation, structure and likely role in the tropical circulation. Meteorology over the Tropical Oceans, D. B. Shaw, Ed., Roy. Meteor. Soc., 155–218.

Gray, W. M., and R. W. Jacobson Jr., 1977: Diurnal variation of deep cumulus convection. Mon. Wea. Rev., 105, 1171–1188.

Gregory, D. and M. J. Miller, 1989: A numerical study of the parameterization of deep tropical convection. Q. J. R. Meteorol. Soc., 115, 1209–1242.

Graham, N. E. and T. P. Barnett, 1987: Sea surface temperature, surface wind divergence, and convection over tropical oceans. Science, 238, 657-9, https://doi.org/10.1126/science.238.4827.657.

Grimm, A. M., and P. L. Silva Dias, 1995: Analysis of tropical–extratropical interactions with influence functions of a barotropic model. J. Atmos. Sci., 52, 3538–3555.

Grise, K.M., S.M. Davis, I.R. Simpson, D.W. Waugh, Q. Fu, R.J. Allen, K.H. Rosenlof, C.C. Ummenhofer, K.B. Karnauskas, A.C. Maycock, X. Quan, T. Birner, and P.W. Staten, 2019: Recent tropical expansion: Natural variability or forced response? J. Climate, 32, 1551–1571.

Guinn, T. A., and W. H. Schubert, 1993: Hurricane spiral bands. J. Atmos. Sci.,50, 3380–3403.

Hack, J. J., W. H. Schubert, D. E. Stevens, and H.-C. Kuo, 1989: Response of the Hadley circulation to convective forcing in the ITCZ. J. Atmos. Sci.,46, 2957–2973.

Hadley, G., 1735: Concerning the cause of the general trade-winds. Philos. Trans., 29, 58–62.

Hardwick-Jones, R., S. Westra, and A. Sharma, 2010: Observed relationships between extreme sub-daily precipitation, surface temperature and relative humidity, Geophys. Res. Lett., 37, L22805, https://doi.org/10.1029/2010GL045081.

Haertel, P.T. and Kiladis, G.N., 2004: Dynamics of 2-day equatorial waves. Journal of the atmospheric sciences, 61(22), pp. 2707–2721.

Halpern, D., 1987: Observations of annual and El Nino thermal and flow variations at 0°, 110°W and 0°, 95°W during 1980–1985. J. Geophys. Res., 92, 8197–8212.

Hanley, D. E., M. A. Bourassa, J. J. O'Brien, S. R. Smith, and E. R. Spade, 2003: A quantitative evaluation of ENSO indices. J. Climate, 16, 1249–1258.

Han, W., J. P. McCreary, D. L. T. Anderson, and A. J. Mariano (1999), Dynamics of the eastern surface jets in the equatorial Indian Ocean, J. Phys. Oceanogr., 29, 2191–2209, https://doi.org/10.1175/1520-0485(1999)029<2191:DOTESJ>2.0.CO;2.

Hann, J., 1901: Lehrbuch der Meteorologie. C. H. Tauchnitz, 805 pp.

Hartmann, D.L. and M. L. Michelsen, 1993: Large-scale effects on the regulation of tropical sea surface temperature. J. Climate, 6, 2049–2062.

Hartmann, D. L., 2007: The Atmospheric General Circulation and its Variability. J. Meteor. Soc. Japan, 85B, 123–143.

Hastenrath, S., and L. Heller, 1977: Dynamics of climate hazards in Northeast Brazil, Q. J. R. Meteorol. Soc., 103, 77–92.

Hastenrath, S., and L. Greischar, 1993: Further work on the prediction of northeast Brazil rainfall anomalies, J. Clim., 6, 743–758.

Hayes, S. P., L. J. Mangum, J. Picaut, A. Sumi, and K. Takeuchi, 1991: TOGA-TAO: A moored array for real-time measurements in the tropical Pacific Ocean. Bull. Amer. Meteor. Soc., 72, 339–347.

He, J. and B. J. Soden, 2015: Anthropogenic weakening of the tropical circulation: the relative roles of direct CO_2 forcing and sea surface temperature change. J. Climate, 28, 8728–8742.

Held, I. and B. Soden, 2006: Robust responses of the hydrological cycle to global warming. J. Climate, 19, 5686–5699.

Held, I. and A. Y. Hou, 1980: Non-Linear Axially-Symmetric Circulations in a Nearly Inviscid Atmosphere. J. Atmos. Sci., 37, 515–533.

Hendon, H. H., and M. L. Salby, 1994: The life cycle of the Madden-Julian Oscillation, J. Atmos. Sci., 51, 2225–2237.

Hendon, H.H., E. Lim, G. Wang, O. Alves, and D. Husdon, 2009: Prospects for predicting two flavors of El Niño. Geophys. Res. Lett., 36, L19713, https://doi.org/10.1029/2009GL040100.

Herbertson, A. J., 1901: Outlines of Physiography: An Introduction to the Study of the Earth. Edward Arnold, 312 pp.

Hess, P. G., D. S. Battisti, P. J. Rasch, 1993: Maintenance of the intertropical convergence zones and the tropical circulation on a water-covered earth. J Atmos Sci 50: 691–713.

Hirst, A. C., 1986: Unstable and damped equatorial modes in simple coupled ocean-atmosphere models. J. Atmos. Sci., 43, 606–630.

Hirst, A. C. and S. Hastenrath, 1983: Atmosphere-ocean mechanisms of climate anomalies in the Angola-tropical Atlantic sector. J. Phys. Oceanogr., 13, 1146–1157.

Hoffert, M. I., B. P. Flannery, A. J. Callegari, C. T. Hsieh and W. Wiscombe, 1983: Evaporation-limited tropical temperatures as a constraint on climate sensitivity, J. Atmos. Sci., 40, 1659–1668.

Hoerling, M. P., A. Kumar, and T. Xu, 2001: Robustness of the nonlinear climate response to ENSO's extreme phases. J. Climate, 14, 1277–1293.

Hoerling, M., and A. Kumar, 2003: The perfect ocean for drought, Science, 299, 691–694.

Hoerling M. P., J. W. Hurrell, T. Xu, G. T. Bates, A. Phillips, 2004: Twentieth century North Atlantic climate change. Part II: Understanding the effect of Indian Ocean warming. Climate Dynamics 23 (3–4), 391–405.

Holton, J. R., 1972: Waves in the equatorial stratosphere generated by tropospheric heat sources, J. Atmos. Sci., 29, 368–375, https://doi.org/10.1175/1520-0469(1972)029<0368:WITESG>2.0.CO;2.

Holton, J. R., 1973: On the frequency distribution of atmospheric Kelvin waves, J. Atmos. Sci., 30, 499–501, https://doi.org/10.1175/1520-0469(1973)030<0499:OTFDOA>2.0.CO;2.

Hopsch, S. B., C. D. Thorncroft, and K. R. Tyle, 2010: Analysis of African easterly wave structures and their role in influencing tropical cyclogenesis. Mon. Wea. Rev., 138, 1399–1419, https://doi.org/10.1175/2009MWR2760.1.

Horii, T., and K. Hanawa, 2004: A relationship between timing of El Nino onset and subsequent evolution. Geophys. Res. Lett., 31, L06304, https://doi.org/10.1029/2003gl019239.

Hoskins, B. J. and D. J. Karoly, 1981: The steady linear response of a spherical atmospheric thermal and orographic forcing. J. Atmos. Sci., 38, 1179–1196.

Houghton, R. W., 1983: Seasonal variations of the subsurface thermal structure in the Gulf of Guinea. J. Phys. Oceanogr., 13, 2070–2081.

Hseih, J. -S. and K. H. Cook, 2007: A study of the energetics of African easterly waves using a regional climate model. J. Atmos. Sci., 64, 421–440, https://doi.org/10.1175/JAS3851.1

Huang, B., Peter W. Thorne, et. al., 2017: Extended Reconstructed Sea Surface Temperature version 5 (ERSSTv5), Upgrades, validations, and intercomparisons. J. Climate, https://doi.org/10.1175/JCLI-D-16-0836.1.

Huber, M. and R. Caballero, 2003: Eocene El Niño: evidence for robust tropical dynamics in the "hothouse". Science, 299, 877–881.

Im, S.-H., S.-I. An, S. T. Kim, and F.-F. Jin, 2015: Feedback processes responsible for El Niño-La Niña amplitude asymmetry, Geophys. Res. Lett., 42, 5556–5563, https://doi.org/10.1002/2015GL064853.

Janicot, S., V. Moron, and B. Fontaine, 1996: Sahel droughts and ENSO dynamics. Geophys. Res. Lett., 23, 515–518, https://doi.org/10.1029/96GL00246.

Janicot, S., F. Mounier, N. M. J. Hall, S. Leroux, B. Sultan, and G. N. Kiladis, 2009: Dynamics of the West African monsoon. Part IV: Analysis of 25–90-day variability of convection and the role of the Indian monsoon. J. Climate, 22, 1541–1565.

Janicot, S., and Coauthors, 2011: Intraseasonal variability of the West African monsoon. Atmos. Sci. Lett., 12, 58–66.

Janowiak, J., 1988: An investigation of interannual rainfall variability in Africa. J. Climate, 1, 240–255.

Janowiak, J. E. and P. Xie, 2003: A global-scale examination of monsoon related precipitation. J. Climate, 16, 4121–4133.

Jin, F. F., 1997a: An equatorial ocean recharge paradigm for ENSO. Part I: Cocneptual model. J. Atmos. Sci., 54, 811–829.

Jin, F. F., 1997b: An equatorial ocean recharge paradigm for ENSO. Part II: A stripped-down coupled model. J. Atmos. Sci., 54, 830–847.

Johanson C. M., Q. Fu, 2009: Hadley cell widening: model simulations versus observations. J Clim, 22: 2713–2725. https://doi.org/10.1175/2008JCLI2620.1.

Johnson, R. H., T.M. Rickenbach, S. A. Rutledge, P. E. Ciesielski, and W. H. Schubert, 1999: Trimodal characteristics of tropical convection. J. Climate, 12, 2397–2418, https://doi.org/10.1175/1520-0442(1999)012,2397:TCOTC.2.0.CO;2.

Jones, C. and B. C. Weare, 1996: The role of the low-level moisture convergence and ocean latent heat fluxes in the Madden and Julian Oscillation: An observational analysis using ISCCB data and ECMWF analyses. J. Climate, 9, 3086–3104.

Jones, C., D. E. Waliser, K. M. Lau, and W. Stern, 2004: Global occurrences of extreme precipitation events and the Madden–Julian oscillation: Observations and predictability. J. Climate, 17, 4575–4589.

Kalnay, E., and Coauthors, 1996: The NCEP/NCAR 40-Year Reanalysis Project. Bull. Amer. Meteor. Soc., 77, 437–471.

Karmakar, N. and V. Misra, 2019: The relation of intraseasonal variations with local onset and demise of the Indian summer monsoon. J. Geophys. Res., https://doi.org/10.1029/2018JD029642.

Karmakar, N. and V. Misra, 2020: Differences in northward propagation of convection over the Arabian Sea and Bay of Bengal during boreal summer. Journal of Geophysical Research: Atmospheres, 125, e2019JD031648. https://doi.org/10.1029/2019JD031648

Karmakar, N., W. R. Boos, and V. Misra, 2020: Influence of intraseasonal variability on the development of monsoon depressions. Geophys. Res. Lett., 48, e2020GL090425, https://doi.org/10.1029/2020GL090425.

Keenlyside, N. S., and M. Latif, 2007: Understanding equatorial Atlantic interannual variability. Journal of Climate, 20, 131–142. https://doi.org/10.1175/JCLI3992.1.

Kemball-Cook, S. and B. Wang, 2001: Equatorial waves and air-sea interaction in the Boreal summer intraseasonal oscillation. J. Climate, 14, 2923–2942.

Kenyon, J. and G. C. Hegerl, 2008: Influence of modes of climate variability on global temperature extremes. J. Climate, 21, 3872–3889, https://doi.org/10.1175/2008JCLI2125.1.

Kessler, W. S., 1990: Observations of long Rossby waves in the northern tropical Pacific, J. Geophys. Res., 95, 5183–5217, https://doi.org/10.1029/JC095iC04p05183

Kessler, W. S., 2002: Is ENSO a cycle or a series of events? Geophys. Res. Lett., 29, 2125, https://doi.org/10.1029/2002GL015924.

Kidston, J., A. A. Scaife, S. C. Hardiman, D. M. Mitchell, N. Butchart, M. P. Baldwin, and L. J. Gray, 2015: Stratospheric influence on tropospheric jet streams, storm tracks and surface weather. Nat. Geosci., 8, 433–440, https://doi.org/10.1038/ngeo2424.

Kikuchi, K. and B. Wang, 2008: Diurnal precipitation regimes in the global tropics. J. Climate, 21, 2680–2696.

Kiladis, G. N. and K. M. Weickmann, 1992a: Circulation anomalies associated with tropical convection during northern winter. Mon. Wea. Rev., 120, 1900–1923.

Kiladis, G. N. and K. M. Weickmann, 1992b: Extratropical forcing of tropical Pacific convection during northern winter. Mon. Wea. Rev, 120, 1924–1938.

Kiladis, G. N., 1998: Observation of Rossby waves linked to convection over the eastern tropical Pacific. J. Atmos. Sci., 55, 321–339.

Kiladis, G. N., K. H. Straub, and P. T. Haertel, 2005: Zonal and vertical structure of the Madden-Julian oscillation. J. Atmos. Sci., 62, 2790–2809, https://doi.org/10.1175/JAS3520.1.

Kiladis G N, Wheeler M C, Haertel P T, Straub K H and Roundy P E 2009 Convectively coupled equatorial waves Rev. Geophys. 47 RG2003.

Kim, K.-M., and K.-M. Lau, 2001: Dynamics of monsoon-induced biennial variability in ENSO. Geophys. Res. Lett., 28, 315–318.

Kikuchi, K., B. Wang, and Y. Kajikawa, 2012: Bimodal representation of the tropical intraseasonal oscillation. *Climate Dyn.*, **38**, 1989–2000, https://doi.org/10.1007/s00382-011-1159-1.

Kirtman, B. P., 1997: Oceanic Rossby Wave Dynamics and the ENSO period in a coupled model. J. Climate, 10, 1690–1704.

Kirtman, B. P., and E. K. Schneider, 2000: A spontaneously generated tropical atmospheric general circulation. J. Atmos. Sci., 57, 2080–2093.

Kirtman, B. P. and J. Shukla, 2000: Influence of the Indian summer monsoon on ENSO. Quart. Roy. Met. Soc., 126, 213–239.

Kleeman, R. and A. M. Moore, 1997: A theory for the limitations of ENSO predictability due to stochastic atmospheric transients. J. Atmos. Sci, 54, 753–767.

Knutson, T. R., and S. Manabe, 1995: Time-mean response over the tropical Pacific to increased CO_2 in a coupled ocean-atmosphere model, J. Clim., 8, 2181–2199.

Knutson, T.R., J. L. McBride, J. Chan, K. Emanuel, G. Holland, C. Landsea, I. Held, J. P. Kossin, A. K. Srivatsava, M. Sugi, 2010: Tropical cyclones and climate change. Nat. Geosci. 3, 157–163. https://doi.org/10.1038/ngeo779.

Kohyama, T., D. L. Hartmann, and D. S. Battisti, 2017: La Niña like mean state response to global warming and potential oceanic roles. J. Climate, 30, 4207–4225.

Korty, R.L., K. A. Emanuel, M. Huber, R. A. Zamora, 2017: Tropical cyclones downscaled from simulations with very high carbon dioxide levels. J. Clim. 30, 649–667.

Koster, R.D., and others 2010: Contribution of land surface initialization to subseasonal forecast skill: First results from a multi-model experiment. Geophysical Research Letters, 37, L02402, https://doi.org/10.1029/2009GL041677.

Krishnan, K., T. P. Sabin, D. C. Ayantika, A. Kitoh, M. Sugi, H. Murakami, A. G. Turner, M. Slingo, K. Rajendran, 2012: Will the South Asian monsoon overturning circulation stabilize any further ? Clim Dyn (online). https://doi.org/10.1007/s00382-012-1317-0.

Krishnan, R., K. V. Ramesh, B. K. Samala, G. Meyers, M. Slingo, M. J. Fennessy, 2006: Indian Ocean monsoon coupled interactions and impending monsoon droughts. Geophys Res Lett 33:L08711.

Krishnamurti, T. N. and H. Bhalme, 1976: Oscillations of a monsoon system. Part I: Observational aspects. J. Atmos. Sci., 33(10), 1937–1954.

Krishnamurti, T. N, and Y. Ramanathan, 1982: Sensitivity of the monsoon onset to differential heating. J. Atmos. Sci., 39, 1290–1306.

Krishnamurti, T. N. and D. Subrahmanyam, 1982: The 30–50 day mode at 850 mb during MONEX. J. Atmos. Sci., 39, 2088–2095.

Krishnamurti, T. N. and L. Bounoua, 1996: An Introduction to Numerical Weather Prediction. CRC Press, Boca Raton, Florida, 286pp.

Krishnamurti, T. N., H. L. Pan, C. B. Chang, J. Ploshay, D. Walker, A. W. Oodally, 1979: Numerical weather prediction for GATE. Quart. Roy. Met. Soc., 105, 979–1010, https://doi.org/10.1002/qj.49710544617.

Krishnamurti, T. N., H. S. Bedi, D. Oosterhof, and V. Hardiker, 1994: The formation of Hurricane Frederic of 1979. Mon. Wea. Rev., 122, 1050–1074.

Krishnamurti, T. N., S. Pattnaik, L. Stefanova, T. S. V. VijayaKumar, B. P. Mackey, O'Shay, and R. J. Pasch, 2005: The hurricane intensity issue. Mon. Wea. Rev., 133, 1886–1912.

Krishnamurti, T. N., L. Stefanova, and V. Misra, 2013: Tropical Meteorology: An Introduction. Springer, New York, 423 pp.

Krishnamurti, T. N., R. Krishnamurti, A. Simon, A. Thomas, and V. Kumar, 2016: A mechanism of the MJO invoking scale interactions. Meteorol Monogr 56:5.1–5.16. https://doi.org/10.1175/AMSMONOGRAPHS-D-15-0009.1

Krishnamurthy, V., and J. L. Kinter III, 2003: The Indian monsoon and its relation to global climate variability. *Global Climate, X.* Rodó and F. A. Comín, Eds., Springer-Verlag, 186–236.

Krishnamurthy, V. K. and B. P. Kirtman, 2009: Relation between Indian monsoon variability and SST. J. Climate, 22, 4437–4458.

Ksepka, D. T. and D. B. Thomas, 2012: Multiple Cenozoic invasions of Africa by Penguis (Aves, Sphenisciformes). Proc. R. Soc. B., 279, 1027–1032, https://doi.org/10.1098/rspb.2011.1592.

Kucharski, F., A. Bracco, J. H. Yoo, A. M. Tompkins, L. Feudale, P. Ruti, and A. Dell'Aquila, 2009: A Gill–Matsuno-type mechanism explains the tropical Atlantic influence on African and Indian monsoon rainfall. Quarterly Journal of the Royal Meteorological Society, 135(640), 569–579.

Kug, J.-S., F.-F. Jin, and S.-I. An, 2009: Two types of El Niño events: Cold tongue El Niño and warm pool El Niño. J. Climate, 22, 1499–1515, https://doi.org/10.1175/2008JCLI2624.1.

Kug, J. -S. and Y. -G. Ham, 2011: Are ther two types of La Niña? Geophys. Res. Lett., 38, L16704, https://doi.org/10.1029/2011/GL048237.

Kuo, H. L., 1965: On formation and intensification of tropical cyclones through latent heat release by cumulus convection. J. Atmos. Sci., 22, 40–63.

Kuo, H. L., 1974: Further studies of the parameterization of the influence of cumulus convection on large-scale flow. J. Atmos. Sci., 31, 1232–1240.

Kushnir, Y, R. Seager, J. Miller, and J. C. H. Chiang, 2002: A simple coupled model of tropical Atlantic decadal climate variability, Geophys. Res. Lett. 29(23), 2133, https://doi.org/10.1029/2002GL015874.

Lackmann, G. M., 2015: Hurricane Sandy before 1900 and 2100, Bull. Amer. Soc., 96(4), 547–560, https://doi.org/10.1175/BAMS-D-14-00123.1

Landsea, C., 1993: A climatology of intense (or major) Atlantic hurricanes. Mon. Wea. Rev., 121, 1703–1713.

Landsea, C. W., G. A. Vecchi, L. Bengtsson, and T. R. Knutson, 2010: Impact of duration threshold on Atlantic tropical cyclone counts. Journal of Climate, 23, 2508–2519.

Landsea, C. W. and J. L. Franklin, 2013: Atlantic Hurricane Database Uncertainty and Presentation of a New Database Format. Mon. Wea. Rev., 141, 3576–3592.

Latif, M. and T. P. Barnett, 1995: Interactions of the tropical oceans. J. Climate, 8, 952–964.

Lau, K.-M., and L. Peng. 1987: Origin of low-frequency (intraseasonal) oscillations in the tropical atmosphere. Part I: Basic theory, J. Atmos. Sci., 44, 950–972.

Lau, K. -M., H. T. Wu, and S. Bony, 1997: The role of large-scale atmospheric circulation in the relationship between tropical convection and sea surface temperature. J. Climate, 10, 381–392.

Lau, K.-M., and H.-T. Wu, 2011: Climatology and changes in tropical oceanic rainfall characteristics inferred from Tropical Rainfall Measuring Mission (TRMM) data (1998–2009), J. Geophys. Res., 116, D17111, https://doi.org/10.1029/2011JD015827.

Lawrence, D. M. and P. J. Webster, 2002: The boreal summer intraseasonal oscillation: relationship between northward and eastward movement of convection. J Atmos Sci., 59(9):1593–1606

Lee, S.-K., Wang, C., Mapes, B.E., 2009. A simple atmospheric model of the local and teleconnection responses to tropical heating anomalies. J. Clim. 22 (2), 227–284.

Lengaigne, M., E. Guilyardi, J. P. Boulanger, C. Menkes, P. Delecluse, P. Inness, 2004: Triggering of El Niño by westerly wind events in a coupled general circulation model. Climate Dyn., 23, 601–620.

L'Heureux, M. L., and R. W. Higgins, 2008: Boreal winter links between the Madden–Julian oscillation and the Arctic Oscillation. J. Climate, 21, 3040–3050.

Liebmann, B., and C. A. Smith, 1996: Description of a complete (interpolated) outgoing longwave radiation dataset. Bull. Amer. Meteor. Soc., 77, 1275–1277.

Lin, J., B. Mapes, M. Zhang, and M. Newman, 2004: Stratiform precipitation, vertical heating profiles, and the Madden-Julian Oscillation. J. Atmos. Sci., 61, 296–309.

Lin, H., and G. Brunet, 2009: The influence of the Madden–Julian oscillation on Canadian wintertime surface air temperature. Mon. Wea. Rev., 137, 2250–2262.

Lin, I. –I, S. J. Camargo, C. M. Patricola, J. Boucharel, S. Chand, P. Klotzbach, J. C. L. Chan, B. Wang, P. Chang, T. Li and F. -F. Jin, 2020: ENSO and Tropical Cyclones. In El Niño Southern Oscillation in a Changing Climate, M. J. McPhaden, A. Santoso and W. Cai, Eds., Geophysical Monograph Series, 253, 377–408.

Lindzen, R. S., and S. Nigam, 1987: On the role of sea surface temperature gradients in forcing low-level winds and convergence in the tropics. J. Atmos. Sci., 44, 2418–2436.

Lindzen, R. S. and A. Y. Hou, 1988: Hadley Circulation for zonally-averaged heating centered off the equator. J. Atmos. Sci., 45, 2416–2427.

Link, J., Tolman, H. & Robinson, K. NOAA's strategy for unified modelling. Nature 549, 458 (2017). https://doi.org/10.1038/549458b.

Liu, P., 2014: MJO structure associated with the higher-order CEOF modes. *Climate Dyn.*, **43**, 1939–1950, https://doi.org/10.1007/s00382-013-2017-0.

Liu, C. S. Shige, Y. N. Takayabu, E. Zipser, 2015: Latent heating contribution from precipitation systems with different sizes, depths, and intensities in the tropics. J. Climate, 28, 186–203.

Liu, N., C. Liu, B. Chen, and E. Zipser, 2020: What are the favorable large-scale environments for the highest flash rate thunderstorms on Earth? J. Atmos. Sci., https://doi.org/10.1175/JAS-D-19-0235.1.

Liu, P., Q. Zhang, C. Zhang, Y. Zhu, M. Khairoutdinov, H. -M. Kim, C. Schumacher, and M. Zhang, 2016: A revised real-time multivariate MJO index. Mon. Wea. Rev., **144**, 627–642, https://doi.org/10.1175/MWR-D-15-0237.1.

Lorenz, E. N., 1955: Available potential energy and the maintenance of the general circulation. Tellus, 7, 157–167, https://doi.org/10.1111/j.2153-3490.1955.tb01148.x

Lorenz, E. N., 1963: Deterministic nonperiodic flow. J. Atmos. Sci., 20, 130–141, https://doi.org/10.1175/1520-0469(1963)020<0130:DNF>2.0.CO;2.

Lorenz, E. N., 1967: The nature and theory of the general circulation of the atmosphere. World Meteorological Organization, Tech Note No. 218, T. P. 115, 161 pp.

Lorenz, E. N., 1969: The predictability of a flow which possesses many scales of motion. Tellus, 21, 289–307. https://doi.org/10.3402/tellusa.v21i3.10086.

Losada, T., B. Rodríguez-Fonseca, S. Janicot, S. Gervois, F. Chauvin, and P. Ruti, 2010: A multi-model approach to the Atlantic equatorial mode: Impact on the West African monsoon. Climate Dynamics, 35(1), 29–43.

Losada, T., B. Rodriguez-Fonseca, E. Mohino, J. Bader, S. Janicot, and C. R. Mechoso, 2012: Tropical SST and Sahel rainfall: A non-stationary relationship. Geophysical Research Letters, 39, L12705. https://doi.org/10.1029/2012GL052423.

Losada, T., and B. Rodríguez-Fonseca, 2016: Tropical atmospheric response to decadal changes in the Atlantic equatorial mode. Climate Dynamics, 47(3–4), 1211–1224.

Loschnigg, J. and P. J. Webster, 2000: A coupled ocean-atmosphere system of SST modulation for the Indian Ocean. J. Climate, 13, 3342–3360.

Lübbecke, J. F., and M. J. McPhaden, 2017: Symmetry of the Atlantic Niño mode, Geophys. Res. Lett., 44, 965–973, https://doi.org/10.1002/2016GL071829.

Lübbecke, J. F. and Coauthors, 2018: Equatorial Atlantic variability-modes, mechanisms, and global teleconnections. Wiley Interdiscip. Rev.: Climate Change, 9, e527, https://doi.org/10.1002/wcc.527.

Lucas, C., M. A. LeMone, and E. J. Zipser, 1994: Vertical velocity in oceanic convection off tropical Australia. Atmos. Sci., 51, 3183–319.

Lucas, C., B. Timbal, and H. Nguyen, 2014: The expanding tropics: a critical assessment of the observational and modeling studies. WIREs Clim Change, 5, 89–112, https://doi.org/10.1002/wcc.251.

Madden, R. A., and P. R. Julian, 1971: Detection of a 40- 50 day oscillation in the zonal wind in the tropical Pacific. J. Atmos. Sci., 28, 702–708.

Madden, R. A., and P. R. Julian, 1972: Description of global-scale circulation cells in the tropics with a 40–50 day period. J. Atmos. Sci., 29, 1109–1123.

Mapes, B. E., T. T. Warner, and M. Xu, 2003: Diurnal patterns of rainfall in northwestern South America. Part III: Diurnal gravity waves and nocturnal convection offshore. Mon. Wea. Rev., 131, 830–844.

Marengo, J. A., C. A. Nobre, and J. Tomasella, 2008: The drought of Amazonia in 2005. J. Climate, 21, 495–516.

Marengo, J. A., and Coauthors, 2012: Recent developments on the South American Monsoon System. Int. J. Climatol., 32(1), 1–21, https://doi.org/10.1002/joc.2254.

Marshall, J., A. Donohoe, D. Ferreira, and D. McGee, 2014: The ocean's role in setting the mean position of the Inter-Tropical Convergence Zone. Clim. Dyn., 42, 1967–1979.

Matsumoto, J., 1990: Withdrawal of the Indian summer monsoon and its relation to the seasonal transition from summer to autumn over East Asia. Mausam, 41, 196–202.

Matsuno, T., 1966: Quasi-Geostrophic Motions in the Equatorial Area. J. Meteor. Soc. Jpn., 44, 25–43.

Matthews, A. J., 2012: A multiscale framework for the origin and variability of the South Pacific Convergence zone. Quart. Roy. Met. Soc., 138, 1165–1178.

Markham, C. G., and D. R. McLain, 1977: Sea surface temperature related to rain in Ceara, Northeastern Brazil, Nature, 265, 320–323.

Maycock, A. C. and P. Hitchcock, 2015: Do split and displacement sudden stratospheric warmings have different annular mode signatures? Geophys. Res. Lett., 42, 10,943-10,951. https://doi.org/10.1002/2015GL066754

McPhaden, M. J., 1982: Variability in the central equatorial Indian Ocean. Part I: Ocean dynamics. J. Mar. Res., 40, 157–176.

McPhaden, M.J., 2012: A 21st century shift in the relationship between ENSO SST and Warm Water Volume anomalies. Geophys. Res. Lett., 39, L09706, https://doi.org/10.1029/2012GL051826

McPhaden, M. J., A. J. Busalacchi, R. Cheney, and Coauthors, 1998: The Tropical Ocean-Global Atmosphere observing system. A decade of progress. J. Geophys. Res. (Oceans), https://doi.org/10.1029/97JC02906.

Meehl, G. A., 1994: Coupled land–ocean–atmosphere processes and south Asian monsoon variability. Science, 266, 263–267.

Meehl, G. A., and W. M. Washington, 1996: El Niño like climate change in a model with increased atmospheric CO2 concentration, Nature, 382, 56–60.

Meehl, G. A., and J. M. Arblaster, 2002: The tropospheric biennial oscillation and Asian-Australian monsoon rainfall. J. Climate, 15, 722–744.

Meehl, G. A., P. R. Gent, J. M. Arblaster, B. L. Otto-Bliesner, E. C. Brady, and A. Craig, 2001: Factors that affect the amplitude of El Niño in global coupled climate models. Clim. Dyn., 17, 515–526.

Meehl, G. A., L. Goddard, G. Boer, R. Burgman, G. Brantstator, C. Cassou, S. Corti, G. Danabasoglu, F. Doblas-Reyes, and Co-Authors, 2014: Decadal climate prediction: An update from the trenches. Bull. Amer. Soc., 95(2), 243–267, https://doi.org/10.1175/BAMS-D-12-00241.1.

Mehta, V. M., 1998: Variability of the tropical ocean surface temperatures at decadal–multidecadal timescales. Part I: The Atlantic ocean. J. Climate, 11, 2351–2375.

Meinen, C. S., and M. J. McPhaden, 2000: Observations of warm water volume changes in the Equatorial Pacific and their relationship to El Niño and La Niña. J. Climate, 13, 3551–3559.

Merle, J., 1980: Variabilite´ thermique annuelle et interannuelle de l'oce´an Atlantique e´quatorial Est. L'hypothe´se d'un "El Niño" Atlantique. Oceanol. Acta 3, 209–220.

Merrill, R. T., 1988: Environmental influences on hurricane intensification. J. Atmos. Sci., 45, 1678–1687, https://doi.org/10.1175/1520-0469(1988)0452.0.CO;2.

Merryfield, W. J., and Coauthors, 2020: Current and emerging developments in subseasonal to decadal prediction. Bull. Amer. Soc., https://doi.org/10.1175/BAMS-D-19-0037.1.

Mo, K. C., C. Jones, and J. N. Paegle, 2012: Pan-America. Intraseasonal Variability of the Atmosphere–Ocean Climate System, 2nd ed. W. K.-M. Lau and D. E. Waliser, Eds., Springer, 111–146.

Mohino, E., B. Rodríguez-Fonseca, T. Losada, S. Gervois, S. Janicot, J. Bader, and F. Chauvin, 2011: Changes in the interannual SST-forced signals on West African rainfall. AGCM intercomparison. Climate Dynamics, 37, 1707–1725.

Molcard, R., M. Fieux, and A. G. Ilahude, 1996: The Indo–Pacific throughflow in the Timor Passage. J. Geophys. Res., 101 (C5), 12 411–12 420.

Molinari, J., D. Knight, M. Dickinson, D. Vollaro, and S. Skubis, 1997: Potential vorticity, easterly waves, and eastern Pacific tropical cyclogenesis. Mon. Wea. Rev., 125, 2699–2708.

Moncrieff, M. W., 2004: Analytic representation of the large-scale organization of tropical convection. J. Atmos. Sci., 61, 1521–1538.

Moore, A. M. and R. Kleeman, 1999a: Stochastic forcing of ENSO by the Intraseasonal Oscillation. J. Climate, 12, 1199–1220.

Mori, S., and Coauthors, 2004: Diurnal land–sea rainfall peak migration over Sumatera Island, Indonesian Maritime Continent, observed by TRMM satellite and intensive rawinsonde soundings. Mon. Wea. Rev., 132, 2021–2039.

Moura, A., and J. Shukla, 1981: On the dynamics of droughts in northeast Brazil: Observations, theory, and numerical experiments with a general circulation model. J. Atmos. Sci., 38, 2653–2675.

Meyers, G., 1996: Variation of Indonesian Throughflow and El Nin~o-Southern Oscillation. Journal of Geophysical Research, 101,12255–12263.

Mishra, V., A. D. Tiwari, S. Aadhar, R. Shah, M. Xiao, D. S. Pai, D. Lettenmaier, 2019: Drought and famine in India, 1870–2016. Geophys. Res. Lett., https://doi.org/10.1029/2018GL081477.

Misra, V., 2006: Understanding the predictability of seasonal precipitation over northeast Brazil. Tellus, 58A, 307–319.

Misra, V., 2008: Coupled interactions of the Monsoons. Geophys. Res. Lett., 35, L12705, https://doi.org/10.1029/2008GL033562.

Misra, V. and S. DiNapoli, 2014: The variability of the southeast Asian summer monsoon. Int. J. Climatology, 34(3), 831–840, https://doi.org/10.1002/joc.3735.

Misra, V., and Coauthors, 2007: Validating and understanding the ENSO simulation in two coupled climate models. Tellus, 59A, 292–308.

Moore, A. M. and R. Kleeman, 1999b: The Nonnormal Nature of El Niño and Intraseasonal variability. J. Climate, 12, 2965–2982.

Moy, C.M., G.O. Seltzer, D.T. Rodbell, and D.M. Anderson, 2002: Variability of El Niño/Southern Oscillation activity at millennial timescales during the Holocene epoch. Nature, 420, 162–165.

Munnich, M, M.A. Cane and S.E. Zebiak, 1991: A study of self excited oscillations of the tropical ocean-atmosphere system. II. Nonlinear cases, J. Atmos. Sci., 43, 1238–1248.

Murakami, M.,1976: Analysis of summer monsoon fluctuations over India. J. Meteor. Soc. Japan, 54 (1), 15–31.

Murakami, T., 1951: On the study of the change of the upper westerlies in the last stage of the Baiu season (rainy season in Japan). J Meteorol. Soc. Japan, 29, 162–175.

Murakami, T., 1958: The sudden change of upper westerlies near the Tibetan Plateau at the beginning of summer season. J. Meteorol. Soc. Japan, 36, 239–247.

Murakami, H., Levin, E., Delworth, T.L., Gudgel, R., Hsu, P.-C., 2018. Dominant effect of relative tropical Atlantic warming on major hurricane occurrence. Science 362, 794–799. https://doi.org/10.1126/science.aat6711.

Nakazawa, T., 1986: Mean features of 30–60 day variations as inferred from 8-year OLR data, J. Meteorol. Soc. Jpn., 64, 777–786.

Nakazawa, T., 1988: Tropical super clusters within intraseasonal variations over the western Pacific, J. Meteorol. Soc. Jpn., 66, 823–836.

Neelin, J. D. and I. M. Held, 1987: Modeling tropical convergence based on the moist static energy budget. Mon. Wea. Rev., 115, 3–12.

Neelin, J. D. and J. Y. Yu, 1994: Modes of tropical variability under convective adjustment and the Madeen-Julian oscillation. 1: Analytical theory. J. Atmos. Sci., 51, 1876–1894.

Neelin, J., C. Chou, and H. Su, 2003: Tropical drought regions in global warming and El Nino teleconnections. Geophys Res Lett. https://doi.org/10.1029/2003GLO018625.

Newell, R., 1979: Climate and the ocean. Amer. Sci., 67, 405–416.

Newell, R. E., 1986: An approach towards equilibrium temperature in the tropical eastern Pacific, J. Phys. Oceanogr., 16, 1338–1342.

Newman, M., S.-I. Shin, and M. A. Alexander, 2011a: Natural variation in ENSO flavors. Geophys. Res. Lett., L14705, https://doi.org/10.1029/2011GL047658.

Nicholls, M. E., Pielke R. A., and W. R. Cotton, 1991: Thermally forced gravity waves in an atmosphere at rest. J Atmos Sci, 48(16):1869–1884.

Newman, M., M. A. Alexander, and J. D. Scott, 2011b: An empirical model of tropical ocean dynamics. Climate Dynamics, 37, 1823–1841, https://doi.org/10.1007/s00382-011-1034-0

Nigam, S., 1994: On the dynamical basis for the Asian monsoon rainfall-El Niño relationship. J. Climate, 7, 1750–1771.

Nigam, S. and C. Chung, 2000: ENSO surface winds in CCM3 simulation: diagnosis of errors. J. Climate, 13, 3172–3186.

Nigam, S., C. Chung, and E. DeWeaver, 2000: ENSO diabatic heating in ECMWF and NCEP reanalyses, and NCAR CCM3 simulation. J. Climate, 13, 3152–3171.

Nnamchi, H. C., J. Li, F. Kucharski, I. -S. Kang, N. S. Keenlyside, P. Chang, and R. Farneti, 2015: Thermodynamic controls of the Atlantic Niño. Nature Communications, 6, 8895. https://doi.org/10.1038/ncomms9895

Nobre, P., and J. Shukla, 1996: Variations of sea surface temperature, wind stress, and rainfall over the tropical Atlantic and South America. Journal of Climate, 9(10), 2464–2479.

Noska, R. and V. Misra, 2016: Characterizing the onset and the demise of the Indian summer monsoon. Geophys. Res. Lett., https://doi.org/10.1002/2016GL068409.

Nouri, N., N. Devineni, V. Were, and R. Khanbilvardi, 2021: Explaining the trends and variability in the United States tornado records using climate teleconnections and shifts in observational practices, Sci Rep, 11, 1741, https://doi.org/10.1038/s41598-021-81143-5.

Okumura, Y., and Xie, S.P., 2004: Interaction of the Atlantic Equatorial Cold Tongue and the African Monsoon. J. Clim. 17, 3589–3602.

Okumura, Y. M. and C. Deser, 2010: Asymmetry in the duration of El Niño and La Niña. J. Climate, 23, 5826–5843.

Ooyama, K., 1964: A dynamical model for the study of tropical cyclone development. Geofisc. Internacional Mexico, 4, 187–198.

Ooyama, K., 1969: Numerical simulation of the life cycle of tropical cyclones. J. Atmos. Sci., 26, 3–40.

Orsolini, Y. J., R. Senan, G. Balsamo, F. J. Doblas-Reyes, F. Vitart, A. Weisheimer, A. Carrasco, and R. E. Benestad, 2013: Impact of snow initialization on sub-seasonal forecasts. Climate Dyn., 41, 1969–1982, https://doi.org/10.1007/s00382-013-1782-0.

Palmer, T., and D. Mansfield, 1984. Response of two atmospheric general circulation models to sea surface temperature anomalies in the tropical east and west Pacific, Nature, 310, 483–485.

Parhi, P., A. Giannini, P. Gentine, and U. Lall, 2016: Resolving contrasting regional rainfall responses to El Niño over Tropical Africa. J. Climate, 29, 1461–1476.

Park, H.-S. S., Chiang, J. C. H., & Bordoni, S. (2012). The mechanical impact of the Tibetan Plateau on the seasonal evolution of the South Asian monsoon. Journal of Climate, 25(7), 2394–2407.

Park, W., N. Keenlyside, M. Latif, A. Stroh, R. Redler, E. Roeckner, G. Madec, 2009: Tropical Pacific climate and its response to global warming in the Kiel Climate model. J. Climate, 22, 71–92.

Penland, C. and P. D. Sardeshmukh, 1995: The optimal growth of tropical sea surface temperature anomalies. J. Climate, 8, 1999–2024.

Perigaud, C. and P. Delecluse, 1992: Annual sea level variations in the southern tropical Indian Ocean from geosat and shallow water simulations. Journal of Geophysical Research, 97, 20169–20178.

Philander, S. G., 1983: El Niño-Southern Oscillation phenomena. Nature, 302, 295–301.

Philander, S. G. H., D. Gu, D. Halpern, G. Lambert, N. -C. Lau, T. Li and R. C. Pacanowski, 1996: Why the ITCZ is mostly north of the equator. J. Climate, 9, 2958–2972.

Picaut, J., F. Masia, Y. du Penhoat, 1997: An advective-reflective conceptual model for the oscillatory nature of the ENSO. Science, 277, 663–666.

Potemra, J. T., R. Lukas., and G. T. Mitchum, 1997: Large-scale estimation of transport from the Pacific to the Indian Ocean. Journal of Geophysical Research, 102, 27795–27812.

Prive, N. C., and R. A. Plumb, 2007: Monsoon dynamics with interactive forcing. Part II: Impact of eddies and asymmetric geometries. J. Atmos. Sci., 64, 1431–1442.

Qu, T., and J.-Y. Yu, 2014: ENSO indices from sea surface salinity observed by Aquarius and Argo. J. Oceanogr., 70, 367–375, https://doi.org/10.1007/s10872-014-0238-4.

Rahmstorf, S. and D. Coumou, 2011: Increase of extreme events in a warming world. Proc. Natl. Acad. Sci. USA, 108, 17 905–17 909, https://doi.org/10.1073/pnas.1101766108.

Rajagopalan, B., Y. Kushnir, and Y. M. Tourre, 1998: Observed decadal midlatitude and tropical Atlantic climate variability. Geophys.Res. Lett., 25, 3967–3970.

Rahul, S., and C. Gnanaseelan, 2016: Can large scale surface circulation changes modulate the sea surface warming pattern in the Tropical Indian Ocean? Clim Dyn 46(11):3617–3632. https://doi.org/10.1007/s00382-015-2790-z

Ramanathan, V., and W. Collins, 1991: Thermodynamic regulation of ocean warming by cirrus clouds deduced from observations of the 1987 El Nin~o. Nature, 351, 27–32.

Ramanathan, V., P. J. Crutzen, J. T. Kiehl, and D. Rosenfield, 2001: Aerosols, climate, and hydrological cycle. Science, 294, 2119–2124.

Ramesh Kumar, M. R., R. Krishnan, S. Syam, A. S. Unnikrishnan, and D. S. Pai, 2009: Increasing trend of "Break-Monsoon" conditions over India—role of ocean-atmosphere processes in the Indian Ocean. IEEE Geosci Rem Sens Lett 6:332–336.

Ramage, C., 1971: Monsoon Meteorology, Int. Geophys. Ser., vol. 15, 296 pp., Academic Press, San Diego, Calif.

Randall, D. A., Harshvardhan, and D. A. Dazlich, 1991: Diurnal variability of the hydrologic cycle in a general circulation model. J. Atmos. Sci., 48, 40–62.

Rao, S. A., J.-J. Luo, S. K. Behera, and T. Yamagata, 2008: Generation and termination of Indian Ocean dipole events in 2003, 2006 and 2007. Climate Dyn., 33, 751–767, https://doi.org/10.1007/s00382-008-0498-z.

Rappin, E. D., M. C. Morgan, and G. J. Tripoli, 2009: The impact of outflow environment on tropical cyclone intensification and structure. J. Atmos. Sci., 68, 177–194.

Rasmusson, E. M., and T. H. Carpenter, 1982: Variations in tropical sea surface temperatures and surface wind fields associated with the Southern Oscillation/El Niño. Mon. Wea. Rev., 110, 354–384.

Raymond, D. J., 1995: Regulation of moist convection over the West Pacific warm pool. J. Atmos. Sci., 52, 3945–3959.

Raymond, D. J. and K. A. Emanuel, 1993: The Kuo Cumulus Parameterization. In: Emanuel, K. A. and D. J. Raymond (eds) The representation of convection in numerical models. Meteorological Monographs. American Meteorological Society, Boston, MA. https://doi.org/10.1007/978-1-935704-13-3_12

Rayner, N. A., D. E. Parker, E. B. Horton, C. K. Folland, L. V. Alexander, D. P. Rowell, E. C. Kent, and A. Kaplan (2003), Global analyses of sea surface temperature, sea ice, and night marine air temperature since the late nineteenth century, J. Geophys. Res., 108(D14), 4407, https://doi.org/10.1029/2002JD002670.

Raymond, D. J., 2001: A new model of the Madden-Julian oscillation. J. Atmos. Sci., 58, 2807–2819.

Raymond, D. J. and Z. Fuchs, 2009: Moisture modes and the Madden–Julian oscillation. J Clim., 22:3031–3046.

Raymond, D. J., S. K. Esbensen, C. Paulson, M. Gregg, C. S. Bretherton, W. A. Petersen, R. Cifelli, L. K. Shay, C. Ohlmann and P. Zuidema, 2004: EPIC2001 and the Coupled Ocean-Atmosphere System of the Tropical East Pacific. Bull. of the Amer. Meteor. Soc., 85, 1341–1354.

Rennick, M. A., 1976: The generation of African waves. J. Atmos. Sci., 33, 1955–1969.

Richter, I., S. K. Behera, Y. Masumoto, B. Taguchi, H. Sasaki, and T. Yamagata, 2013: Multiple causes of interannual sea surface temperature variability in the equatorial Atlantic Ocean. Nature Geoscience, 6, 43–47. https://doi.org/10.1038/ngeo1660.

Richter, I., S. P. Xie, S. K. Behera, and Coauthors, 2014: Equatorial Atlantic variability and its relation to mean state biases in CMIP5. Clim Dyn 42:171–188. https://doi.org/10.1007/s00382-012-1624-5

Riehl, H., 1945: Waves in the easterlies and the polar front in the tropics. Misc. Rep. No. 17, Department of Meteorology, University of Chicago, 79 pp.

Riehl, H. and J. S. Malkus, 1958: On the heat balance of the equatorial trough zone. Geophysica, Helsinki, 503–537.

Romps, D. M., 2014: An analytical model for tropical relative humidity, *J. Clim.*, 27(19), 7432–7449, https://doi.org/10.1175/JCLI-D-14-00255.1.

Ropelewski, C. F., and M. S. Halpert, 1989: Precipitation patterns associated with the high index phase of the Southern Oscillation. J. Climate, 2, 268–284.

Rouault, M., P. Florenchie, N. Faucherau, and C. J. C. Reason, 2003: South East tropical Atlantic warm events and southern African rainfall. Geophys. Res. Lett., 30, 8009, https://doi.org/10.1029/2003GL014840.

Roxy, M.K., Ghosh, S., Pathak, A. et al., 2017: A threefold rise in widespread extreme rain events over central India. Nat Commun 8, 708, https://doi.org/10.1038/s41467-017-00744-9

Roy, T., 2016: Were Indian famines 'Natural' or 'Manmade'? LSE working Paper No. 243. Available from: https://www.lse.ac.uk/Economic-History/Assets/Documents/WorkingPapers/Economic-History/2016/WP243.pdf.

Rui, H., and B. Wang, 1990: Development characteristics and dynamic structure of tropical intraseasonal convection anomalies. J. Atmos. Sci., 47, 357–379.

Ruiz-Barradas, A., J. A. Carton, and S. Nigam, 2000: Structure of interannual-to-decadal climate variability in the tropical Atlantic sector. J. Clim., 13, 3285–3297.

Ruppert, J. H., 2016: Diurnal timescale feedbacks in the tropical cumulus regime. J. Adv. Model. Earth Sys., https://doi.org/10.1002/2016MS000713.

Sabeerali, C., S. A. Rao, G. George, D. N. Rao, S. Mahapatra, A. Kulkarni, R. Murtugudde, 2014: Modulation of monsoon intraseasonal oscillations in the recent warming period. Journal of Geophysical Research: Atmospheres 119 (9):5185–5203.

Saha, K., F. Sanders, and J. Shukla, 1981: Westward propagating predecessors of monsoon depressions. Mon. Wea. Rev.,109, 330–343.

Saha, S. and Coauthors, 2010: The NCEP climate forecast system reanalysis. Bull Am Soc., 91, 1015–1058, https://doi.org/10.1175/2010BAMS3001.1.

Sahami, K., 2003: Aspects of the heat balance of the Indian Ocean on intra-annual and interannual time scales. Ph.D. dissertation, The University of Colorado, 191 pp.

Saji, N. H., B. N. Goswami, P.N. Vinayachandran, and T. Yamagata, 1999: A dipole mode in the tropical Indian Ocean, Nature, 401, 360–363.

Saravanan, R. and P. Chang, 2000: Interaction between tropical Atlantic variability and El Niño-Southern Oscillation. J. Climate, 13, 2277–2292.

Schneider, E. K. and R. S. Lindzen, 1977: Axially symmetric steady-state models of the basic state for instability and climate studies. Part I: Linearized calculations. J. Atmos. Sci., 34(2), 263–279.

Schneider, E. K., B. Huang, and J. Shukla, 1995: Ocean wave dynamics and El Niño. J. Climate, 8, 2415–2439.

Schneider, N., 1998: The Indonesian Throughflow and the Global Climate System. J. Climate, 676-689. https://doi.org/10.1175/1520-0442(1998)011<0676:TITATG>2.0.CO;2

Schott, F., and D. Quadfasel, 1982: Variability of the Somali Current and associated upwelling. Progress in Oceanography, 12, 357–381.

Schott, F. A. and J. P. McCreary, 2001: The monsoon circulation of the Indian Ocean. Progress in Oceanography, 51, 1–123.

Scroxton, N., S. G. Bonham, R. E. M. Rickaby, S. H. F. Lawrence, M. Hermoso, and A. M. Haywood, 2011: Persistent El Niño-Southern Oscillation variation during the Pliocene Epoch Paleoceanography, 26, PA2215, https://doi.org/10.1029/2010pa002097.

Seager, R. and R. Murtugudde, 1997: Ocean dynamics, thermocline adjustment, and regulation of tropical SST. J. Climate, 10, 521–534.

Seeley, J. T. and D. M. Romps, 2016: Tropical cloud buoyancy is the same in a world with or without ice. Geophys. Res. Lett., https://doi.org/10.1002/2016GL068583.

Shannon, L. V., A. J. Boyd, G. B. Bundrit, J. Taunton-Clark, 1986: On the existence of El Niño type phenomenon in the Benguela System. J. Mar. Sci., 44, 495–520.

Shapiro, L. J., 1986: The three-dimensional structure of synoptic-scale disturbances over the tropical Atlantic. Mon. Wea. Rev., 114, 1876–1891.

Sigmond, M., J. F. Scinocca, V. V. Kharin, and T. G. Shepherd, 2013: Enhanced seasonal forecast skill following stratospheric sudden warmings. Nat. Geosci., 6, 98–102, https://doi.org/10.1038/ngeo1698.

Sikka, D. and S. Gadgil, 1980: On the maximum cloud zone and the ITCZ over Indian, longitudes during the southwest monsoon. Mon Weather Rev 108(11):1840–1853. https://doi.org/10.1175/1520-0493(1980)1082.0.CO;2.

Singh, A., T. Delcroix, and S.Cravatte, 2011: Contrasting the flavors of El Niño and Southern Oscillation using sea surface salinity observations. J. Geophys. Res., 116, C06016, https://doi.org/10.1029/2010JC006862.

Smith, N., W. S. Kessler, K. Hil, and D. Carlson, 2015: Progress in observing and predicting ENSO. WMO Bulletin, 62 (1), Available from: https://public.wmo.int/en/resources/bulletin/progress-observing-and-predicting-enso-0

Smith, T. M., and R. W. Reynolds, 2004: Improved extended reconstruction of SST (1854–1997), J. Clim., 17, 2466–2477.

Sobel, A. H., and C. S. Bretherton, 2000: Modeling tropical precipitation in a single column. J. Climate, 13, 4378–4392.

Sobel, A. H. and E. Maloney, 2013: Moisture modes and the eastward propagation of the MJO. J Atmos Sci., 70, 187–192.

Sobel, A. H., J. Nilsson, and L. M. Polvani, 2001: The weak temperature gradient approximation and balanced tropical moisture waves. J. Atmos. Sci., 58, 3650–3665.

Sobel, A. H., S. E. Yuter, C. S. Bretherton, and G. N. Kiladis, 2004: Large-scale meteorology and deep convection during TRMM KWAJEX. Mon. Wea. Rev., 132, 422–444.

Song, K. and S. -W. Son, 2018: Revisiting the ENSO-SSW relationship. J. Climate, 31, 2133–2143.

Song, X. and L. Yu, 2013: How much net surface heat flux should go into the Western Pacific Warm Pool. J. Geophys. Res., 118, 3569–3585, https://doi.org/10.1002/jgrc.20246.

Sooraj, K. P., J.-S.Kug, T.Li, and I.-S. Kang, 2009: Impact of El Niño onset timing on the Indian Ocean: Pacific coupling and subsequent El Niño evolution. Theor. Appl. Climatol., 97, 17–27.

Sprintall, J., A. Gordon, R. Murtugudde, and D. Susanto, 2000: A semi-annual Indian Ocean forced Kelvin wave observed in the Indonesian Seas in May 1997. Journal of Geophysical Research, 105, 17217–17230.

Stevens, B., D. A. Randall, X. Lin, and M. T. Montgomery, 1997: On large-scale circulations in convecting atmospheres by Kerry A. Emanuel, J. David Neelin and Cristopher S. Bretherton. Quart. J. Roy. Meteor. Soc., 123, 1771– 1778.

Sprintall, J., A. L. Gordon., A. Koch-Larrouy, T. Lee, J. T. Potemra, K. Pujiana, S. E. Wijffels, 2014: The Indonesian seas and their role in the coupled ocean climate system. Nature Geosci 7:487–492.

Stachnik J. P., C. Schumacher, 2011: A comparison of the Hadley circulation in modern reanalyses. J Geophys Res, 116: D22102. https://doi.org/10.1029/2011JD016677.

Straub, K. H., 2013: MJO initiation in the real-time multivariate MJO index. J. Climate, 26, 1130–1151, https://doi.org/10.1175/JCLI-D-12-00074.1.

Su, J., R. Zhang, T. Li, X. Rong, J. S. Ug, and C. C. Hong (2010), Causes of the El Niño and La Niña amplitude asymmetry in the equatorial eastern Pacific, J. Clim., 23, 605–617, https://doi.org/10.1175/2009JCLI2894.1.

Sumi, A., 1992: Pattern formation of convective activity over the aqua-planet with globally uniform sea-surface temperature (SST). J Meteor Soc Jpn 70, 855–876

Sui, C.-H., and K.-M. Lau (1989), Origin of low-frequency (intraseasonal) oscillations in the tropical atmosphere. Part II: Structure and propagation of mobile wave-CISK modes and their modification by lower boundary forcings, J. Atmos. Sci., 46, 37–56.

Sun, J. and Z. Wu, 2020: Isolating spatiotemporally local mixed Rossby-gravity waves using multi-dimensional ensemble empirical mode decomposition. Clim. Dyn., 54, 1383–1405.

Suarez, M. J., and P. S. Schopf, 1988: A delayed action oscillator for ENSO. J. Atmos. Sci., 45, 3283–3287.

Swapna, P., R. Krishnan, and J. Wallace, 2014: Indian Ocean and monsoon coupled interactions in a warming environment. Climate Dynamics 42 (9–10):2439–2454.

Tai, K. S., and Y. Ogura, 1987: An observational study of easterly waves over the eastern Pacific in the northern summer using FGGE data. J. Atmos. Sci.,44, 339–361.

Takayabu, Y. N., 1994: Large-scale cloud disturbances associated with equatorial waves. part II: Westward-propagating inertiogravity waves, J. Meteorol. Soc. Jpn., 72, 451–465.

Tam, C. -Y. and N. -C. Lau, 2005: Modulation of the Madden-Julian Oscillation by ENSO: Inferences from observations and GCM simulations. J. Met. Soc. Japan, 83, 727–743.

Tanimoto, Y., and S.-P. Xie, 1999: Ocean–atmosphere variability over the Pan–Atlantic basin. J. Meteor. Soc. Japan, 77, 31–46.

Tanimoto, Y., and S.-P. Xie, 2002: Inter-hemispheric decadal variations in SST, surface wind, heat flux and cloud cover over the Atlantic Ocean. J. Meteorol. Soc. Jpn., 80, 1199–1219.

Teng, H., G. Branstator, A. B. Tawfik, and P. Callaghan, 2019: Circumglobal response to prescribed soil moisture over North America. J. Climate, 32, 4525–4545, https://doi.org/10.1175/JCLI-D-18-0823.1.

Thompson, D. B., and P. E. Roundy, 2013: The relationship between the Madden–Julian oscillation and U.S. violent tornado outbreaks in the spring. Mon. Wea. Rev., 141, 2087–2095.

Thorncroft, C. D. and K. Hodges, 2000: African Easterly Wave Variability and its Relationship to Atlantic Tropical cyclone activity. J. Climate, 14, 1166–1179.

Thorncroft, C. D. and M. Blackburn 1999: Maintenance of the African easterly jet. Quart. J. Roy. Meteor. Soc., 125, 763–786.

Thorne, P. W., J. R. Lanzante, T. C. Peterson, D. J. Seidel, and K. P. Shine, 2011: Tropospheric temperature trends: history of an ongoing controversy. WIREs Climate Change, 2, 66–88, https://doi.org/10.1029/2007GL029875.

Timmermann, A., et al. 2018: El Niño-Southern Oscillation complexity. Nature, 559, 535–545, https://doi.org/10.1038/s41586-018-0252-6.

Tobin, I., S. Bony, R. Roca, 2012: Observational evidence for relationships between the degree of aggregation of deep convection, water vapor, surface fluxes, and radiation. J Clim., 25, 6885–6904.

Torrence, C., and P. J. Webster, 1998: The annual cycle of persistence in the El Niño/Southern Oscillation. Quart. J. Roy. Meteorol. Soc., 125, 1985–2004.

Torrence, C., and P. J. Webster, 1999: Interdecadal changes in the ENSO-monsoon system. J. Climate, 12, 2679–2690.

TPOS2014: Report of the Tropical Pacific Observation System 2020 (TPOS 2020) Workshop, Vol. 1. Workshop report and recommendations, La Jolla, United States, 27–30 January 2014, pp 66.

Trenberth, K. E., 1984: Signal versus noise in the Southern Oscillation. Mon. Wea. Rev., 112, 326–332.

Trenberth K. E., D. P. Stepaniak, 2001: Indices of El Niño evolution. J. Climate 14: 1697–1701.

Trenberth K. E., D. P. Stepaniak, 2004: The flow of energy through the earth's climate system. Quart. Roy. Met. Soc., 130, 2677–2701.

Trenberth, K. E., G. Branstator, D. Karoly, A. Kumar, N.-C. Lau, and C. Ropelewski, 1998: Progress during TOGA in understanding and modeling global teleconnections associated with tropical sea surface temperatures. J. Geophys. Res., 103 (C7), 14 291–14 324.

Trenberth, K. E., D. P. Stepaniak, and J. M. Caron, 2000: The global monsoon as seen through the divergent atmospheric circulation. J. Climate, 13, 3969–3993.

Trenberth, K. E., J. W. Hurrell, and D. P. Stepaniak, 2006: The Asian Monsoon: Global Perspectives. The Asian Monsoon, B. Wang (Ed.), Heidelberg, Springer, pp 67–87.

Troup, A. J., 1965. The Southern Oscillation. Quart. J. Roy. Meteor. Soc., 91, 490–506.

Tziperman, E., S. E. Zebiak, and M. A. Cane, 1997: Mechanisms of seasonal-ENSO interaction. J. Atmos. Sci., 54, 61–71.

Uehling, J., V. Misra, A. Bhardwaj, and N. Karmakar, 2021: Characterizing the local variations of the Northern Australian Rainy Season. Mon. Wea. Rev. In review.

Utsumi, N., S. Seto, S. Kanae, E. E. Maeda, and T. Oki, 2011: Does higher surface temperature intensify extreme precipitation?, Geophys. Res. Lett., 38, L16708, https://doi.org/10.1029/2011GL048426.

Van der Wiel K., A. J. Matthews, D. P. Stevens, and M. M. Joshi, 2015: A dynamical framework for the origin of the diagonal South Pacific and South Atlantic Convergence Zones. Quar. Roy. Met. Soc. https://doi.org/10.1002/qj.2508.

Van der Wiel, K., A. J. Matthews, M. M. Joshi, and D. P. Stevens, 2016: Why the South Pacific Convergence Zone is diagonal. Clim. Dyn., 46, 1683–1698. https://doi.org/10.1007/s00382-015-2668-0

Van Oldenborgh, G. J., K. van der Wiel, A. Sebastian, R. Singh, J. Arrighi, F. Otto, K. Haustein, S. Li, G. Vecchi, H. Cullen, 2017: Attribution of extreme rainfall from Hurricane Harvey August 2017: Environmental Research Letters, 12, 12, https://doi.org/10.1088/1748-9326/aa9ef2.

Vecchi, G. A., and B. J. Soden, 2007: Global warming and the weakening of the tropical circulation. J. Clim., 20, 4316-4340. https://doi.org/10.1175/JCLI4258.1.

Vecchi, G. A., et al., 2006: Weakening of tropical Pacific atmospheric circulation due to anthropogenic forcing, Nature, 441(7089), 73–76, https://doi.org/10.1038/nature04744.

Vecchi, G. A., A. Clement, and B. J. Soden, 2008: Examining the tropical Pacific response to global warming. EOS, Trans. Amer. Geophys. Union, v(89), 81–83.

Ventrice, M. J., M. C. Wheeler, H. H. Hendon, C. J. Schreck III, C. D. Thorncroft, and G. N. Kiladis, 2013: A modified multivariate Madden–Julian oscillation index using velocity potential. Mon. Wea. Rev., 141, 4197–4210, https://doi.org/10.1175/MWR-D-12-00327.1.

Vitart, F., 2017: Madden–Julian oscillation prediction and teleconnections in the S2S database. Quart. J. Roy. Meteor. Soc., 143, 2210–2220, https://doi.org/10.1002/qj.3079.

Vitart, F. and A. W. Robertson, 2018: The sub-seasonal to seasonal prediction project (S2S) and the prediction of extreme events. NPJ Climate and Atmospheric Science, 1:3, https://doi.org/10.1038/s41612-018-0013-0.

Von der Heydt, A. S., A. Nnafie, and H. A. Dijkstra, 2011: Cold tongue/warm pool and ENSO dynamics in the Pliocene. Clim. Past Discuss, 7, 997–1027, https://doi.org/10.5194/cpd-7-997-2011.

Waliser, D. E., 1996: Some considerations on the thermostat hypothesis. Bull. Amer. Soc., 77, 357–360.

Waliser, D. E. and R. C. J. Somerville,1994: Preferred latitude of the ITCZ. J Atmos Sci 51: 1619–1639.

Waliser D. E., Z. Shi, J. R. Lanzante, A. H. Oort, 1999: The Hadley circulation: assessing NCEP/NCAR reanalysis and sparse in situ estimates. Climate Dyn 15:719–735.

Walker, G. T. and E. W. Bliss, 1932: World Weather V. Memoirs of the Royal Meteorological Society, 4, 53–84.

Wallace, J. M., 1992: Effect of deep convection on the regulation of tropical sea surface temperature. Nature, 357, 230–231.

Wang, B., 2003: Fundamental dynamics of the tropical intraseasonal oscillation. Extended abstract to ECMWF workshop, November 3–6. Available from: https://www.ecmwf.int/sites/default/files/elibrary/2004/12985-fundamental-dynamics-tropical-intraseasonal-oscillation.pdf.

Wang, B., and H. Rui, 1990: Dynamics of the coupled moist Kelvin-Rossby wave on an equatorial beta plane, J. Atmos. Sci., 47, 397–413.

Wang, B. and LinHo, 2002: Rainy season of the Asian-Pacific summer monsoon. J. Clim, 15, 386–398.

Wang, B. and Q. Ding, 2008: Global monsoon: dominant mode of annual variation in the tropics. Dyn. of Atm. and Ocn., 44, 165–183, https://doi.org/10.1016/j.dynatmoce.2007.05.002.

Wang, C., 2001: A unified oscillator model for the El Niño-Southern Oscillation. J. Climate, 14, 98–115.

Wang, W., and M. J. McPhaden, 1999: The surface-layer heat balance in the equatorial Pacific Ocean. Part I: Mean seasonal cycle. J. Phys. Oceanogr., 29, 1812–1831.

Wara, M. W., A. C. Ravelo, M. L. Delaney, 2005: Permanent El Niño-like conditions during the Pliocene warm period. Science, 309, 758–761.

Watanabe, T., A. Suzuki, S. Minobe, T. Kawashima, K. Kameo, K. Minoshima, Y. M. Aguila, R. Wani, H. Kawahata, K. Sowa, T. Nagai, and T. Kase, 2011: Permanent El Niño during the Pliocene warm period not supported by coral evidence. Nature, 471, 209–211, https://doi.org/10.1038/nature09777.

Waugh, D. W., K. M. Grise, W. J. M. Seviour, S. M. Davis, N. Davis, O. Adam, S. -W. Son, I. R. Simpson, P. W. Staten, A. C. Maycock, C. C. Ummenhofer, T. Birner, and A. Ming, 2018: Revisiting the relationship among metrics of tropical expansion. J. Climate, 31, 7565–7581.

Webster, P. J. and R. Lukas, 1992: TOGA COARE: The Coupled Ocean-Atmosphere Response Experiment. Bull. Amer. Soc., 73(9), 1377–1416.

Webster, P. J. and S. Yang, 1992: Monsoon and ENSO: Selectively Interactive Systems. Quart. J. Roy. Meteor. Soc., 118, 877–926

Webster, P. J., V. Magaña, T. Palmer, J. Shukla, R. Tomas, M. Yanai, and T. Yasunari,1998: Monsoon: processes, predictability, and the prospects for prediction, J. Geophys. Res., 103, 14,451–14,510.

Webster, P. J., A. M. Moore, J. P. Loschnigg, and R. R. Leben, 1999: Coupled ocean-atmosphere dynamics in the Indian Ocean during 1997–98, Nature, 401, 356–360.

Webster, P., 2006: The Coupled Monsoon System. The Asian Monsoon, B. Wang (Ed.), Heidelberg, Springer, pp. 3–66.

Weickmann, K. M., G. R. Lussky, and J. E. Kutzbach, 1985: Intraseasonal (30–60 day) fluctuations of outgoing longwave radiation and 250 mb stream function during northern winter. Mon. Wea. Rev., 113, 941–961.

Weisberg, R. H., and Wang, C., 1997: A western Pacific oscillator paradigm for the El Niño Southern Oscillation. Geophys. Res. Lett., 24, 779–782.

Weisman, M.L. and J.B. Klemp, 1982: The Dependence of Numerically Simulated Convective Storms on Vertical Wind Shear and Buoyancy. Mon. Wea. Rev., 110, 504–520, https://doi.org/10.1175/1520-0493.

Westra, S., L. V. Alexander, and F. W. Zwiers, 2013: Global increasing trends in annual maximum daily precipitation, J. Clim., 26, 3904–3918.

Westra, S., H. J. Fowler, J. P. Evans, L. V. Alexander, P. Berg, F. Johnson, E. J. Kendon, G. Lenderink, and N. M. Roberts, 2014: Future changes to the intensity and frequency of short duration extreme rainfall, Rev. Geophys., 52, 522–555, https://doi.org/10.1002/2014RG000464.

Wheeler, M., and G. N. Kiladis, 1999: Convectively coupled equatorial waves: Analysis of clouds and temperature in the wavenumber-frequency domain, J. Atmos. Sci., 56, 374– 399.

Wheeler, M. C., and H. H. Hendon, 2004: An all-season real-time multivariate MJO index: Development of an index for monitoring and prediction. Mon. Wea.Rev., 132, 1917–1932.

Wheeler, M. C. and J. L. McBride, 2012: Australian monsoon. Intraseasonal Variability of the Atmosphere–Ocean Climate System, 2nd ed. W. K.-M. Lau and D. E. Waliser, Eds., Springer, 147–198.

Williams, E. R., and S. Stanfill, 2002: The physical origin of the land-ocean contrast in lightning activity. Comp. Rendus. Phys., 3, 1277–1292.

Williams, I. N., Y. Lu, L. M. Kueppers, W. J. Riley, S. Biraud, J. E. Bagley, and M. S. Torn, 2016: Land–atmosphere coupling and climate prediction over the US southern Great Plains. J. Geophys. Res. Atmos., 121, 12 125–12 144, https://doi.org/10.1002/2016JD025223.

Wing, A. A. and K. A. Emanuel, 2014: Physical mechanisms controlling self-aggregation of convection in idealized numerical modeling simulations. J Adv Model Earth Syst. 2014;6:59–74.

Wing, A. A. and T. W. Cronin, 2016: Self-aggregation of convection in long channel geometry. Q J R Meteorol Soc., 142,1–15. https://doi.org/10.1002/qj.2628.

Wing, A. A., K. Emanuel, C. E. Holloway, C. Muller, 2017: Convective self-aggregation in numerical simulations: a review. Surv Geophys., 38(6):1173–1197. https://doi.org/10.1007/s10712-017-9408-4.

Wing, A. A., 2019: Self-aggregation of deep convection and its implications on climate. Current Climate Change Reports, 5, 1–11, https://doi.org/10.1007/s40641-019-00120-3.

Wing, A.A., C.L. Stauffer, T. Becker, K.A. Reed, M.-S. Ahn, N.P. Arnold, S. Bony, M. Branson G.H. Bryan, J.-P. Chaboureau, S.R. de Roode, K. Gayatri, C. Hohenegger, I.-K. Hu, F. Jansson, T.R. Jones, M. Khairoutdinov, D. Kim, Z.K. Martin, S. Matsugishi, B. Medeiros, H. Miura, Y. Moon, S.K. Müller, T. Ohno, M. Popp, T. Prabhakaran, D. Randall, R. Rios-Berrios, N. Rochetin, R. Roehrig, D.M. Romps, J.H. Ruppert, Jr., M. Satoh, L.G. Silvers, M.S. Singh, B. Stevens, L. Tomassini, C.C. van Heerwaarden, S. Wang, and M. Zhao, 2021: Clouds and convective self-aggregation in a multi-model ensemble of radiative-convective equilibrium simulations, J. Adv. Model. Earth Syst., 12, e2020MS002138. https://doi.org/10.1029/2020MS002138.

Wolding, B. O., and E. D. Maloney, 2015: Objective diagnosis and the Madden–Julian oscillation. Part I: Methodology. *J. Climate*, **28**, 4127–4140, https://doi.org/10.1175/JCLI-D-14-00688.1.

Wolter, K., and M. S. Timlin, 1993: Monitoring ENSO in COADS with a seasonally adjusted principal component index. Proceedings of the 17th Climate Diagnostics Workshop, Norman, OK, NOAA/NMC/CAC, NSSL, Oklahoma Climate Survey, CIMMS and the School of Meteorology, University of Oklahoma, Norman, OK, 52–57.

Wolter, K. and M. S. Timlin, 2011: El Niño/Southern Oscillation behavior since 1871 as diagnosed in an extended multivariate ENSO index (MEI.ext). Int. J. Climatol., https://doi.org/10.1002/joc.2336.

Wu, Z., D. S. Battisti, and E. S. Sarachik, 2000a: Rayleigh friction, Newtonian Cooling, and the Linear Response to Steady Tropical Heating. J. Atmos. Sci., 57, 1937–1957.

Wu, Z., E. S. Sarachik, and D. S. Battisti, 2000b: Vertical structure of convective heating and the three-dimensional structure of the forced circulation on an Equatorial Beta Plane. J. Atmos. Sci., 57, 2169–2187.

Wunsch, C., and A. E. Gill, 1976: Observations of equatorially trapped waves in Pacific sea level variations, Deep Sea Res., 23, 371–390.

Wyrtki, K., 1973: An equatorial jet in the Indian Ocean. Science, 181, 262–264.

Wyrtki, K., 1985: Water displacements in the Pacific and the genesis of El Niño cycles. J. Geophys. Res., 90, 7129–7132.

Xie, S.-P., 1994: On the genesis of the equatorial annual cycle. J. Climate, 7, 2008–2013.

Xie, S.-P., and Y. Tanimoto, 1998: A pan-Atlantic decadal climate oscillation, Geophys. Res. Lett., 25, 2185–2188.

Xie, S. -P. and J. A. Carton, 2004: Tropical Atlantic Variability: Patterns, Mechanisms, and Impacts. In Earth Climate: The Ocean-Atmosphere Interaction, ed. C. Wang, S. -P. Xie, J. A. Carton, pp. 121–42. Washington DC: Amer. Geophys. Union.

Xie, Y.-B., S.-J. Chen, I.-L. Zhang, and Y.-L. Hung, 1963: A preliminarily statistic and synoptic study about the basic currents over southeastern Asia and the initiation of typhoon (in Chinese). Acta Meteor. Sin., 33, 206–217.

Yanai, M., 1961: A detailed analysis of typhoon formation. J. Meteor. Soc. Japan, 39, 187–213.

Yanai, M., and T. Maruyama, 1966: Stratospheric wave disturbances propagating over the equatorial Pacific, J. Meteorol. Soc. Japan., 44, 291–294.

Yanai, M., S. Esbensen, and J. -H. Chu, 1973: Determination of bulk properties of tropical cloud clusters from large-scale heat and moisture budgets. J. Atmos. Sci., 30, 611–627.

Yanai, M. and C. Li, 1994: Mechanism of heating and the boundary layer over the Tibetan Plateau. Mon. Wea. Rev., 122, 305–323.

Yanai, M., C. Li, and Z. Song, 1992: Seasonal heating of the Tibetan Plateau and its effects on the evolution of the Asian summer monsoon, J. Meteorol. Soc. Jpn., 70(1), 319–351.

Yang, G.-Y., and J. Slingo, 2001: The diurnal cycle in the tropics. Mon. Wea. Rev., 129, 784–801.

Yano, J.-I., and K. Emanuel, 1991: An improved model of the equatorial troposphere and its coupling with stratosphere, Atmos. Sci., 48, 377–389.

Yeh, S.-W., J.-S. Kug, B. Dewitte, M.-H. Kwon, B. Kirtman, and F.-F. Jin, 2009: El Niño in a changing climate. Nature, 461, https://doi.org/10.1038/nature08316.

Yoshimori, M., A. Broccoli, 2009: On the link between Hadley circulation and radiative feedback processes. Geophys. Res. Lett. 36:L20701.

Yoshida, K. 1959: Preprints, International Oceanographic Congress, American Association for the Advancement of Science, 789-791. Washington D.C., 1959.

Yu, J. -Y. and J. D. Neelin, 1994: Modes of tropical variability under convective adjustment and the MaddenJulian oscillation. Part II: numerical results. J Atmos Sci, 51(13):1895–1914.

Yulaeva, E., and J. M. Wallace, 1994: The signature of ENSO in global temperature and precipitation fields derived from the microwave sounding unit. J. Climate, 7, 1719–1736.

Zebiak, S. E., 1993: Air–sea interaction in the equatorial Atlantic region. J. Clim. 6, 1567–-1586.

Zebiak, S. E. and M. A. Cane, 1987: A model for El Niño and the Southern Oscillation. Mon. Wea. Rev., 115, 2262–2278.

Zhen, W., P. Braconnot, E. Guilyardi, U. Merkel, and Y. Yu, 2008: ENSO at 6ka and 21ka from ocean-atmosphere coupled model simulations. Clim. Dyn., 30, 745–762.

Zhang, C., 2005: Madden-Julian Oscillation. Rev. Geophys., 43, RG2003, https://doi.org/10.1029/2004RG000158.

Zhang, C., 2013: Madden-Julian oscillation: bridging weather and climate. Bull. Amer. Soc., 1849–1870, https://doi.org/10.1175/BAMS-D-12-00026.1.

Zhang, F., Y. Q. Sun, L. Magnusson, R. Buizza, S. -J. Lin, J. -H. Chen, and K. Emanuel, 2019: What is the predictability limit of midlatitude weather? J. Atmos. Sci., 76(4), 1077–1091, https://doi.org/10.1175/JAS-D-18-0269.1.

Zhang, R., T. Delworth, 2005: Simulated tropical response to a substantial weakening of the atlantic thermohaline circulation. J Clim 18:1853–1860. https://doi.org/10.1175/JCLI3460.1.

Zhou, J., and K.-M. Lau, 1998: Does a monsoon climate exist over South America? J. Climate, 11, 1020–1040.

Zhou, T., A. G. Turner, J. L. Kinter, B. Wang, Y. Qian, X. Chen, B. Wu, B. Wang, B. Liu, L. Zou, and B. He, 2016: GMMIP (v1.0) contribution to CMIP6: Global Monsoons Model Intercomparison Project. Geosci. Model Dev., 9, 3589–3604, https://doi.org/10.5194/gmd-9-3589-2016.

Zipser, 2003: Some views on "Hot Towers" after 50 years of Tropical Field Programs and Two years of TRMM data. Cloud Systems, Hurricanes, and the Tropical Rainfall Measuring Mission (TRMM), Meteor. Monogr., No. 51, Amer. Meteor. Soc., 49–58.

Zipser, E. J., D. J. Cecil, C. Liu, S. W. Nesbitt, and D. P. Yorty, 2006: Where are the most intense thunderstorms on Earth? Bull. Amer. Soc., 1057–1071, https://doi.org/10.1175/BAMS-87-8-1057.

Zheng, W., P. Braconnot, and E. Guilyardi, et al., 2008: ENSO at 6ka and 21ka from ocean–atmosphere coupled model simulations. Climate Dyn., 30, 745–762. https://doi.org/10.1007/s00382-007-0320-3s

Index

© Springer Nature Switzerland AG 2023
V. Misra, *An Introduction to Large-Scale Tropical Meteorology*, Springer Atmospheric Sciences, https://doi.org/10.1007/978-3-031-12887-5

Printed in the United States
by Baker & Taylor Publisher Services